T0237967

Springer Complexity

Springer Complexity is an interdisciplinary program publishing the best research and academic-level teaching on both fundamental and applied aspects of complex systems – cutting across all traditional disciplines of the natural and life sciences, engineering, economics, medicine, neuroscience, social and computer science.

Complex Systems are systems that comprise many interacting parts with the ability to generate a new quality of macroscopic collective behavior the manifestations of which are the spontaneous formation of distinctive temporal, spatial or functional structures. Models of such systems can be successfully mapped onto quite diverse "real-life" situations like the climate, the coherent emission of light from lasers, chemical reaction-diffusion systems, biological cellular networks, the dynamics of stock markets and of the internet, earthquake statistics and prediction, freeway traffic, the human brain, or the formation of opinions in social systems, to name just some of the popular applications.

Although their scope and methodologies overlap somewhat, one can distinguish the following main concepts and tools: self-organization, nonlinear dynamics, synergetics, turbulence, dynamical systems, catastrophes, instabilities, stochastic processes, chaos, graphs and networks, cellular automata, adaptive systems, genetic algorithms and computational intelligence.

The two major book publication platforms of the Springer Complexity program are the monograph series "Understanding Complex Systems" focusing on the various applications of complexity, and the "Springer Series in Synergetics", which is devoted to the quantitative theoretical and methodological foundations. In addition to the books in these two core series, the program also incorporates individual titles ranging from textbooks to major reference works.

Editorial and Programme Advisory Board

Dan Braha, New England Complex Systems Institute and University of Massachusetts Dartmouth, USA

Péter Érdi, Center for Complex Systems Studies, Kalamazoo College, USA and Hungarian Academy of Sciences, Budapest, Hungary

Karl Friston, Institute of Cognitive Neuroscience, University College London, London, UK

Hermann Haken, Center of Synergetics, University of Stuttgart, Stuttgart, Germany

Janusz Kacprzyk, System Research, Polish Academy of Sciences, Warsaw, Poland

Scott Kelso, Center for Complex Systems and Brain Sciences, Florida Atlantic University, Boca Raton, USA

Jürgen Kurths, Nonlinear Dynamics Group, University of Potsdam, Potsdam, Germany

Linda Reichl, Center for Complex Quantum Systems, University of Texas, Austin, USA

Peter Schuster, Theoretical Chemistry and Structural Biology, University of Vienna, Vienna, Austria

Frank Schweitzer, System Design, ETH Zurich, Zurich, Switzerland

Didier Sornette, Entrepreneurial Risk, ETH Zurich, Zurich, Switzerland

Springer Series in Synergetics

Founding Editor: H. Haken

The Springer Series in Synergetics was founded by Herman Haken in 1977. Since then, the series has evolved into a substantial reference library for the quantitative, theoretical and methodological foundations of the science of complex systems.

Through many enduring classic texts, such as Haken's *Synergetics and Information and Self-Organization*, Gardiner's *Handbook of Stochastic Methods*, Risken's *The Fokker Planck-Equation* or Haake's *Quantum Signatures of Chaos*, the series has made, and continues to make, important contributions to shaping the foundations of the field.

The series publishes monographs and graduate-level textbooks of broad and general interest, with a pronounced emphasis on the physico-mathematical approach.

For further volumes:
http://www.springer.com/series/712

Boris P. Bezruchko · Dmitry A. Smirnov

Extracting Knowledge From Time Series

An Introduction to Nonlinear Empirical Modeling

 Springer

Prof. Boris P. Bezruchko
Saratov State University
Astrakhanskaya Street 83
410012 Saratov
Russia
bezruchkobp@gmail.com

Dr. Dmitry A. Smirnov
Russian Academy of Science
V.A. Kotel'nikov Institute of
RadioEngineering and Electronics
Saratov Branch
Zelyonaya Str. 38
410019 Saratov
Russia
smirnovda@yandex.ru

ISSN 0172-7389
ISBN 978-3-642-26482-5   ISBN 978-3-642-12601-7 (eBook)
DOI 10.1007/978-3-642-12601-7
Springer Heidelberg Dordrecht London New York

© Springer-Verlag Berlin Heidelberg 2010
Softcover reprint of the hardcover 1st edition 2010
This work is subject to copyright. All rights are reserved, whether the whole or part of the material is
concerned, specifically the rights of translation, reprinting, reuse of illustrations, recitation, broadcasting,
reproduction on microfilm or in any other way, and storage in data banks. Duplication of this publication
or parts thereof is permitted only under the provisions of the German Copyright Law of September 9,
1965, in its current version, and permission for use must always be obtained from Springer. Violations
are liable to prosecution under the German Copyright Law.
The use of general descriptive names, registered names, trademarks, etc. in this publication does not
imply, even in the absence of a specific statement, that such names are exempt from the relevant protective
laws and regulations and therefore free for general use.

Cover design: Integra Software Services Pvt. Ltd., Pondicherry

Printed on acid-free paper.

Springer is part of Springer Science+Business Media (www.springer.com)

Preface

Mathematical modelling is ubiquitous. Almost every book in exact science touches on mathematical models of a certain class of phenomena, on more or less specific approaches to construction and investigation of models, on their applications, etc. As many textbooks with similar titles, Part I of our book is devoted to general questions of modelling. Part II reflects our professional interests as physicists who spent much time to investigations in the field of non-linear dynamics and mathematical modelling from discrete sequences of experimental measurements (time series). The latter direction of research is known for a long time as "system identification" in the framework of mathematical statistics and automatic control theory. It has its roots in the problem of approximating experimental data points on a plane with a smooth curve.

Currently, researchers aim at the description of complex behaviour (irregular, chaotic, non-stationary and noise-corrupted signals which are typical of real-world objects and phenomena) with relatively simple non-linear differential or difference model equations rather than with cumbersome explicit functions of time. In the second half of the twentieth century, it has become clear that such equations of a sufficiently low order can exhibit non-trivial solutions that promise sufficiently simple modelling of complex processes; according to the concepts of non-linear dynamics, chaotic regimes can be demonstrated already by a third-order non-linear ordinary differential equation, while complex behaviour in a linear model can be induced either by random influence (noise) or by a very high order of equations. Possibility to construct non-linear predictive models and availability of fast computers with huge memory size provides new opportunities in processing of signals encountered in nature and different fields of practice ranging from economy to medicine.

Our book is devoted to mathematical modelling of *processes* (i.e. motions, temporal changes). It is addressed to a reader who aims at the usage of empirical modelling machinery to solve practical tasks. We consider problems and techniques of modelling from observational data and describe possible applications. Moreover, we have located computer practical works and special educational software at our website http://www.nonlinmod.sgu.ru. As well, we touch on world-outlook questions which inevitably arise in "making mathematical models". The contents and style of the book have been developed as a result of our research activity and many years of teaching at different natural science departments of Saratov State University

(Saratov, Russia): Department of Physics, Department of Nonlinear Processes and Department of Nano- and Biomedical Technologies. Several chapters have been used in teaching of future geologists, biologists and economists.

We hope that our book will be useful for a wide readership. Therefore, we present material at different levels of complexity. The first chapters are suited for a reader who has education at the level of a secondary school. However, general questions of modelling and examples of contemporary dynamical models can be interesting for specialists as well. Part II is more specific and formalised. It is addressed to people who plan to construct models from time series and to use them for prediction, validation of ideas about underlying dynamical laws, restoration of hidden variables and so on. We do not give rigorous proofs. Discussion of mathematical results is often presented at the level of vivid illustrations with references to special works. Deeper understanding of the problems can be achieved via information located at the websites of our and other research groups, e.g.

> http://sgtnd.narod.ru/eng/index.htm
> http://www.nonlinmod.sgu.ru/index_en.htm
> http://www.cplire.ru/win/InformChaosLab/index.html
> http://chaos.ssu.runnet.ru/
> http://webber.physik.uni-freiburg.de/~jeti
> http://math.gmu.edu/~tsauer/
> http://www.maths.uwa.edu.au/~kevin/
> http://www.mpipks-dresden.mpg.de/~kantz/
> http://www.dpi.physik.uni-goettingen.de/~ulli/
> http://www.maths.ox.ac.uk/~lenny
> http://www.pik-potsdam.de/members/kurths
> http://www.stat.physik.uni-potsdam.de/
> http://www.agnld.uni-potsdam.de/
> http://inls.ucsd.edu/~hdia/
> http://www.eie.polyu.edu.hk/~ensmall/matlab/

Our bibliography concerning modelling problems is by no means complete. We refer only to some of many research papers and monographs, which were the most influential for our research and teaching activity.

The book contains the results of investigations of our group of "non-linear dynamical modelling" which unites university lecturers, academic researchers, Ph.D. students and undergraduate students from Saratov State University and Saratov Branch of V.A. Kotel'nikov Institute of RadioEngineering and Electronics of Russian Academy of Sciences (SB IRE RAS). Our colleagues Ye.P. Seleznev, V.I. Ponomarenko, M.D. Prokhorov, T.V. Dikanev, M.B. Bodrov, I.V. Sysoev, A.S. Karavaev, V.V. Astakhov, S.A. Astakhov, A.V. Kraskov, A.Yu. Jalnine, V.S. Vlaskin and P.V. Nakonechny are to some extent co-authors of the book.

Moreover, we present the results of our joint research with other groups: I.I. Mokhov (A.M. Obukhov Institute of Atmospheric Physics RAS, Moscow, Russia), P.A. Tass and U.B. Barnikol (Institute of Neuroscience and Biophysics – 3,

Research Centre Juelich, Germany), R. Stoop and A. Kern (University of Zuerich, Switzerland), G. van Luijtelaar (Nijmegen Institute for Cognition and Information, Radboud University of Nijmegen, The Netherlands), G.D. Kuznetsova and E.Yu. Sitnikova (Institute of Higher Nervous Activity and Neurophysiology RAS, Moscow, Russia), J.-L. Perez Velazquez (Hospital for Sick Children and University of Toronto, Canada), R. Wennberg (Toronto Western Hospital, Canada), J. Timmer, B. Schelter and M. Winterhalder (University of Freiburg, Germany) and R.G. Andrzejak (von Neumann Institute for Computing, Research Centre Juelich, Germany).

We are grateful to M. Cencini, J. Timmer, T. Sauer and E. Wan for allowing us to use some results of their investigations along with several figures, which are appropriate for the topics considered in Chaps. 2, 8 and 10. We acknowledge a great help of D.V. Sokolov in the technical preparation of all the figures in the monograph.

Prof. N.G. Makarenko (Pulkovo Observatory RAS, St. Petersburg, Russia) and Prof. S.P. Kuznetsov (SB IRE RAS) have read the first (Russian) edition of the manuscript and made a lot of useful remarks. We are very grateful to Prof. J. Kurths for the attention to our work and many useful advice and recommendations concerning the preparation of the second (English) edition of the book, which would not appear without his help.

Finally, we acknowledge many colleagues whose communications in non-linear dynamics and time series analysis problems have been very useful to us: V.S. Anishchenko, V.N. Belykh, O.Ya. Butkovsky, A.F. Golubentsev, Yu.A. Danilov, A.S. Dmitriev, A.M. Feigin, G.T. Guria, N.B. Janson, M.V. Kapranov, V.B. Kazantsev, A.A. Kipchatov, Yu.A. Kravtsov, A.P. Kuznetsov, P.S. Landa, A.Yu. Loskutov, Yu.L. Maistrenko, V.V. Matrosov, V.I. Nekorkin, G.V. Osipov, A.I. Panas, A.N. Pavlov, A.S. Pikovsky, V.P. Ponomarenko, A.G. Rokakh, M.G. Rosenblum, A.G. Rozhnev, V.D. Shalfeev, A.N. Silchenko, D.I. Trubetskov, and D.A. Usanov.

Our investigations partly reflected in this book were supported by grants of the Russian Foundation for Basic Research, Russian Science Support Foundation, the President of Russia, Ministry of Education and Science of Russia, American Civilian Research and Development Foundation, and programs of Russian Academy of Sciences.

Saratov, Russia   Boris P. Bezruchko
                                                                           Dmitry A. Smirnov

Introduction

Throughout the present book, we consider the problem of mathematical modelling as applied scientists who use mathematics as a tool to obtain practically useful results. Mathematical model is a symbolic construction whose properties must coincide with some relevant properties of an object under investigation. As far as applications are concerned, the main point is to make a construction which allows to achieve some practical purpose. Such a purpose may relate to forecast of future behaviour, automatic control, clustering of data, validation of substantial ideas about an object, diagnostics of causal relationships in a complex system and many other problems discussed in the book.

A model or a way used to obtain it may appear "imperfect" in some strict sense, e.g. formulation of a problem is not completely correct, its solution is not unique. However, is it worth speaking of a unique "true" mathematical model of a real-world object if the mathematics itself has arisen quite recently as compared with the Universe and many objects of modelling? This polemic question and a lot of similar questions, some of them being "eternal", have determined the contents of Chap. 1 where we discuss general problems of modelling, including definitions and systematisations of models, the role of mathematics and causes of its efficiency, and approaches to model construction. The concept of ill-posed problems briefly considered in the second part of the book (Chap. 5) is also closely related to such questions.

Chapter 2 is devoted to the two "world-outlook" approaches to modelling differing by the "degree of optimism" with respect to principal predictability of natural phenomena. The first one is the deterministic (dynamical) approach. It is very optimistic. At the beginning, it "claimed" even practical possibility to predict the future precisely based on the precise knowledge of a present state. Currently, when the concept of "dynamical chaos" has been formulated and non-linear dynamics has become a widely known field of knowledge, the claims for the *practically achievable* accuracy of forecasts are much more moderate.

The second approach is called probabilistic (stochastic). It is less optimistic. One refuses precise forecast and tends to determine only probabilities of different scenarios of the future. In Chap. 2 we discuss the assessment of prediction opportunities and practical motivations to call a process under investigation "random". With the example of the well-known "coin flip", we illustrate necessity of "co-operation" between "deterministic" and "stochastic" viewpoints and narrowness of any viewpoint taken separately.

Throughout the book, we describe mainly deterministic models and popularise approaches of *non-linear dynamics*. The latter is a sufficiently "young" and currently developing scientific field whose terminology is not yet fully established. A person who starts to study its problems can come into troubles since even leading specialists sometimes use the same term with almost contradictory meanings, e.g. a collection of definitions of "dynamical system", which is one of the main concepts in non-linear dynamics, is presented in Sect. 2.1.1. Therefore, we discuss a terminology and present a short review of the basic concepts and illustrative tools of non-linear dynamics. As illustrations, we use numerical examples, data of laboratory experiments and signals from real-world objects which can be of interest for a wide readership.

Chapter 3 presents the main capabilities of the mathematical apparatus, implementing the deterministic approach, and some exemplary models. We focus on the ordinary differential equations and discrete maps since they are the most popular tools for dynamical modelling and lie in the field of our direct professional interests. Chapter 4, the last in Part I, briefly exposes stochastic models and the role of noise.

While the purpose of Part I (Chaps. 1, 2, 3 and 4) is to introduce a general view on the topic of modelling, Part II (Chaps. 5, 6, 7, 8, 9, 10, 11, 12 and 13) is focused on a single approach to model construction which can be called "empirical modelling" or "modelling from data". Previously, it was not considered as "respectable" analogously to a "design" of clothes based on wrapping a client into a piece of material linked from the edges. However, such a modelling is currently being actively used and developed since fast computers have become widely available and the concept of dynamical chaos has been formulated so that it has become clear that a complicated behaviour can be described with sufficiently simple non-linear models. Moreover, this approach is often the only possible one in practice since one often cannot follow the most established and reliable way, i.e. write down model equations from the so-called "first principles" (general laws for a certain range of phenomena such as conservation principles and Newton's laws in mechanics and Maxwell's equations in electrodynamics) taking into account specific features of an object under investigation. In a typical practical situation, the main source of information about an object is the data of measurements represented as a set of values of an observed quantity measured at subsequent time instants – a *time series*.

Construction of models from experimental time series is called "system identification" in mathematical statistics and automatic control theory (Ljung, 1991) and "reconstruction[1] of dynamical systems" (Anishchenko et al., 2002; Gouesbet

---

[1] The term "reconstruction" seems completely appropriate only for the case of restoration of equations from their solutions. In modelling of real-world systems, the term "model construction" is more suitable. However, we use the term "reconstruction" as well since it is already widely used in the literature.

et al., 2003a; Malinetsky and Potapov, 2000) in non-linear dynamics. Predecessors of the contemporary reconstruction problems were the problems of approximation and statistical investigation of dependencies between observed quantities considered already in the middle of the eighteenth century. Originally, an observed process was modelled with an explicit function of time $\eta = f(t)$, which approximated a set of experimental data points on the plane (t, η). The purposes of modelling were prediction of the future and smoothing of noise-corrupted data. At the beginning of the twentieth century, a serious step forward in the development of techniques for empirical modelling of complex processes was done in the framework of mathematical statistics when Yule suggested to use *linear stochastic* models (Yule, 1927). That approach was the main tool during half a century (1920s–1970s) and found multiple applications, especially to prediction and automatic control problems (Box and Jenkins, 1970; Ljung, 1991; Pugachev and Sinitsyn, 1985). Formulation of the concept of dynamical chaos and the rise of computational powers led to a new situation: In the last two decades empirical modelling is performed on the basis of non-linear difference and differential equations including multidimensional models. Among the pioneering works in this field, we would mention Abarbanel et al. (1989), Baake et al. (1992), Bock (1981), Breeden and Hubler (1990), Broomhead and Lowe (1988), Casdagli (1989), Crutchfield and McNamara (1987), Cremers and Hubler (1987), Farmer and Sidorowich (1987); Giona et al. (1991), Gouesbet (1991), Mees (1991) and Smith (1992). The problems considered are topical both from fundamental and applied points of view. Empirical models are demanded in different fields of science and practice including physics, meteorology, seismology, economics, medicine and physiology (Kravtsov, 1997). The tasks which are solved with the aid of reconstruction are very diverse. They include forecast, quantitative validation of physical ideas about an object, restoring time courses of quantities inaccessible to a measuring device and diagnostics of causal relationships between processes.

We give an overview of the problems and methods of modelling from time series in Chaps. 5, 6, 7, 8, 9, 10, 11, 12, and 13. Our consideration supplements previous reviews presented in monographs (Abarbanel, 1996; Anishchenko et al., 2002; Casdagli and Eubank, 1992; Dahlhaus et al., 2008; Gerschenfeld and Weigend, 1993; Gouesbet et al., 2003a; Kantz and Schreiber, 1997; Malinetsky and Potapov, 2000; Mees, 2001; Ott et al., 1994; Small, 2005; Soofi and Cao, 2002; Winterhalder et al., 2006) and papers (Abarbanel et al., 1993; Gouesbet et al., 2003b; Pavlov et al., 1999; Rapp et al., 1999; Shalizi, 2003; Smirnov and Bezruchko, 2006; Voss et al., 2004). As well, our original recent results on non-linear data analysis are presented, especially in Chaps. 9, 12 and 13. Mainly, we speak of finite-dimensional deterministic models in the form of discrete maps or ordinary differential equations. Similar to Part I, we select different examples to illustrate measurement resulting from both laboratory experiments (where experimental conditions allow purposeful selection of the regimes of object functioning, control of external influences and correction of an initial state) and real-world data observed in the past or in the situations where observation conditions cannot be changed (so that one must use the data as it is, including inevitable distortions).

Material is discussed on the basis of a typical scheme of the modelling procedure presented in Chap. 5. Chapter 6 is devoted to acquisition of data and its preliminary analysis, which can serve to extract additional information about an object useful for specifying the structure of model equations. In Chap. 7 many important problems in empirical modelling are discussed with the example of the simplest kind of models – explicit functions of time. Further exposition follows the principle "from simple to complex", namely in the "direction" of decrease in the amount of prior knowledge about an object. We proceed from the case when almost everything is known and only the values of parameters in model equations remain to be calculated (Chap. 8) via an intermediate variant (Chap. 9) to the situation when nothing is known a priori about an appropriate form of model equations (Chap. 10). Then, we give examples of useful applications of empirical models (Chap. 11) including our own results on the detection of coupling between complex processes (Chap. 12). Finally, Chap. 13 presents "outdoor" examples from the fields of electronics, physiology and geophysics and provides a more detailed consideration of different steps of a modelling procedure, which should be of interest for a wider audience.

Throughout the book, we often refer to Internet resources containing useful information on mathematical modelling including research papers, tutorials, training and illustrative computer programs. This information should supplement the contents of the book.

References

Abarbanel, H.D.I.: Analysis of Observed Chaotic Data. Springer, New York (1996)

Abarbanel, H.D.I., Brown, R., Kadtke, J.B.: Prediction and system identification in chaotic nonlinear systems: time series with broadband spectra. Phys. Lett. A. **138**, 401–408 (1989)

Abarbanel, H.D.I., Brown, R., Sidorowich, J.J., Tsimring, L.S.: The analysis of observed chaotic data in physical systems. Rev. Mod. Phys. **65**, 1331–1392 (1993)

Anishchenko, V.S., Astakhov, V.V., A.B.: Neiman, Vadivasova, T.Ye., L.: Schimansky-Geier, Nonlinear effects in chaotic and stochastic systems. Tutorial and Modern Development. Springer, Berlin (2002)

Baake, E., Baake, M., Bock, H.J., Briggs, K.M.: Fitting ordinary differential equations to chaotic data. Phys. Rev. A. **45**, 5524–5529 (1992)

Bock, H.G.: Numerical treatment of inverse problems in chemical reaction kinetics. In: Ebert, K.H., Deuflhard, P., Jaeger, W., et al. (eds.) Modelling of Chemical Reaction Systems, pp. 102–125. Springer, New York (1981)

Box, G.E.P., Jenkins, G.M.: Time series analysis. Forecasting and Control. Holden-Day, San Francisco (1970)

Breeden, J.L., Hubler, A.: Reconstructing equations of motion from experimental data with unobserved variables. Phys. Rev. A. **42**, 5817–5826 (1990)

Broomhead, D.S., Lowe, D.: Multivariable functional interpolation and adaptive networks. Complex Syst. **2**, 321–355 (1988)

Casdagli, M.: Nonlinear prediction of chaotic time series. Physica D. **35**, 335–356 (1989)

Casdagli, M., Eubank, S. (eds.): Nonlinear modeling and forecasting. SFI Studies in the Sciences of Complexity, vol XII. Addison-Wesley, New York (1992)

Cremers, J., Hubler, A.: Construction of differential equations from experimental data. Z. Naturforschung A. **42**, 797–802 (1987)

Crutchfield, J.P., McNamara, B.S.: Equations of motion from a data series. Complex Syst. **1**, 417–452 (1987)

Dahlhaus, R., Kurths, J., Maass, P., Timmer, J. (eds.): Mathematical Methods in Time Series Analysis and Digital Image Processing. Springer, Complexity, Berlin (2008)

Farmer, J.D., Sidorowich, J.J.: Predicting chaotic time series. Phys. Rev. Lett. **59**, 845–848 (1987)

Gerschenfeld, N.A., Weigend, A.S. (eds.): Time series prediction: forecasting the future and understanding the past. SFI Studies in the Science of Complexity, Proceedings V. XV. Addison-Wesley, New York (1993)

Giona, M., Lentini, F., Cimagalli, V.: Functional reconstruction and local prediction of chaotic time series. Phys. Rev. E. **44**, 3496–3502 (1991)

Gouesbet, G., Meunier-Guttin-Cluzel, S., Ménard, O. (eds.): Chaos and Its Reconstructions. Nova Science Publishers, New York (2003a)

Gouesbet, G., Meunier-Guttin-Cluzel, S., Ménard, O.: Global reconstructions of equations of motion from data series, and validation techniques, a review. In: Gouesbet, G., Meunier-Guttin-Cluzel, S., Ménard, O. (eds.) Chaos and Its Reconstructions, pp. 1–160. Nova Science Publishers, New York (2003b)

Gouesbet, G.: Reconstruction of the vector fields of continuous dynamical systems from scalar time series. Phys. Rev. A. **43**, 5321–5331 (1991)

Kantz, H., Schreiber, T.: Nonlinear Time Series Analysis. Cambridge University Press, Cambridge (1997)

Kravtsov, Yu.A. (ed.): Limits of Predictability. TsentrCom, Moscow, (in Russian) (1997)

Ljung, L.: System Identification. Theory for the User. Prentice-Hall, Engle Wood Cliffs, NJ (1991)

Malinetsky, G.G., Potapov, A.B.: Contemporary Problems of Nonlinear Dynamics. Editorial URSS, Moscow (in Russian) (2000)

Mees, A.I.: Dynamical systems and tesselations: Detecting determinism in data. Int. J. Bif. Chaos. **1**, 777–794 (1991)

Mees, A.I. (ed.): Nonlinear Dynamics and Statistics. Birkhaeuser, Boston (2001)

Ott, E., Sauer, T., Yorke J.A. (eds.): Coping with Chaos: Analysis of Chaotic Data and The Exploitation of Chaotic Systems. Wiley-VCH, New York (1994)

Pavlov, A.N., Janson, N.B., Anishchenko, V.S.: Reconstruction of dynamical systems. J. Commun. Technol. Electron. **44**(9), 999–1014 (1999)

Pugachsev, V.S., Sinitsyn, I.N.: Stochastic Differential Systems, 560 p. Nauka, Moscow (in Russian) (1985)

Rapp, P.E., Schmah, T.I., Mees, A.I.: Models of knowing and the investigation of dynamical systems. Phys. D. **132**, 133–149 (1999)

Shalizi, C.R.: Methods and techniques of complex systems science: an overview, vol. 3, arXiv:nlin.AO/0307015 (2003). Available at http://www.arxiv.org/abs/nlin.AO/0307015

Small, M.: Applied Nonlinear Time Series Analysis. World Scientific, Singapore (2005)

Smirnov, D.A., Bezruchko, B.P.: Nonlinear dynamical models from chaotic time series: methods and applications. In: Winterhalder, M., Schelter, B., Timmer, J. (eds.) Handbook of Time Series Analysis, pp. 181–212. Wiley-VCH, Berlin (2006)

Smith, L.A.: Identification and prediction of low-dimensional dynamics. Phys. D. **58**, 50–76 (1992)

Soofi, A.S., Cao, L. (eds.): Modeling and Forecasting Financial Data: Techniques of Nonlinear Dynamics. Kluwer, Dordrecht (2002)

Voss, H.U., Timmer, J., Kurths, J.: Nonlinear dynamical system identification from uncertain and indirect measurements. Int. J. Bif. Chaos. **14**, 1905–1933 (2004)

Winterhalder, M., Schelter, B., Timmer, J. (eds.): Handbook of Time Series Analysis. Wiley-VCH, Berlin (2006)

Yule, G.U.: On a method of investigating periodicities in disturbed series, with special reference to Wolfer's sunspot numbers. Philos. Trans. R. Soc. London A. **226**, 267–298 (1927)

Contents

Some Abbreviations and Notations

ADC	analogue-to-digit converter
DS	dynamical system
DE	differential equation
DDE	delay differential equation
ANN	artificial neural network
ML	maximum likelihood
LS	least squares
ODE	ordinary differential equation
$\mathbf{A}$	matrix (upper-case letter, boldface font)
$\mathbf{x}$	vector (lower-case letter, boldface font)
x	scalar (lower-case letter, italic type)
$\dot{x}$	derivative of x with respect to time $(\mathrm{d}x/\mathrm{d}t)$
$\hat{a}$	estimator of a quantity a from a sample (from a time series)
$E[x]$	expectation of a quantity x
$\mathrm{var}[x]$	variance of a quantity x

Part I
Models and Forecast

Chapter 1
The Concept of Model. What is Remarkable in Mathematical Models

1.1 What is Called "Model" and "Modelling"

Dictionaries tell us that the word "model" originates from the Latin word "modulus" which means "measure, template, norm". This term was used in proceedings on civil engineering several centuries BC. Currently, it relates to an enormously wide range of material objects, symbolic structures and ideal images ranging from models of clothes, small copies of ships and aeroplanes, different pictures and plots to mathematical equations and computational algorithms. Starting to define the concept of "model", we would like to remind about the difficulty to give strict definitions of basic concepts. Thus, when university professors define "oscillations" and "waves" in their lectures on this subject, it is common for many of them to repeat the joke of Russian academician L.I. Mandel'shtam, who illustrated the problem with the example of the term "heap": How many objects, and of which kind, deserve such a name? As well, he compared strict definitions at the beginning of studying any topic to "swaddling oneself with barbed wire". Among classical examples of impossibility to give exhaustive formulations, one can mention the terms "bald spot", "forest", etc. Therefore, we will not consider variety of existing definitions of "model" and "modelling" in detail. Any of them relates to the purposes and subjective preferences of an author and is valid in a certain sense. However, it is restricted since it ignores some objects or properties that deserve attention from other points of view.

We will call "model" *something (ideal images, material things or symbolic constructions) whose set of properties intersects a set of properties of an original (an object) in a domain essential for the purpose of modelling*. We will call "modelling" *the process of creation and usage of a model*. Here, the term "original" stands for an object of modelling, e.g. a material thing, a phenomenon, a process. "Ideal" means "being thought" by a human being, existing in his/her mind. "Symbolic constructions" are some abstractions in the form of formulas, plots, chains of symbols, etc. "Set of properties" is a collection of properties. Their "intersection" means a subset belonging to both sets (coincidence of some properties from the two sets). In other words, a model is something similar to an original in certain respects. A model is created and/or used by a human being for his/her purposes.

The above considerations are illustrated in Fig. 1.1 with the example of modelling of an aeroplane. Each point on the plot corresponds to some property: "white",

B.P. Bezruchko, D.A. Smirnov, *Extracting Knowledge From Time Series*, Springer Series in Synergetics, DOI 10.1007/978-3-642-12601-7_1,
© Springer-Verlag Berlin Heidelberg 2010

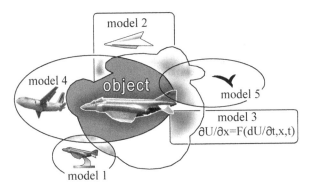

Fig. 1.1 An illustration of the definition of "model". Domains of intersection between the sets of properties are dashed. The set of the object properties is bounded by the thick curve, while the model sets are bounded by thinner ones

"blue", "big", "metallic", "capable of gliding", etc. The lines bound the sets of properties of the object and models. Models may be different and characterised with greater or smaller sets of properties. They can be even more complicated than the object. As well, the domains of intersection between the sets of model and object properties may vary drastically ranging from "microscopic size" to complete inclusion of one of them into the other one. The choice of a model is determined by the purpose of modelling. Thus, a small copy on a pedestal (model 1) is sufficient to model the shape of the aeroplane. In such a case, the domain of intersection includes points corresponding, e.g., to geometrical proportions and colours of the aeroplane elements.

The ability to glide is better modelled with a "paper jackdaw" (model 2), i.e. a properly convoluted sheet of paper. In this case, the domain of intersection includes properties related to flight, while colour and shape become irrelevant. Capability of the aeroplane wings to vibrate can be modelled with the aid of mathematical equations (model 3). Then, oscillatory character of the wings' motion coincides with the solutions to the model equations. Capability to move in the air is common for the aeroplane and a bird (model 5), though the principles of flying are completely different and a biological system is characterised by a higher level of organisation complexity. As sufficiently complete models of the aeroplane, one can use any other aeroplanes (model 4), but the former one is absolutely identical only to itself. Figure 1.1 is also a model; it is a model of our representation.

Everything can be a model of everything under the condition that the properties of a model and an object coincide in part that allows to achieve a purpose. A model can exist in different forms, for example, as certain associations between neurons in our brain, as symbols on a sheet of paper, as a magnetisation profile on a personal computer (PC) hard disk, as a thing made from metal or wood, etc. Thus, a brick is a good model of the geometrical shape of a PC case and vice versa. A computer models many intellectual abilities of a human being and vice versa.

The above definition of a model differs from the definitions of a model as "a selective abstract copying of the properties of an original", "simplified representation of an original", "caricature to an object" in the following respect. Our definition does not imply passive copying of the reality similar to flat photographic images of the "three-dimensional" world. Models play a role of "cameras", "glasses", "filters", i.e. the tools with which we perceive the world. The "active role" of models manifests itself in that the results of our observations (the way how we perceive the facts) depend on the ideas present in our mind at the instant of a shoot. Fixation of each frame is based on the previous frame and is influenced by the latter. Thus, the reflection or perception of the reality depends on the models used by a person observing the reality. Moreover, personality is characterised by a set of models used in practice.

A good analogy to an active cognitive role of a model is a lamp lighting a certain area in a surrounding darkness (in general, the darkness of ignorance). Another convenient image finely illustrating some features of the cognition process is the "dressing" of an object under investigation with existing models of "clothes". Knowledge in the mind of a person is analogous to a wardrobe, a set of dresses and shoes. The richer and more diverse this set, the wider our capabilities to understand and describe various events.

During the cognitive process, a scientist achieves a success if he/she manages to "generate" a fruitful model of something unknown. People perceive only the part of the visible world which is known, i.e. whose model is available. Thousand years of observations of ancient scientists confirmed validity of the geocentric theory. Nowadays, the Sun "rotates" around the Earth and moves from the east to the west in the same way as in the past. However, any schoolchild would explain it now as a result of the revolution of the Earth on its axis and give evidences for that. "An eye cannot see anything unclear to a mind. Anything getting clear to a mind is instantaneously seen by an eye" (M.A. Bedel). Validity of these retranslated words of the Indian poet of seventeenth century is further confirmed by such a modern example as the discovery of the "dynamical chaos" phenomenon. Up to the end of twentieth century, complicated unpredictable behaviour was typically related to complexity, presence of many elements in systems under investigation (a crowd, an ensemble of molecules, etc). However, when contemporary ideas about chaos emerged in the 1960s and simple low-dimensional deterministic models with chaotic behaviour were developed, many researchers quickly understood the situation from the new point of view and started to observe irregular, unrepeatable motions even in pendulums of the grandfather's clocks. To make it possible, it was necessary to include into the "wardrobe" of models new elements such as nonlinear maps and differential equations with chaotic dynamics to be discussed in the next chapters.

To conclude the discussion of the terms "model" and "modelling", we cite several words of specialists in different fields. An opinion of a philosopher M. Wartofsky was the following: contrary to other animals, people create their representations of what we do or what we want, i.e. artefacts or models. A model is not only a simple copy of an existing thing but also a representation of a future practice. We acquire knowledge with the aid of models (Wartofsky, 1979). Russian academician

N.N. Moiseev, who got known due to his investigations of the "nuclear winter" model that affected the mighty of this world in the 1980s, also stressed a role of models in the cognitive process: "We can think only with images approximately reflecting the reality. Any absolute knowledge (absolute truth as philosophers say) is cognised via an infinite asymptotic chain of relative truths approximately reflecting some features of the objective reality. *Those relative truths are called models or model knowledge.* Models can be formulated in any language: Russian, English, French, etc. They can use a language of graphical constructions, language of chemistry, biology, mathematics, etc" (Moiseev, 1979). As an example of a more technical approach, we cite a tutorial in mathematical modelling recently published by N.E. Bauman Moscow State Technical University (Zarubin, 2001): "From a sufficiently general position, mathematical modelling can be considered as one of the techniques for the cognition of the real world during the period when the so-called information-oriented society is being formed. It is an intellectual nucleus of quickly developing informational technologies. In engineering, mathematical modelling is understood as an adequate replacement of a technical device or process under investigation by a corresponding mathematical model and further investigation of the latter with modern computers and methods of computational mathematics. Since such investigation of a mathematical model can be considered as performing an experiment on a computer with the use of computational algorithms, the term 'numerical experiment' is often used in scientific and technical literature as a synonym of the term 'mathematical modelling'. The meaning of these terms is typically regarded as intuitively clear and not discussed in detail". We would agree with the author in that almost any student of a technical university has intuitive concepts of "model" and "modelling". This is also the case for any specialist in exact science or just for a well-educated person since the entire knowledge of a human being about the real world is a model in its essence and in its form. For the sake of definiteness, when speaking of models and modelling, we follow the definition given at the beginning of this section. Let us finish it again with the words of M. Wartofsky: Any model fixes a certain relationship to the world or to a modelled object and involves its creator or user into this relationship. Therefore, it is always possible to reconstruct a subject of modelling from a model. The subject of modelling is an individual bearing such a relationship to the world which is expressed in the given model (Wartofsky, 1979). In other words, we are what we operate with (what models we use).

1.2 Science, Scientific Knowledge, Systematisation of Scientific Models

The entire human knowledge is a model, but we confine our consideration only with models expressing scientific knowledge. Science is a sphere of human activity whose function is the development and systematisation of objective knowledge about the reality. It uses specific methods and relies upon definite ethical norms. According to materialistic point of view, the term "objective" means such contents of human representations that "do not depend on a subject, either on a single person or mankind". Whether it is possible in principle is the question for philosophical discussions. Let us mention just two more concrete criteria of scientific knowledge (of its objectivity and validity).

The first one is the *principle of verification*: confirmation of ideas by experiment. This approach is regarded as insufficient by the school of the famous philosopher Karl Popper (1959). Their reasoning is that it is always possible to find confirming examples. Therefore, it is necessary to use also the opposite approach, i.e. the *principle of falsification*. From this point of view, a scientific system of knowledge must allow experimental refutation. In other words, scientific theories must allow "risky" experiments. According to this principle, astronomy is a science, because it is possible to carry out experiments capable of giving negative (i.e. inconsistent with theoretical predictions) results and refuse some theoretical statements, while astrology is not a science. However, Popper's ideas are also criticised by other philosophers who state that "the principle of falsification itself does not stand the test for falsifiability". Even this polemics demonstrates complexity of the problem since it relates to one of the eternal philosophical questions, to the problem of the objectivity of the truth. In what follows, we consider only examples of modelling of the objects studied by classical physics, biology and other fields of knowledge traditionally accepted as scientific disciplines. We avoid situations whose modelling requires preliminary philosophical discussions.

A characteristic of science is the intention to get maximally generalised, impersonal knowledge and to use corresponding methods such as measurement, mathematical discourse and calculations. In some sense, the most impersonal one is the knowledge expressed with the language of mathematics which is based on the most formalised notion, a "number". The expression $1 + 1 = 2$ in decimal system is understood in the same way by both a scientist studying oscillations of a pendulum and a fruit trader on a market. There are even extreme statements of some scientists that there is as much science in some activity as much mathematics is involved in it (Feynman, 1965).

Contrary to science, art is characterised by an individual, personal perception of the reality. Hence, the same object, e.g. a picture, seems gloomy to one group of people, light and refined to another group and just uninteresting to the third one. However, there is no sharp boundary between science and art. Thus, in scientific constructions (structure of a theory, mathematical formulas, a scheme, an idea of an experiment) an essential role is often played by an aesthetical element that was specially noted by many outstanding scientists. For example, Einstein's criterion of internal perfection (Sect. 1.7) is close to an aesthetical one. Science also goes to the territories previously belonging exclusively to art. Thus, contemporary nonlinear dynamics approaches to quantitative characterisation of complex objects with such special measures as fractal dimensions are applied to musical notations and literary texts to evaluate "beauty of music" and "interest in text".

One often speaks of an art when it is impossible to give clear recommendations (an algorithm) on how to solve a problem. It implies the necessity to call a master whose intuition and natural capabilities allow to achieve a purpose. Such sense is implied in the words "art of modelling" that are often used in many books on mathematical modelling. However, time goes on, knowledge increases and, typically, something previously considered as "an art" becomes a kind of technology, a handicraft available for many people. For example, with the development of semiconductor technologies, the art of creating good transistors previously available only to selected masters has become a routine work for industrial robots.

A very important feature of the scientific approach is the systematisation[1] of knowledge. Among many approaches to *systematisation of scientific models*, we note four kinds based on their degree of generality, their law of functioning, reasons for the transfer of the modelling results to an original and their "origin". The latter case is considered in the next chapter.

In respect of the degree of generality, models differ from each other by the width of the covered range of real-world objects and phenomena. According to such a classification, models can be ordered in the form of a pyramid (Neuymin, 1984). Its vertex is occupied by the scientific picture of world. The level below includes physical, biological and other partial pictures of world. Lower, one finds theories of highest level of generality including the theory of relativity, the quantum theory, the theory of solids and the theory of continuous media. One more level below is for theories of medium level of generality including, e.g., thermodynamics, theory of elasticity, theory of oscillations and theory of stability. The lowest levels of the pyramid are occupied by partial theories such as theory of heat engines, theory of resistance of materials, automatic control theory and scientific laws such as Newton's laws, Kirchhoff's laws and Coulomb's law. The base of the pyramid consists of specific models of objects and phenomena (including technical processes), continuous and discrete models of evolution processes, etc. The higher the place occupied by a model in the pyramid, the wider the range of objects described by it. However, each level is relevant for certain class of problems. Thus, knowledge of the quantum theory does not guarantee making a good laser since the latter requires specific models[2] as well.

In respect of the law of functioning, models are divided into two classes: *logical* (or ideal (Glinsky et al., 1965)) and *material* ones. The former models function according to the laws of logic in human consciousness, while the latter ones "live" according to the laws of nature.[3] In their turn, logical models are divided into *iconic*, *symbolic* and mixed ones. Iconic models express properties of an original with the

---

[1] Along with the concept of systematisation, one uses the concept of classification. The latter is stricter since it implies that there are strict boundaries distinguishing different classes. The base for both concepts is some set of properties. For example, buttons can be classified based on their colour, shape, number of holes, way of attaching to clothes, etc.

[2] There is an opinion that less general meaning of the term "model" is more reasonable. Namely, it is suggested to call "model" only such things that are not covered by the terms "theory", "hypotheses", "formalism".

[3] Does logical model reflect the "rules" of nature? Based on that the species "*Homo sapiens*" has successfully competed with other biological species depleted of mind and logic, extended over all continents and reached oceanic depths and cosmos, one can believe that *H. sapiens* correctly

aid of vivid elements having their prototypes in the material world. Based on the definition of "clear" as "habitual and perceptible" (Mandel'shtam 1955), one can say that iconic models are the "clearest" ones to people.[4] Thus, particles are vividly modelled with elastic balls in the kinetic theory of gases. To vividly model an electro-capacity (an ability of a charged body to produce an electric field: $C = q/\varphi$, where q is the charge, φ is the potential of a charged body), it is convenient to use a bucket with water as a prototype since everyone has a practical experience with it. Then, the level of water in a vessel h is used as an analogue of the potential φ while the volume of water V_w serves as an analogue of the charge q. An analogue of electro-capacity is the quantity $C_w = V_w/h = S$ equal to the cross-sectional area of the vessel, rather than its volume as common sense would interpret the word "capacity" at the first glance. The value of C_w rises with the width of the bottom of a cylindrical vessel (a bucket). Maximum amount of liquid that can be stored in a vessel is limited only by a maximum pressure which can be withstood by the walls of a vessel. Analogously, a maximum charge of a capacitor is determined by the breakdown voltage for the surrounding dielectric. $C_w = $ const for a constant cross-sectional area of a vessel. If the cross-sectional area changes with height $S = S(h)$, then C_w is a function of the water level h. A similar property is exhibited by electro-capacity of a varactor, which is a semiconductor capacitor serving us, for example, to switch between TV channels. Electro-capacity of this popular element is a function of the voltage applied to it. In such a way, a vivid iconic model in the form of a vessel filled with a liquid can be used to form concepts of nonlinearity (Bezruchko et al., 1999b).

Symbolic models express properties of an original with the use of conventional signs and symbols. In particular, this class of models includes mathematical expressions and equations, physical and chemical formulas. Mixed (iconic – symbolic) models refer to schemes, diagrams, layouts, graphs, etc.

In their turn, material models can be *physical*, if they consist of the same material substance as an original, or *formal*. Further, models are divided into the following subgroups: *functional* models reflect functions of an original (paper "jackdaw" as a model of an aeroplane); *geometrical* models reproduce geometrical properties of an original (table copy of an aeroplane); *functional–geometrical* models combine both abilities (e.g. a flying model reproducing simultaneously the shape of an aeroplane). As well, one can encounter a subdivision into functional and structural models (Glinsky et al., 1965) and other classifications (Horafas, 1967; Myshkis, 1994).

In respect of *reasons for the transfer of the modelling results to an original*, models are divided into the following groups:

(1) *Conventional* models express properties of an original based on a convention, agreement about the meaning ascribed to the model elements. Thus, all

---

assesses the rules of evolution and interrelations among natural objects. Hence, logical models make objective sense.

[4] This statement can be attributed more readily to people whose "left" hemisphere of the brain is developed better, i.e. perception dominates over logic.

symbolic models, including mathematical ones, are conventional. For example, R. May suggested a one-dimensional map $x_{n+1} = x_n \exp(r(1 - x_n))$ in 1976 as a model for a population rise controlled by an epidemic disease. Here, the quantity x_n stands for the number of individuals at the nth time instant and the parameter r relates to the conditions of infection by convention underlying the model.

(2) *Analogous* models exhibit similarity to an original sufficient to transfer the modelling results based on the reasoning "by analogy". Thus, if an object O_1 possesses the properties $c_1, c_2, \ldots, c_{N-1}, c_N$ and an object O_2 possesses the properties $c_1, c_2, \ldots, c_{N-1}$, then one can assume that the second object also possesses the property c_N. Such a reasoning is hypothetical. It may lead to both true and false results. To mention an example of a failure, the analogy between the motion of a liquid (O_1) and the process of heat conduction (O_2) led in its time to the incorrect conclusion about the existence of "caloric fluid". A positive example is a successful replacement of human organism with animal organisms for studying the effect of drugs.

(3) (Strictly) *similar* models allow rigorous recalculation of model characteristics into characteristics of an original (Barenblatt, 1982). Here, one speaks of full mathematical analogy, proportionality between the corresponding variables in the entire range of their values. Two objects are similar if the following two conditions are fulfilled:

(a) The objects can be mathematically described in the same form. For example among the expressions $z = x \cos y$, $u = 2v \cos 3w$ and $p = \varphi s \cos (2p - 1)$, the first two expressions take the same form, while the third one differs from them. Of the same form are the equations of motion for small-amplitude oscillations of a spring pendulum $d^2x/dt^2 + (k/m)x = 0$ and a mathematical pendulum $d^2x/dt^2 + (g/l)x = 0$.

(b) The corresponding variables entering the mathematical expressions are linearly related with a constant coefficient of proportionality (constant of similarity). For example the formula $x^2 + y^2 = R^2$ for different R defines circles which are similar to each other (concentric).

1.3 Delusion and Intuition: Rescue via Mathematics

People perceive information about an observed object coming from the organs of sense rather than an object directly, i.e. they get the picture of their relationships to the reality rather than the picture of the reality itself. Iconic models are formed on the basis of sensory perception. However, speaking of iconic models, one should not identify them completely with images produced by the senses of a person unburdened with scientific knowledge, e.g. images of archaic consciousness. Images can be formed during an educational process (in a family, school, university, company) or practical experience (scientific activity, production process). Correspondence between images and the real world is to be checked taking into account sensory perception errors, possible teacher's mistakes, false ideas entering scientific knowledge at the current historical stage, etc.

To what extent one can rely on the organs of sense is demonstrated by the following examples borrowed from the book of M. Kline (1985):

(1) Two equal segments in Fig. 1.2a seem to be of different length; the lower segment seems shorter than the upper one.
(2) Equal segments in Fig. 1.2b visually "change" the length for different orientations of arrows at their vertices.
(3) The way of shading of the two parallel lines (Fig. 1.2c) affects their seeming slope.
(4) When temperature of water is determined by a hand, the result depends on where the hand has been before, namely in warm or cold conditions.
(5) Receptors on a tongue get tired and adapt in such a way that gustatory sense depends on the prehistory as well, something sweet seems less sweet after some time.
(6) Perception of the motion speed by a car driver becomes blunt after the car has gathered the speed and maintained it for some time.

Above-mentioned and many other examples demonstrate unreliability of information obtained from the organs of sense and "sensory intuition", an intuition based on previous experience, sensory perception and rough guesses. Intuition relying on scientific knowledge and, first of all, on mathematics differs qualitatively. For example, scientific analysis of motion on the basis of the concepts of dynamics allows to answer more correctly such questions as (i) How should a gun be directed if a target starts to fall down at the instant of a shot? (ii) Where does a key dropped out of a hand during a walk fall down? Sensory intuition prompts to bend the gun barrel down and to find a key somewhere behind. The scientific analysis says that (i) the barrel should be directed to the starting position of the target and (ii) a key falls down near a leg as if a person would stand. Experience in scientific analysis of different facts changes one's intuition. Being based on scientific knowledge including mathematical approaches, rather than on sensory perception, it becomes a tool helping to move forward, to search for new knowledge.

Inconceivable efficiency of mathematics deserves a special discussion. Mathematics is a science studying quantitative relationships and space forms of the real world. It appeared as *a set of useful rules and formulas* for the solution of practical tasks encountered by people in their everyday life. Already civilisations of Ancient

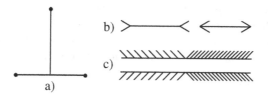

Fig. 1.2 Examples of optical illusion (Kline, 1985)

Egypt and Babylon started to create it about the third millennium BC (Kline, 1985; Poizner, 2000). However, only in the sixth century BC, ancient Greeks caught an opportunity to use mathematics as a tool to get new knowledge. In particular, scientific practice knows a lot of cases where a result was predicted "on the tip of a pen" and after that specially organised experiments managed to find something new, previously unknown to human beings. Thus, unknown planets and their satellites were revealed from the calculations of celestial body trajectories, contortion of a light beam when it passes near a big mass was predicted on the tip of a pen, etc.

No reliable documents remained which would be capable of telling what caused the Greeks to come to a new understanding of mathematics and its role. There exist only more or less plausible guesses of historians. According to one of them, the Greeks detected contradictions in the results concerning the determination of the area of a circle obtained in Babylon and started to investigate which of them is true. According to another one, the new deductive mathematics originates from Aristotelian logic arisen during hot discussions on the social and political topics. Seemingly, mathematics as *a logical deduction and a cognitive tool* emerged in connection with a new world view formed in the sixth century: the nature is made rationally, its phenomena proceed according to an exact plan which is mathematical in essence, human mind is omnipotent so that the plan can be cognised. Reasons for such an optimism were based, e.g., on the realisation of similarity between the geometric forms of the Moon, a ball, etc.; discovery of the dependence of the pitch of tone produced by a string on the length of the latter and that harmonic accords are produced by strings whose lengths relate to each other as some integers.[5] As a result of such observations, two fundamental statements were formulated: (i) the nature is made according to mathematical principles and (ii) numerical relationships are the basis, general essence, and a tool to cognise the order in nature.

Centuries passed, the Greeks' civilisation died under the rush of conquerors, but mathematics remained. Chiefs and peoples appeared at the historical scene and left it, but mathematics developed together with the mankind and the views on its role and importance for the human society changed. It progressed the most strongly during the last centuries and took its especial place in science among the tools for cognition of the world. Finally, a mathematical method has been formed which possesses the following characteristic features:

(1) *Definition of basic concepts*, some of them being suggested directly by the real world (a point, a line, an integer, etc.) while the others are produced by a human mind (functions, equations, matrices, etc.). Interestingly, a part of concepts is depleted of a direct intuitive basis (of an analogue in the nature), e.g. negative numbers. Such concepts were accepted by the scientific community with difficulty, only after convincing demonstration of their usefulness.

(2) *Abstractness*. Mathematical concepts capture essential features of various objects by abstracting from the concrete nature of the latter. Thus, a right line

---

[5] In particular, when two equally strong stretched strings, one of them being twice as long as the other one, oscillate, the interval between their tones is equal to an octave.

reflects a property of all stretched strings, ropes, edges of rulers and trajectories of light beams in a room.

(3) *Idealisation.* Speaking of a line, a mathematician abstracts from the thickness of a chalk line. He/she considers the Earth as an ideal sphere, etc.

(4) *The way of reasoning.* It is the most essential peculiarity which relies mainly upon formulation of axioms (true statements which do not need to be proven) and deductive way of proving (based on several rules of logic) allowing one to get conclusions which are as reliable as original premises.

(5) *The use of special symbols.*

There are many mathematical systems. A system including the smallest number of axioms is regarded as the most perfect of them. Such mathematical "games" appear very useful and lead to findings which allow better understanding of the real world. Mathematics is especially useful when basic laws are already established and the details of complex phenomena are of interest. For example when compared to chess, the laws are the rules of the game governing movements of the pieces, while mathematics manifests itself in the analysis of variants. In chess, one can formulate the laws in Russian, English, etc., while in physics one needs mathematics for such a formulation.

R. Feynman notes that it is impossible to explain honestly all the laws of nature in such a way that people could perceive them only with senses, without deep understanding of mathematics. However regrettable it is, but this is seemingly a fact (Feynman, 1965). He sees the reason in that mathematics is not simply a language but a language plus discourse, a language plus logic. He claims that guessing the equations is seemingly a very good way to discover new laws.

What is the cause of the exceptional efficiency of mathematics? Why is such an excellent agreement between mathematics and real-world objects and phenomena possible, if the former is, in essence, a product of human thought? Can a human mind understand properties of the real-world things without any experiments, just via discourse? Does the nature accord with the human logic? If a range of phenomena are well understood and the corresponding axioms are formulated, why do dozens of our corollaries appear as applicable to the real world as the axioms? These questions are in the "list" of eternal questions of the philosophy of science. All thinkers who tried to address them can be divided into two groups according to their answers (Kline, 1985), though there were many of them since ancient times till now. The first group believes that mathematicians select axioms so as to provide the agreement between their corollaries and experiments, i.e. mathematics is adjusted to the nature. In other words, general and necessary laws do not belong to the nature itself but to the mind who inserts them into the nature, i.e. the scientific truth is created rather than discovered. The second group thinks that the world is based on mathematical principles, in a religious variant the Creator has made the world based on mathematical principles. There are many great names in both groups since the questions mentioned cannot be avoided in the investigations of the nature. As well, it is natural that the discussion is not completed; the questions discussed are non-random in the list of eternal problems of epistemology (Poincare, 1982).

1.4 How Many Models for a Single Object Can Exist?

The above-mentioned questions about the role of mathematics, its exceptional efficiency and the status of mathematical statements lead to a new question about the number of possible models for a single object. If a real-world object possesses an infinite set of properties, then one could speak of infinitely many models taking into account that any model typically describes only a finite number of selected characteristics of an object. Yet, if the world is made according to mathematical laws, one could expect the existence of the best, "true" model. However, from the viewpoint that the strength of mathematics is determined by the efforts of the human mind, there are no reasons to count on the existence of a "true model". It follows from such considerations that a simple answer to the question formulated in the title of this section is lacking.

In respect of the considered question, an important place in epistemology is occupied by the statements of N. Bohr known as the "principle of complementarity". He wrote that difficulties encountered in adjustment of our notions borrowed from sensations to gradually deepening knowledge of the laws of nature originate mainly from the fact that an every word in a language relates to our common notions. He hoped that the idea of complementarity can characterise the existing situation which has a far-reaching analogy with general difficulties in creation of human concepts coming from the division into a subject and an object (Bohr, 1971). He thought that it is impossible in principle to create a theoretical model, which would be useful in practice, without empirical elements. Thus, according to the principle of uncertainty, it is impossible in micro-world to specify precisely the location of a particle and its momentum: the variables in this pair complement each other.[6] If one wants to know coordinates of a particle precisely, then he/she loses precise value of its velocity. According to Bohr's ideas, similar relationships exist between the accuracy of a model and its clarity,[7] possibility of its practical usage. He stated that our ability to analyse harmony of the surrounding world and the breadth of our perception will always be in a mutually exclusive, complementary relationship (Bohr, 1971).

Description of the main statements of the principle of complementarity adapted to the problem of mathematical modelling is given in the introduction to the monograph (Gribov et al., 1997) devoted to modelling of complex molecules. Briefly, it is as follows. A molecule as a single stable and electroneutral system of atomic nuclei and electrons can be adequately described with an equation of state in the form $(T_e + T_n + V)\Psi = E\Psi$, where T_e and T_n are electron and nuclei kinetic operators, respectively, V is an operator of all kinds of interactions between electrons and

---

[6] The variables are complementary if each of them can be specified more precisely at the expense of lower certainty in the value of the other one.

[7] It is appropriate to recall here the words of L.I. Mandel'shtam about the concept of "clear" as something habitual and perceptible.

nuclei, Ψ is a wave function. However, even if one manages to solve this equation, the results cannot be compared to a spectral experiment due to the existence of a large number of isomers and overlaps between their spectra. Selection of a single isomer is envisaged neither in the original definition of the object nor in the form of the equation of state. Specification of only the number of electrons and nuclei with their masses provides the truth, but the clarity is completely lost. Clarity can be provided if one uses representations of molecules such as hard and elastic spatial figures, ideas about charges on atoms, potential surfaces, etc. Such "clear" models are mapped to the class of measured quantities. But the price for their clarity is the truth. A model allowing comparison to measurements can provide only reasonably satisfactory coincidence with the experiment. A good agreement between relatively simple models and an experiment can be achieved only via fitting parameters, i.e. on a semi-empirical basis. The requirement of clarity leads to the necessity of using different models. "Even though it is no longer possible to give a single definition to a molecule as an object under investigation, one can however answer quite clearly the question what 'to investigate a molecule' means. It means to construct a sufficient number of molecular models agreeing with experimental observations and staying in mutually complementary (and often mutually exclusive) relationships and to specify numerical values of the corresponding parameters. The greater the number of different models obtained as a result, the greater the amount of knowledge and clearer the notion about the object under investigation."

The authors of Gribov et al. (1997) note that each of the obtained empirical models (they usually rely on spectral measurements and diffraction pictures for molecules) is "complementary" to the truth and the results obtained with different models cannot be averaged as, for example in multiple weighting of a body. The approach itself and the method of measurements do not meet the requirements justifying the use of averaged characteristics. Averaging of the results coming from experiments, which differ from each other in essence, may not be considered as approaching the truth. Numerical values of the parameters obtained in such a way may not be considered as more precise.

We are close to the views of N. Bohr and the authors of the cited monograph. However, for the sake of objectivity and completeness, we should mention the existence of alternatives to the concept of complementarity. The above ideas are not supported by all scientists thinking of the epistemological problems. There are many famous names among their opponents. The most famous one is, probably, A. Einstein, who believed that the main purpose of physics is a complete and objective description of the real world, independent of the act of observation and the existence of an observer (Einstein, 1976). Therefore, he could not admit a theoretical description if it depended on the observation conditions as required by the conception of complementarity.

Here we finish the discussion of the very important and interesting question remaining in the sphere of professional interests for philosophers of science and scientists working in different fields of research including natural and humanitarian sciences. Let us now consider the ways of mathematical model construction and corresponding historical experience.

1.5 How the Models are Born

We have already noted in the previous sections that the process of modelling cannot be fully reduced to an algorithm and has currently something in common with art. Nevertheless, it is possible to distinguish basic (the most typical and fruitful) ways of model creation and even try to use them as an additional sign for systematisation of the existing variety of models. Developing the systematisation presented in a popular work of N.N. Moiseev (1979), we can single out four ways of model creation:

(1) Intuitive way is based on a guess.
(2) Construction of a model as a result of direct observation of a phenomenon and its study. Models obtained directly from experiments or with the use of experimental data are often called *phenomenological* or *empirical models*. Phenomenological models include those obtained via reconstruction from time series described in Part II. The models of Newton's mechanics are phenomenological in respect of their historical origin.
(3) Construction of a model as a result of deduction process when a new model is obtained as a partial case of a more general model, in particular, from the laws of nature (the *first principles*). Such models are called *asymptotic models* in Moiseev (1979). For example, after creation of special theory of relativity, Newton's mechanics appeared to be its limit for $c \to \infty$. Thus, accumulation of knowledge leads to the conversion of phenomenological models into asymptotic ones. The number of asymptotic models reflects to some extent the maturity of a science (Moiseev, 1979, 1981).
(4) Construction of a model as a result of an induction process when a new model is a generalisation of "elementary" models. An example is *ensemble models* (Moiseev, 1979) that allow to describe behaviour of a system of objects based on the information about the behaviour of the elements (subsystems) and strengths of their interaction. Naturally, a qualitatively new kind of behaviour can be observed as a result of integration of elementary models into an ensemble. Popular ensemble models are sets of coupled oscillators and coupled maps lattices. An example of a new quality is the arousal of oscillatory regimes in a population described with the Lotka–Volterra model after joining "predators" and "preys" up into an ensemble (Volterra, 1976).

Drawing an analogy between scientific models and models of clothes, one can note the following parallels in the processes of their making. Obtaining intuitive models can be compared to the work of a couturier who realises his ideal images in a picture. Creation of an asymptotic model can be compared to making a business suit via simplification of a template produced by an outstanding couturier. Construction of an ensemble model reminds gathering a suit from a white jacket and black trousers. Empirical modelling corresponds to wrapping a client body into a piece of material and lacing the latter along the edges. Common experience makes one doubt in the elegance of such "phenomenological" clothes but its mathematical analogues appear quite efficient due to the development of computers and special techniques

(Part II). Moreover, the empirical way turns out to be the only available one in many cases when processes in such complex systems as astrophysical, biological and others are considered.

1.6 Structural Scheme of Mathematical Modelling Procedure

Variety of situations, objects and purposes lead to infinite multitude of specific modelling problem settings and ways of their solution. Nevertheless, one can single out something common and necessary, the stages which are realised to a certain extent in the construction of almost any mathematical model. They are presented in Fig. 1.3 as blocks which are restricted with more or less straight lines. The straighter lines are used for a stage which can be more readily reduced to an algorithm. The entire procedure of modelling is not typically a direct road to a purpose but rather represents multiple returns to already passed stages, repetitions and corrections, i.e. a step-by-step approach to a satisfactory result. In general, it starts with the assessment of a real situation from the viewpoint of an existing prior model and a purpose (stage 1). As a result, a *conceptual model* reflecting the problem setting is formed (stage 2). The conceptual model is formulated in terms relevant for the problem considered: mechanical, physical, biological, sociological, etc. Then, one specifies the structure of a model, i.e. the most appropriate mathematical tool, kind and number of equations, kind of function entering the equations (stage 3). At the next stage (number 4), one concretises the details of a model if necessary, i.e. introduces additional approximations and estimates parameters in the equations. Finally, at the stage 5, one validates the obtained model using criteria dictated by the purposes of modelling. If the model is unsatisfactory, the procedure is repeated from the beginning or from some intermediate stage and a new model is produced.

Opportunities of a modeller differ for laboratory systems (e.g. electronic circuits and lasers), where active experiments can be performed, and real-world processes (e.g. temporal variations in climate characteristics and physiological processes in living organism), where only the data of passive observations and measurements are available. In the former case, one can get much more information about an object, e.g., due to purposeful variation of its parameters and study of different dynamical regimes. Therefore, the return to the stage 1 in Fig. 1.3 is much more reasonable and more often used in modelling of such systems. In the latter case, getting a model adequately describing an object within a wide range of parameter values is much more difficult so that one formulates model-based conclusions with less confidence

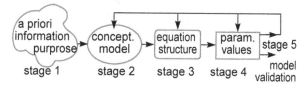

Fig. 1.3 Structural scheme for a typical procedure of mathematical modelling

than for the laboratory systems. This difference is illustrated with several practical examples in Chap. 13.

Let us illustrate the scheme of Fig. 1.3 with the example of modelling a spring pendulum (a load on a spring). This object can be considered, e.g., as a model of a suspension in a car with spring dampers, of an atom in a crystal lattice, of a gene in DNA and of many other systems whose inertial elements are influenced by a restoring force when they deviate from an equilibrium state. In some situations and devices the spring pendulum is of interest by itself and becomes an object of modelling, e.g. as an element of a car suspension. At the stage 1, one easily chooses both the language of description (physics) and the way of model construction (an asymptotic model from Newton's laws with the account of specific properties of the object) since the nature and mechanisms of the object behaviour are well known. If the purpose of modelling is the quantitative description of the pendulum deviations from an equilibrium state, then we can mention two popular problem settings (the stage 2):

> To describe free motion of the object, which can be regarded repetitious, a conceptual model is typically a load of some mass m exhibiting friction-free motion under the influence of an elastic force arising from the deformation of the spring (Fig. 1.4a).
> If damping is essential, then a viscous friction force is incorporated into a conceptual model (it is shown symbolically with a damper in Fig. 1.4b).

The second problem setting is more realistic but even in that case one does not take into account, e.g., peculiarities of the full stop of the load that would require the use of the third, more complicated, conceptual model involving a dry friction force, etc.

The stage 3 is quite simple for the above settings. It reduces to writing the second law of Newton, i.e. second-order ordinary differential equations $\mathbf{F} = m\mathrm{d}^2\mathbf{r}/\mathrm{d}t^2$, where $\mathbf{F}$ is the resultant of forces, $\mathbf{r}$ is radius vector of the centre of mass, $\mathrm{d}^2\mathbf{r}/\mathrm{d}t^2$ is acceleration, t is time. For both conceptual models, the original equation ("the first principle") is the same. However, after one takes into account the above-mentioned assumptions, different models arise (the stage 4). If an elastic force is proportional to the value of the spring deformation ($F_{\mathrm{el}} = -kx$ where the coefficient of elasticity k is constant), then one gets a conservative linear oscillator equation $\mathrm{d}^2x/\mathrm{d}t^2 + (k/m)x = 0$ for the case of Fig. 1.4a. If the friction force is proportional to the

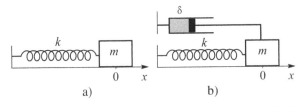

Fig. 1.4 Conceptual models of a load on a spring both with (**a**) and without (**b**) friction

velocity $F_{\text{fr}} = -2\delta \cdot dx/dt$, then a dissipative linear oscillator equation $d^2x/dt^2 + 2\delta \cdot dx/dt + (k/m)x = 0$ is a model for the case of Fig. 1.4b. Different models are obtained for other definitions of the functions entering the equation. Thus, if the elasticity of the spring depends on its deformation ($k = k_1 + k_2x$) or the coefficient of friction depends on the velocity ($\delta = \delta_1 - \delta_2 dx/dt$), the oscillator equation becomes nonlinear and gets essentially richer in its properties.

Differential equations are not always the most appropriate tool to describe motions. Thus, if the purpose of modelling would be a qualitative description of decaying oscillations without nuances of their waveform, one could use a difference equation $x_{n+1} = f(x_n)$, where n is the discrete time, rather than a differential one. A model discrete map can be obtained if one assumes, e.g., exponential decay and expresses the next local maximum value via the previous one as $x_{n+1} = ax_n$, where $a = e^{-\delta T}$, T is the "quasi-period" of the oscillations (Fig. 3.2c). If one constructs a model as an explicit function of time $x = f(t)$, the description reduces to the consideration of the time course of x, deviation of the load from an equilibrium state (cf. Fig. 3.2b). The corresponding model takes the form $x = c\, e^{-\alpha t} \cos(\beta t)$.

At the final stage 5, criteria of model quality are chosen according to the purpose of modelling. It can be qualitative coincidence of oscillatory regimes of a model and an object, accuracy of prediction of the future states based on the current one, etc. If the result of checking is unsatisfactory, then the problem setting is corrected and the entire procedure is repeated again or one just returns to an intermediate stage. Usually, a model "evolves" from a simple version to a more complex one, but movement in the opposite direction is also possible. Construction of a model is finished when it describes a phenomenon accurately enough within the necessary range of parameter values or fulfils other modelling purposes.

1.7 Conclusions from Historical Practice of Modelling: Indicative Destiny of Mechanics Models

Analysis of the history of science allows to formulate (Krasnoshchekov and Petrov, 1983) some basic principles, "solid residual" of the historical practice in modelling in the form of several theses.

Thesis 1. Usually, a stimulus to the creation of a new model is *a few* basic facts. Amount of experimental data is seemingly not of principal importance. Moreover, experimental material by itself is insufficient to create a fruitful theory (a model), no matter how good the former is.

There is an opinion that theoretical constructions can appear only on the basis of a reliable experimental material lighting an object under investigation in detail. However, the history of natural science is full of counterexamples. For example the general theory of relativity resulted from the generalisation of the fact that inertial and gravitational masses are identical, while experimental confirmation came later via specially organised experiments. Another edifying example is the story with establishing the law of gravity which in fact relied on the only result – the third law of Kepler. The rest of the huge body of experimental material obtained by this scientist and other astronomers (in particular, Tycho Brage) played its role mainly for the practical validation of the law.

Thesis 2. A theory relies basically on experimental data. Similarly, an experiment carries useful information if it is performed in the framework of some theoretical conception. An experiment as a simple collection of observed facts with an incorrect conception (or without conception at all) can mislead a researcher. There are many confirmatory examples in the history of science:

(1) Direct observation of the heavenly bodies led to the geocentric theory.
(2) Copernicus proclaiming the heliocentric system was accused by his contemporaries in that the theory did not agree with an experiment. But he resisted "believing his mind more than his senses".
(3) Direct observations of motions led Aristotle to the formulation of mechanics which reigned over the minds during 2000 years but finally appeared incorrect.

The entire history of natural sciences development relates to the settlement of contradiction between a direct experiment as a collection of observed facts and a multitude of formal logical schemes (models) appealed to explain them. Thus, Einstein came to dualistic criterion for the "correctness" of a theory and formulated it as in the following thesis.

Thesis 3. Agreement between a theory and an experiment (*criterion of external justification*) is a necessary but insufficient condition for its "correctness". The second criterion (*criterion of internal perfection*) is "naturalness" and "logical simplicity" of premises.

Thesis 4. When constructing a model, one should use the existing experience in mathematical modelling from such sciences as physics, mechanics and astronomy.

Thesis 5. The choice of mathematical tools is a key point in modelling. In some sense, mathematical modelling represents a search for mathematical constructions whose abstract notions are the most suitable to be "filled in" with a concrete matter of the investigated reality. If there are no such available mathematical tools, one creates new ones. To mention a few examples, (i) mathematical analysis was born from the requirements to describe mechanical motions; (ii) difference equations have been adapted and used for a long time to describe population dynamics.

Thus, in modelling a researcher

- should not be confused with the absence of complete experimental material since "intellectual" material is lacking more often;
- needs a prior conception which turns a collection of experimental facts into "objective" information. Complete absence of such a conception prejudices even the possibility of modelling;
- should possess appropriate mathematical tools and techniques, which can be often borrowed from the previous scientific experience.

An indicative example showing validity of the above ideas is a history of the development of the mechanics laws. Mechanics appeared the first system of models adequately describing a wide range of real-world processes and phenomena. From a long list of the creators of mechanics, one commonly singles out the names of Aristotle (384–322 BC), Galilei (1564–1642) and Einstein (1879–1955). Periods of their activity are separated with centuries and even millennia, i.e. intervals of time which can be considered as big but still finite life times for model conceptions inevitably replacing and complementing each other.

Based on the knowledge and experiment existing in his time, Aristotle classified surrounding bodies and their motions as follows. He divided all bodies into three types: (i) always motionless; (ii) always moving; (iii) able to be both moving and motionless. Motion itself was regarded eternal. He distinguished two kinds of motions: (i) "natural" motions when a body moves to its natural place; (ii) "forced" motions whose cause is a force which permanently supports them. If the force disappears, the motion stops too.

Do Aristotle's ideas agree with an experiment? If one means purely contemplative experience of that time, the answer is "Yes"! Indeed, the Earth and houses on it are always steady; the Sun, the Moon, and river water always move; a thrown stone, a cart, people moves to a steady state and finally stop if a driving effort disappears. Load thrown from some height and water in a river move "naturally". A man exerts physical strength to move a cart. When he leaves it, the "forced" motion ends and the cart stops. But why a stone still moves after being thrown if a force has stopped acting? It finds an explanation as well: air flows the stone and pushes it from behind after a hand has stopped acting. Ether can do it as well: Aristotle thought that there is no emptiness.

Let us use historical experience in modelling formulated in the thesis 3 where we mentioned the criteria of "external justification" and "internal perfection" intended for assessing the quality of theories. Let us consider "internal perfection" of Aristotelian mechanics following the monograph (Krasnoshchekov and Petrov, 1983), where the authors applied the language of contemporary mathematics to set forth the ancient ideas. Let us express the main law of motion following from those observations and prior conceptions in the form $F = pv$, where p is the coefficient of proportionality (the notion of mass did not exist at that times), F is the force and v is the velocity. Contradiction inherent in this law is demonstrated by the following example. A man pulls a cart along a river bank by making certain efforts (Fig. 1.5). The cart performs forced motion along the bank with a velocity v relative to an observer standing on land (i.e. relative to a motionless reference system $x–y$). The cart does not move relative to an observer going on a boat with a parallel course and the same velocity, i.e. no force is applied to the cart. It definitely contradicts an intuitive notion about a force as some objective (independent of an observer) influence on the cart. Even the sweat on the man's forehead confirms such a notion. A way out of the paradoxical situation was found via claiming the reference system connected to the Earth as an absolute one. Thus, the above law of motion is valid only in that system. For all other systems moving relatively to the absolute system with a constant velocity v', one suggested a new rule $F = pv' + pv$. Strikingly, one found experimental confirmations of the Aristotelian approach for almost 2000 years. In part, such a longevity was provided by the support from the Church relying on the geocentric system in its dogmas. However, science and social consciousness

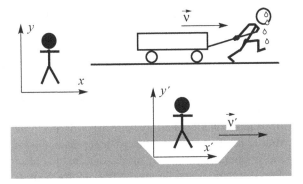

Fig. 1.5 Mental experiment with motions in different reference systems

once became ripe to perceive new ideas. Despite the risk to find himself in a fire of inquisition, Galilei was one of the first persons who encroached on the absolute. "Retire with some of your friend in a spacious room under the hatches of a ship, stock up flies and butterflies. Let you have also a vessel with a small fish swimming in it. Further, hang up somewhere above a bucket from which water will drip drop-by-drop into another vessel with a narrow neck located below and observe. Jump in the direction of the ship motion and vice versa. For $v = 0$ and in a moving ship, you will not be able to detect whether the ship is moving or standing from any of the mentioned phenomena." We do not know whether Galilei performed the above experiments or they were mental. Accuracy of the observation tools and experimental base of those times were significantly worse than modern fast movie cameras, laser interferometers, magnetic pendants and laboratory tables on an air cushion. Tools that were used and described by him could hardly provide convincing accuracy but there was confidence in the result of the experiment. The confidence was supported by the dissatisfaction in the geocentric conception.

Without the absolute reference system, the entire harmony of the Aristotelian world collapsed and an opportunity for a new conception to strengthen itself arose. Laws of motion had to be expressed in a form invariant under the transition from "a bank" to "a ship" and back. If a conception exists, then one can purposefully perform experiments to check it and to reject inappropriate variants. Finally, a new model was born: the law of motion $F = p\mathrm{d}v/\mathrm{d}t$ according to which the force applied by a man to a cart is the same in all reference systems moving evenly and rectilinearly relatively to each other. Acceptance of this idea takes away all contradictions in the considered common experiments and leads to a number of new discoveries: a cart will stop if it is left since it interacts with a land (friction force), ether is not needed to explain the motion of a thrown stone, etc. A concept of inertial reference systems and the Galilei transform arose. Naturalness and logical simplicity of the theory of motion based on the law $F = p\mathrm{d}v/\mathrm{d}t$ do not make one introduce additional hypotheses and, consequently, possess higher "internal perfection".

Three centuries passed and classical mechanics of Galilei and Newton faced insuperable difficulties. A danger came at the turn of nineteenth and twentieth centuries from electrodynamics. Maxwell's theory which generalised basic empirical notions of electromagnetism and allowed to determine the velocity of light appeared non-invariant under the Galilei transform. It caused theoreticians to suspect either the achievements of electromagnetism or postulates of classical mechanics of incorrectness. Scientific sagacity of A. Einstein and his contemporaries allowed to solve the dispute in favour of the theory of electromagnetism; the Galilei transform was declared inappropriate for the world of high velocities. Special theory of relativity became an even more perfect mechanical model, more adequately reflecting the reality and involving classical mechanics as a particular case.

References

Barenblatt, G.I.: Similarity, Self-similarity, Intermediate Asymptotics, Gidrometeoizdat, Leningrad (1982). Translated into English: Scaling, self-similarity, and intermediate asymptotics. Cambridge University Press, Cambridge (1996)

Bezruhcko, B.P., Kuznetsov, S.P., Pikovsky, A.S., et al.: Dynamics of quasi-periodically driven nonlinear systems close to the ending point of a torus doubling bifurcation line. Izvestiya VUZ. Applied Nonlinear Dynamics (ISSN 0869-6632). **5**(6), 3–20 (in Russian) (1997)

Bohr, N.: Selected Scientific Works, vol. 2. Nauka, Moscow (in Russian) (1971)

Einstein, A.: Collection of Scientific Works, vol. 4. Nauka, Moscow (in Russian) (1976)

Feynman, R.: The Character of Physical Law. Cox and Wyman Ltd, London (1965)

Glinsky, B.A., Gryaznov, B.S., Dynin, B.S., Nikitin Ye.P.: Modelling as a Method of Scientific Learning. Moscow State University, Moscow (in Russian) (1965)

Gribov, L.A., Baranov, V.I., Zelentsov, D.Yu.: Electronic-Vibrational Spectra of Polyatomic Molecules. Nauka, Moscow (in Russian) (1997)

Horafas, D.N.: Systems and Modelling, 419p. Mir, Moscow (in Russian) (1967)

Kline, M.: Mathematics and the Search for Knowledge. Oxford University Press, New York (1985)

Krasnoshchekov, P.S., Petrov, A.A.: Principles of Model Construction. Moscow State University, Moscow (in Russian) (1983)

Mandel'shtam L.I.: Lectures on Oscillations. Academy of Science USSR, Moscow (in Russian) (1955)

Moiseev, N.N.: Mathematics Performs Experiment. Nauka, Moscow (in Russian) (1979)

Moiseev, N.N.: Mathematical Problems in System Analysis. Nauka, Moscow (in Russian) (1981)

Myshkis, A.D.: Elements of Theory of Mathematical Models. 192p. Nauka, Moscow (in Russian) (1994)

Neuymin Ya.G.: Models in Science and Technology. Nauka, Moscow (in Russian) (1984)

Poincare, H.: Science and Method. Dover, New York (1982)

Poizner, B.N.: Big bifurcation: birth of mathematical modelling. Izvestiya VUZ. Applied Nonlinear Dynamics (ISSN 0869-6632). **8**(5), 82–96 (in Russian) (2000)

Popper, K.: The Logic of Scientific Discovery. Hutchinson London (1959)

Volterra, V.: Mathematical Theory of the Struggle for Existence. Nauka, Moscow (Russian translation) (1976)

Wartofsky, M.: Models, Representation, and the Scientific Understanding. D. Reidel Publishers, Dordrecht (1979)

Zarubin, V.S.: Mathematical Modelling in Technology. N.E. Bauman MGTU, Moscow (in Russian) (2001)

Chapter 2
Two Approaches to Modelling and Forecast

Before creation of a model, one should specify one's *intentions* in respect of its *predictive ability*. Such a choice determines which mathematical tools are appropriate. If one does not pretend to a precise and unique forecast of future states, then a *probabilistic approach* is traditionally used. Then, some quantities describing an object under investigation are *declared random*, i.e. fundamentally unpredictable, stochastic.[1] Such a "verdict" may be based on different reasoning (Sect. 2.2) but if it is accepted, one uses a body of the theory of probability and mathematical statistics. At that, to characterise dependence between a condition S and an event A, one speaks only of a probability P of A if S has occurred, i.e. of a conditional probability $P(A|S)$.

A *dynamical approach*, which is an alternative to the probabilistic one, relies on the conception of determinism. *Determinism* is a doctrine about regularity and causation of all phenomena in nature and society. Therein, one assumes that each occurrence of an event S (a cause) inevitably leads to an occurrence of an event A (a consequence). Famous French astronomer, mathematician and physicist Pierre Simon de Laplace (1749–1827) was reputed as the brightest proponent of determinism. In respect of his scientific views, he showed solidity which seemed surprising in view of his inconsistency in everyday attachments[2] (Mathematical dictionary, 1988, p. 117). It was Laplace who told Napoleon that he did not need "a hypothesis about the existence of God" in his theory of the Solar system origin. He saw an etalon of a complete system of scientific knowledge in celestial mechanics and tried to explain the entire world including physiological, psychological, and social phenomena, from the viewpoint of mechanistic determinism.

---

[1] "Stochastic" originates from a Greek word which means "capable of guessing, acute". However, it is currently used in a somewhat different sense to denote uncertainty, randomness.

[2] Several words about the picturesque personality of Laplace. Consistency of his materialistic world view stands in a sharp contrast to his political instability; he took a victor's side at each political upheaval. Initially, he was a republican. After Napoleon came to power, he became a Minister of the Interior and, then, was appointed as a member and vice-president of Senate. In the time of Napoleon, he got the title of count of the empire. He voted for dethronement of Napoleon in 1814. After restoration of Bourbons, he got peerage and a title of marquis.

B.P. Bezruchko, D.A. Smirnov, *Extracting Knowledge From Time Series*, Springer Series in Synergetics, DOI 10.1007/978-3-642-12601-7_2,
© Springer-Verlag Berlin Heidelberg 2010

Mathematical realisation of the dynamical (deterministic) approach was provided by the apparatus of infinitesimals which appeared in the seventeenth century due to the efforts of Newton and Leibniz. An arsenal of researchers got a powerful tool for the description of temporal evolution: ordinary differential equations (ODEs). A theorem about unique existence of their solution at fixed initial conditions made differential equations an etalon for deterministic mathematical description of various phenomena: "a unique future corresponds to a given present!". Currently, apart from ODEs one widely uses other mathematical tools for construction of deterministic models (Chap. 3) including difference equations, discrete maps and integro-differential equations. All those models regardless of their concrete meaning, which may be far from mechanics (dynamics), are often called *dynamical models*. In general, the term "dynamical" is currently often used to denote "deterministic" rather than "force" or "mobile".

2.1 Basic Concepts and Peculiarities of Dynamical Modelling

2.1.1 Definition of Dynamical System

The basis of deterministic description is an idea that the entire future behaviour of an object is *uniquely* determined by its state at an initial time instant. A rule determining an evolution from an initial state is called *evolution operator*.[3] *State* or *state vector* is a collection of D quantities $\mathbf{x} = (x_1, x_2, \ldots, x_D)$, where D is called dimension. The quantities x_k are called *dynamical variables*. A state may be both finite dimensional (D is a finite number) and infinite dimensional. The latter is the case, e.g., when a state is a spatial distribution of some quantity, i.e. a smooth function of a spatial coordinate.[4]

Evolution operator Φ_t determines a state at any future time instant $t_0 + t$ based on an initial state $\mathbf{x}(t_0)$: $\mathbf{x}(t_0 + t) = \Phi_t(\mathbf{x}(t_0))$. Mathematically, it can be specified with equations, maps, matrices, graphs and any other means (Chap. 3) under the only condition of a *unique* forecast.

The concept of a *dynamical system* (DS) is a key one in the deterministic approach. It was used already by Poincare at the beginning of the twentieth century but its meaning is still not completely established. The term DS is often understood in different ways. Therefore, it is useful to discuss it in more detail. The word

---

[3] In general, operator is the same as mapping, i.e. a law which relates some element x of a certain given set X to a uniquely determined element y of another given set Y. The term "operator" is often used in functional analysis and linear algebra, especially for mappings in vector spaces. For instance, operator of differentiation relates each differentiable function to its derivative (Mathematical dictionary, 1988).

[4] In this case the state is also called "state vector". The term "vector" is understood in a general sense as an element of some *space* (Lusternik and Sobolev, 1965).

"system"[5] is typically used in a traditional sense as "a collection of elements being in some relations to each other and forming a certain integrity" (Philosophic dictionary, 1983, p. 610). Alternative interpretations relate mainly to the understanding of the word "dynamical" and to what elements and systems are implied (real-world objects, mathematical constructions or both). Some authors even take the term "dynamical" out of the deterministic framework and combine it with randomness. For the sake of illustration, we cite below several selected definitions formulated by known specialists in the field (see also Alligood et al., 2000; Guckenheimer and Holmes, 1983; Katok and Hasselblat, 1995; Loskutov and Mikhailov, 2007):

> The concept of a DS appeared as a generalisation of the concept of a mechanical system whose motion is described with Newton's differential equations. In its historical development, the concept of a DS similarly to any other concept gradually changed getting new and deeper contents... Nowadays, the concept of a DS is quite broad. It covers systems of any nature (physical, chemical, biological, economical, etc) *both deterministic and stochastic.*[6] Description of a DS is very diverse. It can be done with differential equations, functions from algebra of logic, graphs, Markov chains, etc. (Butenin et al., 1987, p. 8).
>
> When speaking of a DS, we imply *a system* of any nature which can take different mathematical forms including ordinary differential equations (autonomous and non-autonomous), partial differential equations, maps on a straight line or a plane (Berger et al., 1984).

In the section "What is a dynamical system?" of the monograph Malinetsky and Potapov (2000), the authors note: "In general, in different books one can find different interpretations of the term DS, e.g. like the following ones:

- a synonym of the term "a set of ODEs $d\mathbf{x}/dt = \mathbf{f}(\mathbf{x}, t)$";
- a synonym of the term "a set of autonomous ODEs $d\mathbf{x}/dt = \mathbf{g}(\mathbf{x})$";
- a mathematical model of some mechanical system.

We[7] will adhere to the viewpoint according to which the concept of a DS is *a generalisation of the concept of a set of autonomous differential equations* and includes two main components: phase space **P** (metric space or manifold) and continuous or discrete one-parametric group (semigroup) $\varphi^t(\mathbf{x})$ or $\varphi(\mathbf{x}, t)$ of its transforms. A parameter t of the group is time."

Another formalised definition is as follows: "A DS is a quadruple (X, B, μ, Φ), where X is a topological space or a manifold, i.e. an abstract image of a state space, B are some interesting subsets in X, e.g. closed orbits or fixed points. They form an algebra in the sense that they include not only separate elements but also their unions and intersections. They are necessary to introduce a measure, since X itself can be immeasurable. μ is a measure, e.g. a volume of some domain or a frequency of an orbit visitations to it. μ is desired to be ergodic, unique, and invariant under the group of transforms Φ_t which defines an evolution. Sometimes, one adds also a

---

[5] From the Greek word "συστημα", i.e. "a whole consisting of parts" (Philosophic dictionary, 1983, p. 610).

[6] Highlighting with italic is ours in all the cited definitions.

[7] The authors G.G. Malinetsky and A.B. Potapov.

typical (in the sense of the measure μ) initial point. For example, the point $x_0 = 0$ is not typical for the operator $\Phi_t, t \in Z : x_{t+1} = \Phi_1(x_t) \equiv x_t(1+x_t)$, since it does not lead to an evolution" (Makarenko, 2002).

> One settled to understand a DS as a system of any nature (physical, chemical, biological, social, economical, etc.) *whose state changes* discretely or continuously in time (Danilov, 2001, p. 6).
> By abstracting from a concrete physical nature of an object, one speaks of it as of DS if it is *possible to specify such a set of quantities* called dynamical variables and *characterising a system state* whose *values at subsequent time instant are obtained from an initial set according to some rule*. This rule is said to determine an evolution operator for the system (Kuznetsov, 2001, p. 7).
> A DS can be thought of as an object of any nature whose state changes in time according to some dynamical law, i.e. as a result of a *deterministic* evolution operator action. Thus, the concept of DS is a consequence of a certain idealisation when one neglects influences of random perturbations inevitably present in any real-world system... Each DS corresponds to some mathematical model... (Anishchenko et al., 2002, pp. 1–2).
> A DS is a system whose behaviour is specified by a certain set of rules (an algorithm). A DS *represents only a model of some real-world system*. Any real-world system is prone to fluctuations and, therefore, cannot be dynamical (Landa, 1996).

The last definition is the closest one to the considerations in our book. It does not lead to difficulties in classification of possible situations. Thus, many real-world phenomena and objects can be successfully considered both with the aid of "probabilistic" (random) and "deterministic" mathematical tools. To illustrate that dynamical ideas can be fruitful under certain conditions and meaningless under different ones in modelling of the same object, we refer to the well-known "coin flips" (Sect. 2.6). There are no contradictions if the name of DS is related only to deterministic models and perceived as a kind of scientific jargon in application to real-world systems.

Further, we call DS *a mathematical evolution model for which one specifies* (i) a state $\mathbf{x}$ and (ii) an evolution operator Φ_t allowing a unique prediction of future states based on an initial one: $\mathbf{x}(t_0 + t) = \Phi_t(\mathbf{x}(t_0))$. In relation to real-world systems, we understand the term DS as a brief version of a statement "a system whose description with a dynamical model is possible and reasonable".

2.1.2 Non-rigorous Example: Variables and Parameters

Let us consider different dynamical systems which could describe an object which is familiar to many people – a usual cat (Fig. 2.1). The choice of quantities playing a role of dynamical variables or parameters of a model is determined by the purpose of modelling. If the purpose is to describe an evolution of the state of the cat's health, one can use its mass $M = x_3$, height $H = x_2$ and hair density $N = x_1$ (number of strands per a unit area) as dynamical variables. The collection $\mathbf{x} = (x_1, x_2, x_3)$ is a state vector of a dimension $D = 3$. Of course, one can imagine a number of other variables, such as blood haemoglobin concentration (x_4) and pulse rate (x_5). It would increase a model dimension D and make an investigation of the model

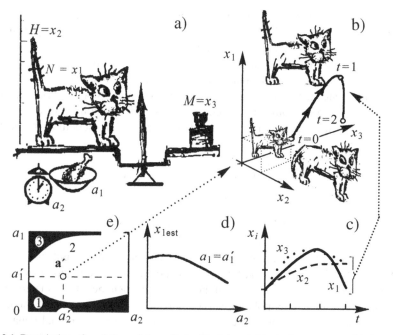

Fig. 2.1 Description of evolution of the cat's health: (**a**) variables and parameters; (**b**) the phase space and a phase orbit at fixed values of parameters $\mathbf{a}' = \left(a_1', a_2'\right)$; (**c**) time realisations $x_i(t)$, i.e. projections of the phase orbit onto the phase space axes; (**d**) a combined space of parameters and states presenting an established value of x_1 versus a_2 at fixed $a_1 = a_1'$; (**e**) a parameter space, the area 2 corresponds to a normal life of a cat and areas where its prolonged existence is impossible due to either hunger (the area 1) or gluttony (the area 3) are painted over. A point $\mathbf{a}'$ in the parameter space corresponds to a definite structure of the entire phase space

more complicated. For the sake of illustration, it is sufficient for us to use the three dynamical variables and consider dynamics of the object in the three-dimensional phase space (Fig. 2.1b). Each point of the phase space corresponds to a vector $\mathbf{x} = (x_1, x_2, x_3)$ reflecting an object state. Thus, the cat is too young and feeble at the point $t = 0$ (Fig. 2.1b), it is young and strong at the point $t = 1$, and it is already beaten by the life at $t = 2$.

Obviously, a current health state of the cat and its variations depend on the quantities which we can keep constant or change as we want. Such quantities are called *parameters*. For instance, these can be nutrition (the mass of food in a daily ration a_1, kg/day) and life conditions (the duration of walks in fresh air a_2, h/day). The number of model parameters as well as the number of dynamical variables is determined by the problem at hand and by the properties of an original. Thus, the health of a cat depends not only on the mass of food but also on the calorie content of food (a_3), amount of vitamins (a_4), concentration of harmful substances in the air (a_5), etc. For simplicity, we confine ourselves to two parameters and consider behaviour of the object in a two-dimensional parameter space, i.e. on a parameter plane (a_1, a_2), see Fig. 2.1e. Each point of the parameter plane corresponds to a certain kind of the temporal evolution of the object, i.e. to a certain kind of a phase orbit passing through an initial point in the phase space. Regions in the parameter space which correspond to different qualitative behaviour are separated with bifurcational sets of points. Bifurcational sets on the parameter plane (Fig. 2.1e) are boundary curves between white and black areas (bifurcational curves).

Just to illustrate the terms introduced above, without pretensions of strict description of such a complex biological object, one can consider the following set of first-order ordinary differential equations as a dynamical system modelling the health state of the cat:

$$dx_1/dt = f_1(x_1, x_2, x_3, a_1, a_2),$$
$$dx_2/dt = f_2(x_1, x_2, x_3, a_1, a_2),$$
$$dx_3/dt = f_3(x_1, x_2, x_3, a_1, a_2).$$

Relying upon everyday-life experience and imagination, one could suggest different forms for the functions f_k, e.g. algebraic polynomials whose coefficients are expressed via a_1 and a_2. It is a very common situation when model parameters enter evolution equations just as polynomial coefficients. According to the theorem of existence and uniqueness of a solution, the set of ordinary differential equations at fixed values of parameters and initial conditions has a unique solution under some general conditions. It means that the set of ODEs specifies a single phase orbit passing through a given initial point in the phase space.

Division of characterising quantities into dynamical variables and parameters is dictated by a modelling task. If the purpose of the cat modelling were description of its mechanical movements in space (rather than the state of its health as above), it would be reasonable to choose different variables and parameters. Thus, neither the animal mass M nor its "downiness" N and height H (previous dynamical variables) change during a jump of the cat. However, these quantities essentially affect its flight and must be taken into account as parameters $a_1 = M$, $a_2 = N$, $a_3 = H$, along with other quantities which influence mechanical motion (e.g. the shape of the cat's body). As dynamical variables, one can consider coordinates of the centre of the mass of the cat ($x_1 = x$, $x_2 = y$, $x_3 = z$) and angular displacements of its longitudinal axis in relation to coordinate axes ($x_4 = \alpha$, $x_5 = \beta$, $x_6 = \gamma$). Further, one can write down an evolution operator based on Newton's equations for progressive and rotational movements in contrast to the above semi-intuitive invention of model equations for the state of the cat health. Thus, depending on the purpose of modelling, the same physical quantities serve as dynamical variables in one case and play a role of parameters in another one.

Dynamical variables and parameters can be recognised in the evolution equations for a dynamical system. For instance, in the system specified by the classical equation of non-linear oscillator with cubic non-linearity (Duffing oscillator)

$$d^2x/dt^2 + 2\delta\, dx/dt + \omega_0^2(bx^3 - x) = 0, \tag{2.1}$$

one of the dynamical variables is the quantity x and the parameters are δ, ω_0, b, i.e. the *parameter vector* $\mathbf{a} = (\delta, \omega_0, b)$ is three dimensional. The system itself is two dimensional ($D = 2$) since one must specify initial values of x and dx/dt to find a particular solution to equation (2.1). The latter becomes clearer if one rewrites equation (2.1) equivalently as a set of two first-order equations for the variables $x_1 = x$ and $x_2 = dx/dt$:

$$dx_1 / dt = x_2; \quad dx_2 / dt = -2\delta x_2 - \omega_0^2 \left(bx_1^3 - x_1 \right).$$

Thus, the derivative dx/dt serves as the second dynamical variable of the system (2.1).

2.1.3 Phase Space. Conservative and Dissipative Systems. Attractors, Multistability, Basins of Attraction

A significant merit of dynamical modelling is a possibility of a vivid representation, especially in the case of low dimension D and small number of parameters. For such a representation, one uses formal *spaces*[8]: *state space* (or *phase space*), *parameter space* and their *hybrid versions*. Along the axes of a formal space, one indicates the values of dynamical variables or parameters. In a hybrid version, parameters are shown along certain axes and variables along others.

A state vector $\mathbf{x}(t)$ at some time instant t corresponds to a point in a phase space with coordinates $x_1(t), x_2(t), x_3(t)$ called *a representative point* since it represents an instantaneous state. In evolution process, a representative point moves along a certain curve called *a phase orbit*. A set of characteristic phase orbits is called *phase portrait* of a system. Having some experience, one can extract a lot of information about possible motions of a system from its phase portrait. Thus, a phase space is three dimensional in the above example with a cat. A single orbit corresponding to a concrete choice of an initial state at $t = 0$ is shown in Fig. 2.1b. It evidences that the animal developed well at the beginning and achieved excellent conditions at $t = 1$. Then, it grew thin and cast the coat up to an instant $t = 2$. We note that a

---

[8] "Space is a logically conceivable form (structure) serving as a medium where other forms or constructions are realised. For instance, a plane or a space serve in elementary geometry as media where various figures are constructed. ... In contemporary mathematics, a space defines a set of objects called points. ... Relations between points define "geometry". As examples of spaces, one can mention: (1) metric spaces ..., (2) "spaces of events" ..., (3) phase spaces. Phase space of a physical system is a set of all its states which are considered as points in that space. ..." (Mathematical dictionary, 1988). A space can be topological (if a certain non-quantitative concept of "closeness" is defined), metric (closeness is determined by "metrics"), etc. The choice of a phase space is determined by what one wants to use in modelling. For example, one needs "a smooth manifold" (Sect. 10.1.1) to use differential equations as a model. To define a limit behaviour of DS orbits, one needs a "complete" space, i.e. each limit point of a convergent sequence should belong to the same space.

phase orbit itself does not carry information about a time instant when a point visits
a certain area.

Usually, one shows the most characteristic orbits on a phase portrait. For illustration, let us consider elements of the phase space of a system (2.1), which models
oscillations in a potential well with two minima in the presence of friction similarly
to a ball rolling in a double-pit profile shown in Fig. 2.2a. The curves on a phase
plane $(x, dx/dt)$ in Fig. 2.2b are phase orbits starting from points 1 and 2. They
cannot intersect since it would violate a dynamical description: a unique present
must lead to a unique future! Situations resembling intersections can be found at
singular points, i.e. *points of equilibrium* where a state of a DS remains constant for
arbitrarily long time. There are three of them on the portrait: O, A_1, A_2. The first
one corresponds to the location of a resting ball on the top of the hill (Fig. 2.2a),
while the others show the left and right pits. Other points of the phase space correspond to states which are left by a representative point at a future time instant. Each
of them corresponds to a certain phase orbit and time realisations of dynamical
variables $x_k(t)$, Fig. 2.2b. We note that in a typical phase orbit, one can distinguish
between a starting interval (*a transient process*) and a later stage with greater degree
of repeatability (*an established motion*). Established motions are less diverse than
transient processes and correspond to objects called *attractors* in a phase space of
a dissipative system. In our example these are states of stable equilibrium: points
A_1, A_2. Indeed, they seem to attract orbits from certain areas of the phase space.
Starting from different points (1 and 2 in Fig. 2.2b), phase orbits can get to different
attractors.

A set of points in a phase space from which a system gets to a certain attractor
is called *basin of attraction*.[9] If an attractor in a phase space is unique, then its
basin of attraction is the entire phase space. If there are several attractors, one says
that *multistability* takes place. Then, their basins divide a phase space between each
other, e.g. as shown with shading in Fig. 2.2c, d. Attractor can exist only in a phase
space of a *dissipative dynamical system*. This is a system exhibiting phase volume
compression illustrated in Fig. 2.3. A set of initial points occupies a volume $V(0)$.
Starting from $V(0)$, a system gets to a volume $V(t)$ after some time interval t. A
system is called dissipative if a phase volume decreases with time, $V(t) < V(0)$. In
a one-dimensional case, a measure of a phase volume V is an interval length, it is
a surface area in a two-dimensional case and a hyper-volume in a multidimensional
case of $D > 3$. Finally, representative points get from an initial volume to attractors

---

[9] Strict definition of attractor is a subject of multiple discussions. A universally accepted definition
is still lacking. One of the popular ones is given in several steps (Malinetsky and Potapov, 2000,
pp. 76–77). "... a set A is called ... *invariant* ... if $\Phi_t A = A$. *Neighbourhood of a set* A is an
open set U containing the *closure* of A, i.e. A together with all its limit points including boundary
points. ... A closed invariant set A is called an *attracting set* if there exists its neighbourhood U
such that $\Phi_t(\mathbf{x}) \to A$ for all $\mathbf{x} \in U$ and $t \to \infty$. A maximal U satisfying this definition is called
basin of attraction of A. ... An attracting set containing an everywhere dense orbit is called an
attractor A." This definition can be roughly reformulated as follows: an attractor is the least set to
which almost all orbits of a DS from some area of non-zero volume tend.

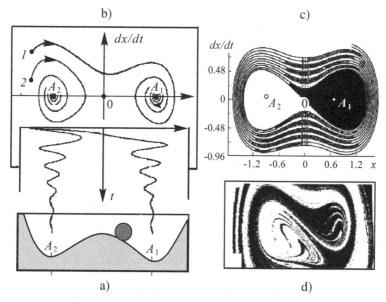

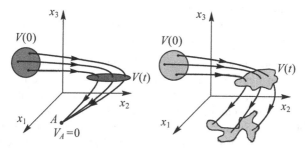

Fig. 2.2 A ball in a double pit: an illustration (**a**); a phase portrait and time realisations (**b**); basins of attraction of the equilibrium points A_1, A_2, i.e. of the two attractors coexisting in the phase space (**c**); basins of attraction in a non-autonomous case at the values of parameters for which two chaotic attractors coexist in the phase space (**d**)

Fig. 2.3 Illustration of some phase volume deformations: (**a**) a dissipative system; (**b**) a conservative system. The curves are phase orbits

whose volume is equal to zero. Such a definition of a dissipative system is broader than that used in physics where a dissipative system is a system with friction in which mechanical energy turns into energy of chaotic molecular motion. In *conservative systems* (friction-free systems in physics) an initial phase volume is preserved and only its form changes, hence attractors are absent.

Some possible kinds of attractors and the character of the corresponding established motions are shown in Fig. 2.4. Apart from equilibrium states represented by points, an attractor can be

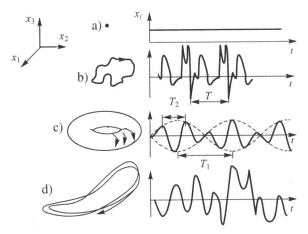

Fig. 2.4 Examples of characteristic sets in a phase space of a continuous-time system and corresponding time realisations

- a *limit cycle*, i.e. a closed curve, an image of a motion repeating itself with some period T (Fig. 2.4b);
- a *torus*, i.e. "an infinitely thin tread winding up on a bagel", an image of a quasi-periodic motion (with two characteristic periods T_1 and T_2 whose ratio is an irrational number) (Fig. 2.4c). A torus can be three- and multidimensional, i.e. represent complex behaviour with three, four, and more incommensurable frequencies of periodic components;
- a fractal set concentrated in a bounded area of a phase space, an image of *chaotic oscillations* called *a strange attractor* (Fig. 2.4d).[10]

Kinds of established motion realised in a DS and corresponding attractors are limited by its dimension. Thus, a phase space of a continuous-time system (e.g. with operators represented by differential equations) can contain only equilibrium points for $D = 1$, equilibrium points and limit cycles for $D = 2$, all the limit sets listed above for $D \geq 3$. Such considerations can help in practice to choose a model dimension. For instance, detection of a chaotic motion indicates that one needs at least three first-order non-linear ordinary differential equations to model an object. A somewhat different situation is found in a discrete-time system. An outlook of an attractor in its phase space can be imagined if one dissects the left pictures in Fig. 2.4 with a plane (a Poincare cross section). A single-turn cycle gives a single point in such a section. More complex cycles give several points. An orbit on a torus "draws" a closed curve in a section representing a quasi-periodic motion in a phase space of a discrete-time system. A chaotic attractor is represented by a set of points structured in a complicated (often self-similar) manner. A chaotic motion can be observed even in a phase space of *one-dimensional* non-invertible maps.

---

[10] "Strange" means here "different from previously known". An overview of kinds of chaotic attractors is given, e.g. in Anishchenko (1997), Katok and Hasselblat (1995) and Kuznetsov (2001).

2.1.4 Characteristics of Attractors

2.1.4.1 Geometrical Characteristics

Apart from visually detected differences, phase portraits are characterised by a number of quantitative measures. The most popular among them are dimensions. An integer-valued *topological dimension* D_T can be defined via an inductive principle (Poincare, 1982): $D_T = 0$ for a point; $D_T + 1$ is the dimension of a set which can be divided into non-intersecting parts with a subset of dimension D_T. According to those rules, a smooth curve has topological dimension $D_T = 1$, a surface $D_T = 2$, a volume $D_T = 3$. In particular, an equilibrium point, a cycle and a torus have topological dimensions 0, 1 and 2, respectively (see, e.g., Malinetsky, 2000, pp. 208–209). Structure of strange attractors differs qualitatively from the above sets. The former are *fractal* (self-similar) so that one needs more complicated measures called *fractal dimensions*. The simplest among them is *capacity* which characterises only geometry of an attractor. One also introduces *generalised dimensions* to take into account a frequency of a representative point visitations to subsets of an attractor. Below, we present only brief information about fractal measures. An educational computer program providing additional illustrations is located at our website (http://www.nonlinmod.sgu.ru). For more detailed study of fractal measures and techniques of their computation, we recommend the lectures 11–13 in the monograph Kuznetsov (2001) and references therein.

To define capacity, one covers a limit set in a D-dimensional phase space with D-dimensional cubes (i.e. line segments, squares, three-dimensional cubes, etc.) with an edge ε. Let a minimal number of cubes sufficient to provide covering be $N(\varepsilon)$.[11] Capacity of a set is

$$D_F = - \lim_{\varepsilon \to 0} \frac{\ln N(\varepsilon)}{\ln \varepsilon}, \qquad (2.2)$$

if the limit exists. One can use D-dimensional balls or sets of another shape instead of cubes (Kuznetsov, 2001, pp. 170–171; Malinetsky and Potapov, 2000, p. 210). Corresponding illustrations are given in Fig. 2.5, where we also present a classical example of a fractal *Cantor set* obtained from a unit segment by subsequent removal of middle thirds. In the latter case, one gets the capacity

$$D_F = - \lim_{\varepsilon \to 0} \frac{\ln 2^N}{\ln\left(1/3\right)^N} = \frac{\ln 2}{\ln 3} \approx 0.63$$

according to the definition (2.2). Majority of fractal sets are of non-integer dimension and can be embedded into spaces whose dimension equals the smallest integer exceeding a fractal dimension. Thus, the Cantor set is not already a finite set of points, but it is still not a line.

---

[11] Covering of a set A is a family of its subsets $\{A_i\}$ such that their union contains A.

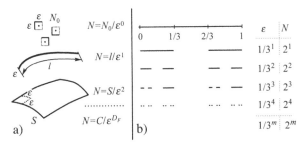

Fig. 2.5 Illustrations to (**a**) definition of capacity; (**b**) the Cantor set

A more subtle characteristic is *Hausdorff dimension*, which generalises capacity to the case of covering with elements of an arbitrary shape and size. Both quantities often coincide, but not always (Kuznetsov, 2001, p. 173; Malinetsky and Potapov, 2000, p. 209). As a rule, accurate numerical estimation of the Hausdorff dimension is impossible (Makarenko, 2002).

Generalised dimensions of Renyi D_q take into account a frequency of a representative point visitation to different attractor areas (Kuznetsov, 2001, pp. 176–190; Malinetsky and Potapov, 2000, pp. 211–214). Let an attractor be partitioned[12] into N non-empty cubes (cells) of size ε. Let us denote a portion of time spent by a representative point at a cell number i as p_i. It is a normalised density of points in a cell, i.e. an estimate of the probability of a visitation to a cell.[13] Then, one defines[14]

$$D_q = \frac{1}{q-1} \lim_{\varepsilon \to 0} \frac{\ln \sum_{i=1}^{N(\varepsilon)} p_i^q}{\ln \varepsilon}. \tag{2.3}$$

One distinguishes special kinds of generalised dimension: capacity at $q = 0$; *information dimension* at $q = 1$ (in the sense of limit for $q \to 1$); *correlation dimension* at $q = 2$. The latter characterises an asymptotic behaviour of *pairs* of points on an attractor. Indeed, a quantity p_i^2 can be interpreted as a probability to find two representative points within an ith cube of size ε. It is this quantity that can be easily estimated. Direct usage of the formula (2.3) leads to computational

---

[12] A partition is a covering with non-overlapping subsets $\{A_i\}$.

[13] It is strictly applicable to attractors supplied with an ergodic measure (Makarenko, 2002).

[14] A mathematical comment (Makarenko, 2002). Let us assume that p_i in each non-empty element of a partition follows an exponential law: $p_i \propto \varepsilon^\alpha$. If we deal with points on a line segment, then $\alpha = 1$ corresponds to a uniform distribution of points. However, $\alpha < 1$ may appear for rarely populated areas. Then, the ratio $p_i/\varepsilon \to \infty$ for $\varepsilon \to 0$. Therefore, such a distribution is called *singular*. For a square, areas with an exponent $\alpha < 2$ support singular distributions. One calculates *a partition function* $\sum_i p_i^q$, where a parameter q allows "to adjust" an estimator to locations with different probability density. If a partition function depends on ε via a power law, one introduces the definition (2.3) and speaks of *multifractal distribution*. If D_q differs for different q, an attractor is called multifractal (Kuznetsov, 2001, p. 182).

difficulties at $D > 3$. Therefore, a number of numerical techniques for the estimation of dimensions from a fragment of a phase orbit sampled discretely in time $(\mathbf{x}(t_1), \mathbf{x}(t_2), \ldots, \mathbf{x}(t_N))$ have been developed. One of the most popular ones is the *algorithm of Grassberger and Procaccia* for the correlation dimension estimation (Grassberger and Procaccia, 1983). It relies on the calculation of the so-called correlation integral

$$C(\varepsilon) = \frac{2}{N(N-1)} \sum_{i=1}^{N} \sum_{j=i+1}^{N} \Theta \left(\varepsilon - \|\mathbf{x}(t_i) - \mathbf{x}(t_j)\| \right),$$

where Θ is the Heavyside function ($\Theta(s) = 0$, $s \le 0$; $\Theta(s) = 1$, $s > 0$) and $\| \cdot \|$ is a norm of a vector (Euclidean or any other). One can easily see that it is an estimate of the probability that two points, arbitrarily chosen on an attractor according to its probability measure, are separated by a distance less than ε. As it follows from equation (2.3), $C(\varepsilon) \approx A\varepsilon^{D_2}$ holds true for $\varepsilon \to 0$. Correlation dimension can be estimated as a slope on the plot $\ln C(\ln \varepsilon)$ at small ε. In practice, the number of orbit points N is limited. Therefore, the size ε of a cell cannot be selected arbitrarily small. Furthermore, the greater the dimension, the greater the number of points required for its reliable estimation. There are different recommendations in respect of the necessary number of points obtained under different assumptions (Eckmann and Ruelle, 1985, 1992; Kipchatov, 1995).

To get integer-valued estimates of dimension of an observed motion, one uses several ideas. One of the most popular is the *false nearest neighbour technique* (Kennel et al., 1992). According to it, one checks the property that a phase orbit restored in a space of a sufficient dimension should not exhibit self-intersections. The technique is applied to reconstruct a phase orbit from a time realisation of a single variable (Sect. 10.1.2).

Another widely known method is the *principal component analysis* (Broomhead and King, 1986), where one distinguishes directions in a phase space along which the motion of a representative point develops more intensively. It is done via the analysis of correlations between state vector components (Sect. 10.1.2).

2.1.4.2 Dynamical Characteristics

The most widely used are *Lyapunov exponents* which characterise a speed of divergence or convergence of initially nearby phase orbits. A weak deviation of a representative point from an orbit on an attractor, i.e. a weak perturbation ε_0, evolves approximately according to an exponential law $\varepsilon(\Delta t) = \varepsilon_0 e^{\lambda \Delta t}$ until it gets large (Fig. 2.6a). As a result, a D-dimensional sphere of initial perturbations transforms into an ellipsoid after some time interval. If one prevents a system from a significant rise of perturbations (from an evolution along the grey arrow in Fig. 2.6a) by limiting an observation time interval τ, it is possible to estimate the exponents via the ratios of an ellipsoid semi-axis length to an initial radius: $\lambda_i = (1/\tau) \ln(\varepsilon_i/\varepsilon_0)$. These values averaged over an entire attractor are called Lyapunov exponents. Let

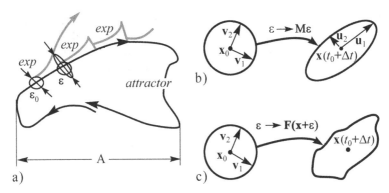

Fig. 2.6 Illustrations of Lyapunov exponents: (**a**) idea of calculation; evolution of a circle with a centre x_0 (**b**) for a linear system; (**c**) for a non-linear system

us denote them $\Lambda_1, \Lambda_2, \ldots, \Lambda_D$. They characterise stability of the motion on an attractor in a linear approximation. The set of values Λ_i in descending order is called *spectrum of Lyapunov exponents*, while sequence of their signs $(+, -$ or $0)$ is called the *spectrum signature*. If all the exponents are negative, i.e. the signature is $\langle -, -, \ldots, - \rangle$, then an attractor is an equilibrium point. The signature of a limit cycle is $\langle 0, -, \ldots, - \rangle$ and that of a two-dimensional torus is $\langle 0, 0, -, \ldots, - \rangle$. Spectrum of Lyapunov exponents for a chaotic attractor contains at least one positive exponent, e.g. $\langle +, 0, -, \ldots, - \rangle$, which determines the speed of divergence of initially close orbits.

Let us now describe some mathematical details. We start with a set of linear ordinary differential equations with variable coefficients:

$$d\varepsilon(t)/dt = \mathbf{A}(t)\varepsilon(t), \tag{2.4}$$

where $\varepsilon \in R^D$ and $\mathbf{A}$ is a matrix of an order D. Let us denote $\varepsilon(t_0) = \varepsilon_0$. Then, a solution to equation (2.4) at a time instant $t_0 + \Delta t$ is

$$\varepsilon(t_0 + \Delta t) = \mathbf{M}(t_0, \Delta t) \cdot \varepsilon_0, \tag{2.5}$$

where $\mathbf{M}(t_0, \Delta t)$ is a matrix of order D which depends on the initial time instant and the interval Δt and takes the form

$$\mathbf{M}(t_0, \Delta t) = \exp\left(\int_{t_0}^{t_0+\Delta t} \mathbf{A}(t')dt' \right), \tag{2.6}$$

where the matrix exponent is understood in the sense of formal expansion in a power series. For example, if $D = 1$ and $\mathbf{A}(t) = a = \text{const}$, then $d\varepsilon(t)/dt = a\varepsilon(t)$ and the solution to equation (2.5) takes a familiar form $\varepsilon(t_0 + \Delta t) = \varepsilon_0 e^{a\Delta t}$. Thus, in the case of constant coefficients, a perturbation evolves according to an exponential law. If coefficients are not constant, then a situation changes to some extent. For instance, one gets $\varepsilon(t_0 + \Delta t) = \varepsilon_0 e^{a\Delta t} e^{b \sin \Delta t}$ for $\mathbf{A}(t) = a + b \cos t$.

To characterise increase (or decrease) in ε in a multidimensional case, one should consider an evolution of a sphere of initial conditions with a centre at the origin and a radius $\|\varepsilon_0\|$. Since the system is linear, a sphere transforms into an ellipsoid. Lengths and orientations of semi-axes of the ellipsoid depend on the matrix $\mathbf{M}$ and, hence, on the value of Δt. An absolute value of ε changes in a different manner depending on the orientation of the initial vector ε_0. To describe it, one can use the so-called *singular value decomposition* of the matrix $\mathbf{M}$. This is a decomposition of the form $\mathbf{M} = \mathbf{U} \cdot \mathbf{\Sigma} \cdot \mathbf{V}^T$, where $\mathbf{U}$ and $\mathbf{V}$ are mutually orthogonal matrices which can be conveniently written in the form of vectors sets $\mathbf{U} = [\mathbf{u}_1, \mathbf{u}_2, \dots, \mathbf{u}_D]$ and $\mathbf{V} = [\mathbf{v}_1, \mathbf{v}_2, .., \mathbf{v}_D]$. If the matrix $\mathbf{M}$ is non-singular, then vectors $\mathbf{u}_1, \mathbf{u}_2, \dots, \mathbf{u}_D$ (called left singular vectors of the matrix $\mathbf{M}$) are mutually orthogonal and of unit length, i.e. they form an orthonormal basis in the space R^D. The same considerations apply to vectors $\mathbf{v}_1, \mathbf{v}_2, \dots, \mathbf{v}_D$ (right singular vectors). The matrix $\mathbf{\Sigma}$ is diagonal. Its diagonal elements $\sigma_1, \dots, \sigma_D$ are listed in descending order. They are called *singular values* of the matrix $\mathbf{M}$. Action of the matrix $\mathbf{M}$ on the vector ε_0 parallel to one of the right singular vectors $\mathbf{v}_i$ multiplies its length by σ_i and transforms it to a vector parallel to an ith left singular vector: $\varepsilon(t_0 + \Delta t) = \sigma_i \|\varepsilon_0\| \mathbf{u}_i$ (Fig. 2.6b). Thus, if at least one singular value of $\mathbf{M}$ exceeds 1 in absolute value, then an initial perturbation rises for some directions (one singular value is greater than 1 and another one is less than 1 in Fig. 2.6b). It rises in the fastest way for the direction of $\mathbf{v}_1$. The quantities showing how a perturbation changes are called *local Lyapunov exponents*:

$$\lambda_i(t_0, \Delta t) = \frac{1}{\Delta t} \ln \sigma_i. \qquad (2.7)$$

They describe an exponential growth in perturbations *averaged over a finite time interval*. According to the definition (2.7), a strict equality $\|\varepsilon(t_0 + \Delta t)\| = \|\varepsilon_0\| \cdot e^{\lambda_i(t_0, \Delta t)\Delta t}$ holds true for respective directions of an initial perturbation. A.M. Lyapunov proved that under certain conditions[15] imposed on a matrix $\mathbf{A}$, there exist finite limits:

$$\Lambda_i = \varlimsup_{\Delta t \to \infty} \frac{1}{\Delta t} \ln \frac{\|\varepsilon(t_0 + \Delta t)\|}{\|\varepsilon_0\|}, i = 1, 2, \dots, D, \qquad (2.8)$$

where the quantities Λ_i are exactly the *Lyapunov exponents*. They show an *efficient* speed of increase (decrease) in perturbations. Which of D exponents is realised for a given ε_0 depends on the direction of the latter. A perturbation changes at a speed determined by the largest Lyapunov exponent Λ_1 for *almost any* direction

---

[15] There exists a number L such that $\frac{1}{\Delta t} \int\limits_{t_0}^{t_0 + \Delta t} |A_{ij}(t')| dt' \leq L$ for all i, j and Δt (Kuznetsov, 2001, p. 140).

of ε_0. If Λ_1 is positive, then a typical perturbation rises. Hence, Λ_1 can be related to a predictability horizon of a system (2.4) in a situation where the equation itself is known precisely but initial conditions are specified at a certain error. A similar analysis can be done for linear difference equations.

Linearised dynamics and Lyapunov exponents. The analysis of stability for non-linear systems is performed via investigation of the linearised equations. Let us consider a non-linear system

$$\mathrm{d}\mathbf{x}/\mathrm{d}t = \mathbf{f}(\mathbf{x}). \tag{2.9}$$

Let $\mathbf{x}(t)$ be one of its orbits with an initial condition $\mathbf{x}(t_0) = \mathbf{x}_0$. Let us call the orbit with $\mathbf{x}(t_0) = \mathbf{x}_0$ a reference orbit and consider an orbit starting at a very close initial condition $\mathbf{x}(t_0) = \mathbf{x}_0 + \varepsilon_0$. Evolution of very small perturbations remaining small over an entire time interval considered is described with a set of equations linearised in a vicinity of the reference orbit:

$$\frac{\mathrm{d}\varepsilon}{\mathrm{d}t} = \frac{\partial \mathbf{f}(\mathbf{x}(\mathbf{x}_0, t))}{\partial \mathbf{x}} \varepsilon. \tag{2.10}$$

This equation coincides with equation (2.4) if one assigns

$$\mathbf{A}(t) = \frac{\partial \mathbf{f}(\mathbf{x}(\mathbf{x}_0, t))}{\partial \mathbf{x}}.$$

One can write down its solution in the form of equation (2.5), where the matrix $\mathbf{M}$ maps an infinitesimal sphere of initial conditions with a centre $\mathbf{x}_0$ to an ellipsoid with a centre $\mathbf{x}(t_0 + \Delta t)$ (Fig. 2.6b). Strictly speaking, if a perturbation is not infinitesimal but finite, an image of a sphere will not be an ellipsoid, but another set. Linearised dynamics only approximately describes an evolution of finite perturbations (Fig. 2.6c). For any reference orbit, there exists a set of Lyapunov exponents characterising linearised dynamics in its vicinity.

In 1968, Oseledets showed that a set of Lyapunov exponents is the same for any generic point $\mathbf{x}_0$ on an attractor. This statement is an essence of multiplicative ergodic theorem (see, e.g., Kornfel'd et al., 1982; Malinetsky and Potapov, 2000; pp. 224–227; Sinai, 1995). Thus, Lyapunov exponents characterise evolution of infinitesimal perturbations not only for a given reference orbit but also for an entire attractor of a DS. The *largest Lyapunov exponent* assesses an efficient speed of growth of infinitesimal perturbations (see also Sect. 2.4).

2.1.5 Parameter Space, Bifurcations, Combined Spaces, Bifurcation Diagrams

Attractors in a phase space evolve (change their shape, size, etc.) under parameter variations and loose stability at certain parameter values. As a result, one observes qualitative changes in a system motion, changes in its phase portrait, e.g. changes in the number of attractors in a phase space. Such a situation has got a name of *bifurcation*[16] (Belykh, 1997; Guckenheimer and Holmes, 1983; Malinetsky, 2000). We stress that a smooth deformation of an attractor and corresponding variations in an oscillation waveform are not regarded as a qualitative change.

To represent vividly the entire picture of possible kinds of established motions and transitions between them, one can use a geometrical representation in a *parameter space* where the values of parameters are shown along the axes. Some special methods are applied for that. The main idea is to single out sets of points separating areas with qualitatively different behaviour, i.e. *bifurcation sets*. In a pictorial example with a cat, a parameter space is the plane a_1, a_2 (Fig. 2.1e), while the boundaries between areas with different shading are bifurcation curves: the area 2 corresponds to a healthy life, while in the areas 1 and 3 the existence tragically and quickly stops due to hunger or gluttony. Parameter spaces shown below (e.g. Figs. 3.6, 3.7, 3.11 and 3.19a) are structured in a much more complicated manner. Bifurcation sets (curves) divide an entire parameter plane into areas where different attractors in a phase space exist. A way to represent vividly a situation of multistability (coexistence of several kinds of motion, several attractors in a phase space) on a parameter plane (e.g. in Figs. 3.6, 3.11 and 3.19a) is to show an area where a certain attractor exists as a separate sheet. Then, overlapping of many sheets at some parameters values is equivalent to multistability. For instance, bistability (coexistence of two attractors) takes place in the domain of intersection of sheets A and B in Fig. 3.6. The third sheet in that domain relates to an unstable cycle. Similarly, multistability in Fig. 3.19a takes place in the domain of intersection of sheets representing different modes of a system under investigation.

It is relevant to note some opportunities provided by the use of *combined spaces*. For instance, one can show a parameter value as an abscissa and a dynamical variable value in an established regime as an ordinate. A *bifurcation diagram* obtained in such a way for a quadratic map (Fig. 3.8e) is widely used to demonstrate universal laws of similarity (scaling) in transition to chaos via the period-doubling cascade. For a map describing a dissipative non-linear oscillator, such a diagram illustrates phenomena of resonance, hysteresis, bistability and bifurcation cascade (Fig. 3.10c–e). Moreover, one can present information in the phase and parameter spaces with colours. Basins of different attractors or areas of existence and evolution of different oscillatory regimes are often shown in such a way (Figs. 2.2 and 3.11).

We have given a very brief introduction to realisation of the dynamical approach. For readers who want to get deep knowledge in the field, we refer to classical works on qualitative theory of differential equations, theory of oscillations and non-linear dynamics (e.g. Andronov et al., 1959, 1967; Arnold, 1971, 1978; Bautin and Leontovich, 1990; Butenin et al., 1987; Guckenheimer and Holmes, 1983; Katok and Hasselblat, 1995; Shil'nikov et al., 2001).

---

[16] Initially, the word "bifurcation" meant division of an evolution pathway into two branches. However, currently any qualitative change is called call "bifurcation".

2.2 Foundations to Claim a Process "Random"

The use of the probabilistic approach is typically related to recognition of some quantity as "random". However, what is "random quantity" and what is its difference from a "non-random" one? Currently, there are several points of view on randomness which allow introduction of quantitative measures. For the most part, they agree with each other, but not always. Sometimes they can even lead to opposite results in the assessment of randomness or non-randomness of some quantity in practice. Here, we consider the problem according to a scheme suggested in Kravtsov (1989, 1997).

2.2.1 Set-Theoretic Approach

Set-theoretic approach underlying contemporary theory of probability (Gnedenko, 1950; Hoel, 1971; Pugachev, 1979, 1984; von Mises, 1964) associates the concept of randomness with possibility to specify a probability distribution law for a given quantity. Absence or presence of regularity is assessed via possible scattering of the values of a quantity: (i) probability distribution density in the form of Dirac δ function corresponds to a deterministic quantity; (ii) a non-zero "width", "smeared character" of distribution corresponds to unpredictable, random quantity.

2.2.1.1 Random Events and Probability

In description of many phenomena, researchers face impossibility to predict a course of events uniquely, even if all controllable conditions are held "the same".[17] To investigate such phenomena, the concepts of random event and probability were introduced in the theory of probability. These concepts are indefinable in theory, only some of their properties are defined via axioms. Their vivid interpretation and connection to practice are the tasks for the users. Below, we remind these basic concepts on an intuitive level, rather than rigorously.

An *event* is an outcome of a *trial*. Let us consider a classical example of the "coin flip" (see also Sect. 2.6). Let a coin be flipped only once. Then, the single flip is a trial. As a result, two events are possible: "a head" (an event A) and "a tail" (an event B). A and B are mutually exclusive events. An event in which either A or B occurs is called a *union of events* A and B and designated as A ∪ B. In our case, it inevitably occurs as a result of any trial. Such an event is called *sure* and its probability is said to be equal to unity: $P\{A \cup B\} = 1$. Since a union of A and B is a sure event, one says that A and B constitute *a complete group* of events. It follows from an idea of symmetry that the events A and B are equiprobable, i.e.

---

[17] We mark the words "the same" with inverted commas to stress their conventional nature. To be realistic, one can speak of sameness only to a degree permitted by the conditions of observations or measurements.

the chances to observe a head or a tail are equal for a usual coin with a uniform density of metal. Equiprobable mutually exclusive events constituting a complete group are called *elementary events*. Probability of a union of mutually exclusive events is equal to the sum of their probabilities. In our case, A and B constitute a complete group, therefore, one can write down: $P\{A \cup B\} = P\{A\} + P\{B\} = 1$. From here and the condition of equiprobability, one gets the individual probabilities as $P\{A\} = P\{B\} = 1/2$.

Elementary events may not be always singled out so easily. Sometimes, geometrical considerations can help (a geometrical definition of probability). Let a trial consist of a random throwing of a point onto an area A of a plane. A point falls into A for sure and all subsets of A are "equal in rights". A point may either fall into a region $B \subset A$ or not. Probability of an event that a point falls into a subset B is defined via the ratio of the areas $\mu(B)/\mu(A)$, where μ stands for the *Lebesgue measure*. The latter is a surface area in our example, but the same formula can be used for a space of any dimension. Such a definition can be interpreted in terms of elementary events if they are introduced as falls of a point into small squares covering A (for a size of squares tending to zero).[18]

The most habitual to physicists is a statistical definition of probability. If an event A is realised M times in a sequence of N independent trials, then the ratio M/N is called a frequency of occurrence of the event A. If a frequency M/N tends to some limit for a number of trials tending to infinity, then such a limit is called a probability of the event A. This is the most vivid (physical) sense of the concept of probability. The property of an event frequency stabilisation is called *statistical stability*. The entire machinery of the theory of probability is appropriate for the phenomena satisfying the condition of statistical stability.

2.2.1.2 Random Quantities and Their Characteristics

Random quantity is any numerical function ξ of a random event. In the case of coin flips the values of a random quantity can be defined as $\xi = 1$ (a head) and $\xi = 0$ (a tail). The probability of $\xi = 1$ is the probability of a head.

For a complete characterisation of a random quantity, one needs to specify probabilities of its possible values. For instance, one uses a *distribution function* $F_\xi(x) \equiv P\{\xi \le x\}$. If ξ is continuous valued and its distribution function is

---

[18] We note that it is important to define clearly what is a "random" point, line or plane for the definition of geometrical probabilities. For instance, let us assess the probability of an event that a "random" chord exceeds in length an edge of an equilateral triangle inscribed into a unit circumference. A chord can be chosen "randomly" in different ways. The first way: let us superpose a vertex of a chord with one of the triangle vertices leaving the other chord vertex free. Then, a portion of favourable outcomes when the length of a chord exceeds the length of an edge is 1/3. The second way: let us select randomly a point in a circle which is the middle of a "random" chord. A chord is longer than a triangle edge if its middle belongs to a circle inscribed into the triangle. Radius of that circle equals half the radius of the circumscribed circle and, hence, a portion of favourable outcomes assessed as the ratio of the areas of the two circles equals 1/4. We get two different answers for two different notions of a random chord.

differentiable, then one defines a *probability density function* $p_\xi(x) \equiv dF_\xi(x)/dx$. Then, a probability for ξ to take a value from an infinitesimal segment $[x, x + dx]$ equals $p_\xi(x)dx$. For the sake of definiteness, we speak of random quantities supplied with probability density functions.

Several often used distributions are the following:

(i) The *normal (Gaussian) law*

$$p_\xi(x) = \left(1 / \sqrt{2\pi\sigma^2}\right) \cdot e^{-\frac{(x-a)^2}{2\sigma^2}}, \qquad (2.11)$$

where a and σ^2 are parameters. This is one of the most often used distributions in the theory of probabilities. The reason is that it possesses many useful theoretical properties and allows obtaining a number of analytical results. Besides, in practice the quantities resulting from influence of multiple factors are often distributed approximately according to the Gaussian law. It finds theoretical justifications: the *central limit theorem* states that a sum of independent identically distributed random quantities is asymptotically normal, i.e. its distribution law tends to the Gaussian one for an increasing number of items.[19]

(ii) The exponential law (*Laplace distribution*):

$$p_\xi(x) = \begin{cases} (1/a)\exp(-x/a), & x \geq 0, \\ 0, & x < 0; \end{cases} \qquad (2.12)$$

(iii) The *uniform distribution* on a segment $[a, b]$

$$p_\xi(x) = \begin{cases} 1/(b-a), & a \leq x \leq b, \\ 0, & x < a, x > b. \end{cases} \qquad (2.13)$$

A random quantity ξ is often characterised by statistical moments of its distribution. An *ordinary moment* of an order n is the quantity

$$E\left[\xi^n\right] \equiv \int_{-\infty}^{\infty} x^n p(x) dx. \qquad (2.14)$$

Here and further, E stands for the mathematical expectation of the quantity in square brackets. The first-order moment is just the expectation of ξ. Its physical meaning is an average over infinitely many independent trials. *Central moments* are defined as ordinary moments for deviations of ξ from its expectation:

---

[19] Some authors mention ironically the frequent use of the normal law in data analysis and the references to the central limit theorem: engineers think that practical applicability of the central limit theorem is a strictly proven statement, while mathematicians believe that it is an experimental fact (see, e.g., Press et al., 1988).

$$E\left[(\xi - E\,[\xi])^n\right] \equiv \int\limits_{-\infty}^{\infty} (x - E\,[\xi])^n p(x)\mathrm{d}x. \tag{2.15}$$

The second-order central moment is called *variance*. This is the most often used measure of scattering. Let us denote it as σ_ξ^2. Then, σ_ξ is called a root-mean-squared (standard) deviation of ξ. The third-order central moment is called skewness (a measure of a distribution asymmetry). The fourth-order central moment is called kurtosis. Skewness is equal to 0 and kurtosis is $3\sigma_\xi^4$ for the normal law (2.11). If all ordinary moments of ξ (for $n = 1, 2, \ldots$) exist, then one can uniquely restore the distribution function from their values. Parameters of a distribution law are related to its moments. For instance, $E[\xi] = a$ and $\sigma_\xi^2 = \sigma^2$ for the normal law (2.11); $E[\xi] = a$ and $\sigma_\xi^2 = a^2$ for the exponential law (2.12); $E[\xi] = (a + b)/2$ and $\sigma_\xi^2 = (b - a)^2/12$ for the uniform law (2.13).

For two random quantities ξ_1 and ξ_2, one considers joint characteristics. The two quantities can be regarded components of a two-dimensional *random vector* ξ. A joint probability density function $p_\xi(x_1, x_2)$ is then defined: a probability that the values of ξ_1 and ξ_2 fall *simultaneously* (in the same trial) into infinitesimal segments $[x_1, x_1 + \mathrm{d}x_1]$ and $[x_2, x_2 + \mathrm{d}x_2]$ equals $p_\xi(x_1, x_2)\mathrm{d}x_1\mathrm{d}x_2$. One also introduces a conditional probability density for one quantity under the condition that the other one takes a certain value, e.g. $p_{\xi_1}(x_1 | x_2 = x^*)$. The quantities ξ_1 and ξ_2 are called *statistically independent* if $p_\xi(x_1, x_2) = p_{\xi_1}(x_1)p_{\xi_2}(x_2)$. In the latter case, the conditional distributions of ξ_1 and ξ_2 coincide with the respective unconditional distributions.

A random quantity depending on time (e.g. one deals with a sequence of values of a quantity ξ) is called a *random process*, see Chap. 4.

2.2.1.3   The Concept of Statistical Estimator

As a rule, in practice one does not know a distribution law and must *estimate* the expectation of an observed quantity or parameters of its distribution from results of several trials. This is a problem of mathematical statistics (Hoel, 1971; Ibragimov and Has'minskii, 1979; Kendall and Stuart, 1979; Pugachev, 1979, 1984; Vapnik, 1979, 1995; von Mises, 1964) which is inverse to problems of the theory of probability where one determines properties of a random quantity, given its distribution law. Let us denote a set of values of a random quantity ξ in N trials as $\{x_1, \ldots, x_N\}$. It is called a *sample*.[20]

A quantity whose value is obtained via processing the data $\{x_1, \ldots, x_N\}$ is called a sample function. An *estimator* of some distribution parameter is a sample function, whose values are in some sense close to the true value of that parameter.[21] We denote estimators with a "hat" like $\hat{a}$.

---

[20] A sample is an N-dimensional random vector with its own distribution law.

[21] Theoretically speaking, any measurable sample function is called estimator. If estimator values are not close to a true parameter value, such an estimator is just a "bad" one.

Let a sample $\{x_1, \ldots, x_N\}$ represent *independent* trials. Let the expectation of ξ be equal to a. Let the value of a be unknown but known to belong to a set A. It is necessary to get an estimator $\hat{a}$, which is as close to a as possible for any true value of a from A. Any estimator is a random quantity since it is a function of random quantities: $\hat{a} = f(x_1, \ldots, x_N)$. One gets a certain value of $\hat{a}$ from a certain sample and another value from another sample, i.e. $\hat{a}$ is characterised by its own probability density function $p_f(\hat{a})$ (Fig. 2.7), which is determined by the distribution law $p_\xi(x)$ and the way how $\hat{a}$ is computed (i.e. by the function f). Different functions f correspond to estimators with different distributions and, hence, with different probabilistic properties.

2.2.1.4 Estimator Bias and Variance

The most important property of an estimator $\hat{a}$ is closeness of its values to a true value of an estimated quantity a. Closeness can be characterised in different ways. The most convenient and widely used one is to define an *estimator error* as the

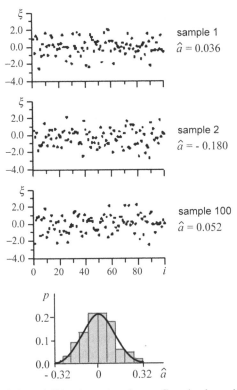

Fig. 2.7 Samples consisting of 100 values taken from a Gaussian law with zero mean and unit variance. The values of an estimator of the expectation for different samples are shown and its probability distribution density (theoretically, it is Gaussian with zero mean and the variance of 0.1) obtained from 100 samples

mean-squared difference between $\hat{a}$ and a:

$$E\left[(\hat{a} - a)^2\right] \equiv \int\limits_{-\infty}^{\infty} (\hat{a} - a)^2 p_f(\hat{a}) \mathrm{d}\hat{a}. \tag{2.16}$$

It can be readily shown that the error is equal to the sum of two items:

$$E\left[(\hat{a} - a)^2\right] = (E[\hat{a}] - a)^2 + \sigma_{\hat{a}}^2. \tag{2.17}$$

An estimator whose bias is equal to zero, i.e. $E[\hat{a}] = a$ for any $a \in A$, is called *unbiased*. If the values of such an estimator are averaged over different samples, one gets a quantity closer to the true value of a since the random errors in $\hat{a}$ compensate each other. One could derive many unbiased estimators of a quantity a, i.e. different functions f. They would differ in their variance. One can show that an unbiased least-variance estimator is unique, i.e. if the least possible value of the variance is σ_{min}^2, then it is exactly achieved only for a single estimator. An unbiased least-variance estimator is an attractive tool, though the least value of the squared error (2.17) may be achieved for another estimator, which is somewhat biased but exhibits significantly smaller variance.

An unbiased estimator of the expectation from a sample of independent values is the *sample mean*. We denote it by angular brackets and a subscript N: $\langle \xi \rangle_N$. This is just an arithmetic mean

$$\langle \xi \rangle_N = f(x_1, \ldots, x_N) = \frac{1}{N} \sum_{i=1}^{N} x_i. \tag{2.18}$$

This is a least-variance estimator of the expectation in the case of the normally distributed quantity ξ. If the distribution of ξ is symmetric and exhibits large kurtosis and/or other deviations form normality, a *sample median*[22] has typically a smaller variance as an estimator of its expectation. As well, the sample median is more stable to variations in the distribution law of ξ. Stability with respect to some perturbations of the distribution law is often called *robustness*. To compute the sample median, one may write down the values in a sample in ascending order: $x_{i_1} < x_{i_2} < \ldots < x_{i_N}$. Then, a sample median is $x_{i_{(N+1)/2}}$ for an uneven N and $\left(x_{i_{N/2}} + x_{i_{N/2+1}}\right)/2$ for an even N. The sample mean (2.18) is an estimator which is unbiased for any distribution of ξ, while a sample median can be biased for asymmetric distribution laws.

The sample moment of an order n can serve as an estimator of the respective ordinary moment $E[\xi^n]$:

---

[22] A median of a distribution is such a number b which divides the x-axis into two equiprobable areas: $P\{\xi < b\} = P\{\xi > b\} = 1/2$. A median coincides with the expectation for a symmetric distribution.

$$\langle \xi^n \rangle_N = \frac{1}{N} \sum_{i=1}^{N} x_i^n. \tag{2.19}$$

A situation with central moments is somewhat different since the value of $E[\xi]$ entering their definition is unknown. Yet, an estimator of variance can be obtained as the *sample variance*

$$\hat{\sigma}_\xi^2 = \frac{1}{N} \sum_{i=1}^{N} \left(x_i - \langle \xi \rangle_N \right)^2. \tag{2.20}$$

It is biased due to the replacement of $E[\xi]$ with a sample mean. Its bias is of the order of $1/N$. One can show that an unbiased estimator is

$$\hat{\sigma}_\xi^2 = \frac{1}{N-1} \sum_{i=1}^{N} \left(x_i - \langle \xi \rangle_N \right)^2. \tag{2.21}$$

2.2.1.5 Estimator Consistency

How do estimator properties change under increase in the sample size N? In general, an estimator distribution law varies with N. Hence, its bias and variance may also change. As a rule, one gets estimator values closer to a true value a at bigger N. If the bias $E[\hat{a}] - a$ tends to zero at $N \to \infty$ for any a from A, then the estimator $\hat{a}$ is called *asymptotically unbiased*. If the estimator $\hat{a}$ converges to a in probability (i.e. the probability that the estimator value differs from a true one more than by ε tends zero for arbitrarily small ε: $\forall \varepsilon > 0 \ P\{|\hat{a} - a| > \varepsilon\} \underset{N \to \infty}{\to} 0$), it is called *consistent*. Consistency is a very important property of an estimator assuring its high goodness for large samples. The sample moments (2.19) are consistent estimators of the ordinary moments (Korn and Korn, 1961; Pugachev, 1979).

2.2.1.6 Method of Statistical Moments

Let us consider the problem of parameter estimation when a functional form of the distribution $p_\xi(x, \mathbf{c})$ is known and $\mathbf{c} = (c_1, \ldots, c_P)$ is a parameter vector taking values from a set $A \subset R^P$. One of the possible approaches is the *method of statistical moments* which is following. The first P theoretical ordinary moments are expressed as functions of parameters. Examples for normal, exponential and uniform distributions are given above, where the first two moments are expressed as simple functions of parameters. Thereby, one obtains a system

$$E\left[\xi\right] = g_1(c_1, \ldots, c_P),$$

$$E\left[\xi^2\right] = g_2(c_1, \ldots, c_P),$$

$$\ldots,$$

$$E\left[\xi^P\right] = g_P(c_1, \ldots, c_P). \tag{2.22}$$

By substituting the sample moments instead of the theoretical ones into equation (2.22), one gets a set of equations for the parameters

$$\langle\xi\rangle_N = g_1(c_1, \ldots, c_P),$$

$$\left\langle\xi^2\right\rangle_N = g_2(c_1, \ldots, c_P),$$

$$\ldots,$$

$$\left\langle\xi^P\right\rangle_N = g_P(c_1, \ldots, c_P), \tag{2.23}$$

whose solution gives estimators $\hat{c}_1, \ldots, \hat{c}_P$. Such moments-based estimators may not possess the best properties for small samples. However, they are asymptotically unbiased and consistent (Korn and Korn, 1961) so that they can be readily used for large samples.

2.2.1.7 Maximum Likelihood Method

Typically, the maximum likelihood method provides estimators with the best properties. According to it, a sample is considered as a random vector $\mathbf{x} = (x_1, \ldots, x_N)$ of dimension N which is characterised by some probability density function depending on a parameter vector $\mathbf{c}$. Let us denote such a conditional probability density as $p_N(\mathbf{x}|\mathbf{c})$. One looks for the parameter values $\mathbf{c} = \hat{\mathbf{c}}$ maximising $p_N(\mathbf{x}|\mathbf{c})$ for an *observed* sample, i.e. an occurrence of the sample $\mathbf{x} = (x_1, \ldots, x_N)$ is the most probable event for the values $\mathbf{c} = \hat{\mathbf{c}}$. They are called *maximum likelihood estimators* (ML estimators).

The function $L(\mathbf{c}) = p_N(\mathbf{x}|\mathbf{c})$ where $\mathbf{x}$ is a fixed vector (an observed sample), is called *likelihood function* or just *likelihood*. It should not be interpreted as a probability density function for parameters $\mathbf{c}$ since the parameters are fixed numbers (not random quantities) according to the problem setting. Therefore, a special term "likelihood" is introduced. ML estimators give a maximal value to the likelihood: $L(\hat{\mathbf{c}}) = \max_{\mathbf{c} \in A} L(\mathbf{c})$. Necessary conditions of the maximum read as

$$\partial L(\mathbf{c})/\partial c_j = 0, \quad j = 1, \ldots, P. \tag{2.24}$$

It is often more convenient to deal with the likelihood logarithm. It gets maximal at the same point as $L(\mathbf{c})$, therefore, ML estimators are found from equation

$$\partial \ln L(\mathbf{c})/\partial c_j = 0, \quad j = 1, \ldots, P, \tag{2.25}$$

which is called *likelihood equation*.

For a sample consisting of independent values, the likelihood function equals the product of probability density functions at each value of x_i and the logarithmic likelihood equals the sum of logarithms:

$$\ln L(\mathbf{c}) = \sum_{i=1}^{N} \ln p(x_i | \mathbf{c}). \tag{2.26}$$

In such a case, ML estimators are consistent and asymptotically unbiased. Asymptotically, they are the least-variance estimators.

For the normal distribution of ξ, the logarithmic likelihood reads as

$$\ln L(a, \sigma^2) = -\frac{N}{2} \ln \left(2\pi\sigma^2\right) - \frac{1}{2\sigma^2} \sum_{i=1}^{N} (x_i - a)^2. \tag{2.27}$$

One can readily see that the ML estimators of the parameters a and σ^2 coincide with the sample mean (2.18) and the sample variance (2.20). Thus, the ML estimator of σ^2 is biased. However, it tends to the unbiased estimator (2.21) with increasing N, i.e. it is asymptotically unbiased. One can show that the ML estimator $\hat{a}$ is distributed here according to the normal law with the expectation a and the variance σ^2/N (see Fig. 2.7 for a concrete illustration). It follows from these observations that the value of the sample mean gets closer to the true value of a for a large number of trials, since the estimator variance decreases with N. In particular, $|\hat{a} - a| < 1.96\sigma/\sqrt{N}$ holds true with a probability of 0.95.

The interval $[\hat{a} - 1.96\sigma/\sqrt{N}, \hat{a} + 1.96\sigma/\sqrt{N}]$ is called 95% confidence interval for the quantity a. The greater the N, the narrower this interval. To estimate it from observations, one can replace the true value of σ with its estimator $\hat{\sigma}$. An estimator $\hat{a}$ is called a *point* estimator since it gives only a single number (a single point). If an interval of the most probable values of an estimated parameter is indicated, then one speaks of an *interval* estimator. Interval estimators are quite desirable, since from a single value of a point estimator one cannot judge to what extent it can differ from the true value.

2.2.1.8 When the ML Technique is Inconsistent

Sometimes, the ML technique can give asymptotically biased estimators. This is encountered, for instance, in the investigation of a dependence between two variables when the values of *both* variables are known with errors. It is studied by the so-called confluent analysis (Aivazian, 1968; Korn and Korn, 1961). As an example, let us consider the following problem. There is a random quantity Z and quantities X and Y related to Z via the following equations (Pisarenko and Sornette, 2004):

$$\begin{aligned} X &= Z + \xi, \\ Y &= Z + \eta, \end{aligned} \tag{2.28}$$

where ξ and η are independent of each other and of Z, and are normally distributed random quantities with zero expectation and the same variance σ^2. One can say that X and Y represent measurements of Z in two independent ways. There is a sample of X and Y values obtained from independent trials: $\{x_i, y_i\}_{i=1}^{N}$. It is necessary to estimate a measurement error variance σ^2.

The simplest way to derive an estimator is to note that a quantity $X - Y = \xi - \eta$ is normally distributed with zero expectation and the variance $2\sigma^2$, since the variance of the sum of two independent quantities is equal to the sum of their variances. Then, one can easily get a consistent estimator of $X - Y$ variance from a sample $\{x_i - y_i\}_{i=1}^{N}$ as follows:

$$\hat{\sigma}_{X-Y}^2 = \frac{1}{N} \sum_{i=1}^{N} (x_i - y_i)^2. \tag{2.29}$$

Hence, the value of σ^2 is estimated via the following equation:

$$\hat{\sigma}^2 = \frac{1}{2N} \sum_{i=1}^{N} (x_i - y_i)^2. \tag{2.30}$$

At the same time, a direct application of the ML technique (without introduction of the above auxiliary variable) gives the likelihood function

$$L(x_1, y_1, \ldots, x_N, y_N \,|z_1, \ldots, z_N, \sigma) = \frac{1}{(2\pi\sigma^2)^N} \exp\left(-\sum_{i=1}^{N} \frac{(x_i - z_i)^2 + (y_i - z_i)^2}{2\sigma^2}\right),$$

which contains unobserved values of Z. By solving the likelihood equations, one then gets estimators:

$$\hat{z}_i = (x_i + y_i)/2, \quad i = 1, \ldots, N, \tag{2.31}$$

$$\hat{\sigma}_{ML}^2 = \frac{1}{4N} \sum_{i=1}^{N} (x_i - y_i)^2. \tag{2.32}$$

Thus, the ML estimator of the variance is twice as small as the unbiased estimators (2.30) at any N, i.e. the former is biased and inconsistent. What is a principal difference of this problem? It is as follows: The number of estimated quantities (equal to $N + 1$ under the ML method) rises with the sample size! In the previous cases, we have considered estimation of a fixed number of parameters.

In general, the less the number of quantities estimated, the better the properties of their estimators.

2.2.1.9 Bayesian Estimation

A very broad branch of the theory of statistical estimation is related to the case when true values of parameters c are also random quantities, i.e. they can vary between different samples according to a probability density function $p(c)$ which is called *prior*. If a prior density is known, then it is reasonable to take it into account in estimation. The corresponding approaches are called *Bayesian.*[23]

In the most widespread version, one tries to find a distribution law for the parameters c under the condition that a sample $x_1, \ldots, x_N$ has been realised. This is a so-called *posterior* probability density function $p(c|x_1, \ldots, x_N)$. It can be derived if a probability density function $p(x_1, \ldots, x_N|c)$ at a given c is known. Then, one finds posterior density via the *Bayesian rule:*[24]

$$p(c|x_1, \ldots, x_N) = \frac{p(c)p(x_1, \ldots, x_N|c)}{\int p(c)p(x_1, \ldots, x_N|c)dc}. \qquad (2.33)$$

We note that the denominator does not depend on the estimated parameters, since integration over them is performed.

If a posterior distribution law is found, then one can get a concrete point estimator $\hat{c}$ in different ways, e.g., as the expectation $\hat{c} = \int c p(c|x_1, \ldots, x_N)dc$ or as its point of maximum (a mode). In the absence of knowledge about a prior density, it is replaced with a constant $p(c)$ that corresponds to a distribution which is uniform over a very broad segment. Then, to a multiplier independent of c, a posterior distribution coincides with the likelihood function. Further, if a Bayesian estimator is defined as a posterior distribution mode, one comes exactly to the ML technique.

As a rule, in practice one sets up a hypothesis: which distribution law an observed quantity follows, whether trials are independent or not, etc. Accepting such assumptions, one applies corresponding techniques. Validity of the assumptions is checked with statistical tools after getting an estimate (Sect. 7.3).

2.2.2 Signs of Randomness Traditional for Physicists

All the signs listed below rely to some extent on the understanding of randomness as a lack of "repeatability" in a process.

(a) *Irregular* (non-periodic) outlook of a time realisation. This is the most primitive sign of randomness. Here, it is directly opposed to periodicity: *absence of strict period means randomness, its presence means determinancy.*

---

[23] From the name of an English priest Thomas Bayes (1702–1761), who suggested the idea in a work published after his death.

[24] In fact, this is a joint probability of two events A and B written down in two versions: $P\{A \cap B\} = P\{A\}P\{B|A\} = P\{B\}P\{A|B\}$. Hence, one deduces $P\{B|A\} = P\{B\}P\{A|B\}/P\{A\}$.

(b) *Decaying correlations*. This is a decrease of an *autocorrelation function* $\rho(\tau)$
(ACF, Sect. 4.1.2) to zero with increasing τ. For a zero-mean stationary process
(Sect. 4.1.3), the ACF reads as $\rho(\tau) = \langle x(t)x(t+\tau)\rangle \Big/ \sqrt{\langle x^2(t)\rangle\langle x^2(t+\tau)\rangle}$.
Angular brackets denote averaging over an ensemble which coincides with tem-
poral averaging for an ergodic process (Sect. 4.1.3). This sign gives, in essence,
a quantitative measure of an observed process deviation from a periodic one.
One cannot reveal periodicity with this approach if a period $T > T_0$, where T_0
is an observation time.

(c) *Continuous spectrum*. According to this sign, a process with a continuous power
spectrum (Sect. 6.4.2) is called random, while a spectrum of a periodic process is discrete. In practice, finiteness of an observation time T_0 limits a spectral resolution: $\Delta\omega_{min} = 2\pi/T_0$. By increasing an observation time T_0, one
would finally establish finiteness of spectral lines for any real-world process
and, strictly speaking, would have to regard any real-world process random
according to any of the signs (a–c).

(d) *Irregularity* of sets of data points in a restored "phase space" (Sect. 10.1):
absence of any signs for a finite dimension and so forth. These are more delicate
characteristics which are not related just to the detection of non-periodicity.

There are also more qualitative criteria: irreproducibility of a process or its
uncontrollability, i.e. impossibility to make conditions under which a process would
occur in the same way or in the way prescribed in advance, respectively.

2.2.3 Algorithmic Approach

An *algorithmic approach* interprets "a lack of regularity" as an excessive complexity of an algorithm required to reproduce a given process in a digital form. The
idea to relate randomness to complexity was put forward for the first time by A.N.
Kolmogorov and independently by Chaitin and Solomonoff.

Any process can be represented as a sequence of 0s and 1s, i.e. written down in a
binary system: $\{y_i\}$, $i = 1, 2, \ldots, N$. Kolmogorov suggested to regard a length l (in
bits) of the shortest program capable of reproducing the sequence $\{y_i\}$ as a measure
of its complexity. For instance, a program reproducing a sequence $1010\ldots 10$ (a
hundred of pairs "10") is very short: *print "10" a hundred times*. If 0s and 1s are
located randomly, a program consists of symbol-wise transmission of a sequence
which appears uncompressible. Thus, $l \sim N$ for random sequences and $l << N$ for
non-random ones.

Unfortunately, there is no generally applicable way to find the minimal length
of a program in practice.[25] New approaches to the concepts of complexity and randomness based on the idea of algorithmic complexity have been developed. A view

---

[25] For the same fundamental reasons that are mentioned in Gödel's theorem stating incompleteness
of any system of axioms as discussed, e.g., in Shalizi (2003) and references therein.

relating those concepts to *predictability* is getting more and more popular during the last years (Badii and Politi, 1997; Kravtsov, 1989; Shalizi, 2003).

2.2.4 Randomness as Unpredictability

Randomness or determinancy of a process is related in Kravtsov (1989, 1997) to possibility of its prediction with the aid of an existing model. The author considers a registered process $x(t)$ and a model process $z(t)$. For the sake of simplicity, it is assumed $\langle x(t) \rangle = \langle z(t) \rangle = 0$. At a current time instant $t = t_0$ the quantities take the values $x = x_0$, $z = z_0$. It is natural to specify a model process so that $z_0 = x_0$ and assess a forecast quality via its error $x(t) - z(t) = \Delta(t)$, $\Delta(t_0) = 0$. The entire approach is based on the statistical description of the pair x, z.

The values of x and z typically diverge as time passes so that the absolute value of $\Delta(t)$ rises. By repeating experiments and comparison of $x(t)$ to $z(t)$, one can form an ensemble of realisations and estimate probability distributions $p(x, z, t, x_0, z_0, t_0)$ and $p(\Delta, t, x_0, z_0, t_0)$. In such a description, the model process $z(t)$ is included into statistical considerations along with the registered process. Measures of predictability can be the following:

(i) The mean-squared error $\sigma_\Delta^2(\tau) \equiv \langle \Delta^2(\tau) \rangle = \langle |x(t) - z(t)|^2 \rangle$, where $t = t_0 + \tau$, $\sigma_\Delta^2(0) = 0$. If the quantities $x(t)$ and $z(t)$ become statistically independent at $\tau \to \infty$, then $\langle x(t)z(t) \rangle = 0$ and $\sigma_\Delta^2(\tau) = \langle x^2(t) \rangle + \langle z^2(t) \rangle$. One assumes that x and z are bounded. Then, a relative error can be reasonably defined as $E(\tau) = \sigma_\Delta^2(\tau) / (\langle x^2(t) \rangle + \langle z^2(t) \rangle)$ so that $E \to 1$ for $t \to \infty$.

(ii) The cross-correlation function between an original and a model processes $D(\tau) = \langle x(t_0 + \tau)z(t_0 + \tau) \rangle / \sqrt{\langle x^2(t_0 + \tau) \rangle \langle z^2(t_0 + \tau) \rangle}$. One has $D(0) = 1$ and $|D(\tau)| \leq 1$ for any τ. From well-known statistical relationships, one can derive

$$D(\tau) = \frac{\langle x^2(t_0 + \tau) \rangle + \langle z^2(t_0 + \tau) \rangle}{2\sqrt{\langle x^2(t_0 + \tau) \rangle \langle z^2(t_0 + \tau) \rangle}} (1 - E(t)).$$

Thus, the degree of predictability can be expressed via different similar quantities. Qualification of a process as random or deterministic is determined by the possibility of its prediction with an available model. Here, random is something that we cannot predict for some reasons: due to the properties of $x(t)$, or due to the kind of a model process $z(t)$, or due to the absence of a model. Such an approach to randomness was developed within a hypothesis distinction theory for the needs of radio-location.

2.3 Conception of Partial Determinancy

Conception of partial determinancy is based on the convention that one chooses unpredictability (predictability) of an observed process $x(t)$ with a certain predictive model $z(t)$ as a sign of randomness (determinancy) of $x(t)$. Randomness and determinancy are not opposed to each other but considered as poles of a single property called partial determinancy.

It is convenient to use cross-correlation between an observed and a model process $D(\tau)$ as a quantity characterising the degree of determinancy (predictability). Its typical plot is shown in Fig. 2.8, where D is an area of full determinancy; DC is an area of partial determinancy; C is an area of random (unpredictable) behaviour. An observed process $x(t)$ appears deterministic (predictable) if $D \approx 1$; unpredictable if $|D| << 1$; partially predictable if $0 < |D| < 1$.

A time interval $\tau = \tau_{det}$ over which the degree of predictability falls down to a certain threshold value, e.g. $D(\tau_{det}) = 1/2$, is called an interval of deterministic behaviour. What affects this quantity? For a real-world system, it is always finite for the following reasons:

- An observed process always differs from an investigated process due to the influence of registering devices, a *measurement noise* $v(t)$.
- There are random and/or non-random unaccounted external influences $\mu(t)$, the so-called "dynamical noise".
- A model does not adequately reflect properties of an object. This is a "noise of ignorance" $\Delta M(t)$ depending on a model structure and parameters values.

Therefore, $\tau_{det} = f(v, \mu, \Delta M)$. Even if one manages to reduce strongly an effect of devices and an error in a deterministic component of a model, unavoidable external fluctuations remain. They can be related to infinite dimension of real-world object microstructure, to noises of different origin, to senescence processes, and so forth and principally limit predictability interval. The limit $\tau_{lim} = \lim_{v \to 0, \Delta M \to 0} \tau_{det} = f(v, \mu, \Delta M)$ is called "a predictability horizon".

As a rule, x and z become statistically independent for $\tau >> \tau_{lim}$ so that $D(\tau) \to 0$. An interval of deterministic behaviour τ_{det} can exceed an autocorrelation

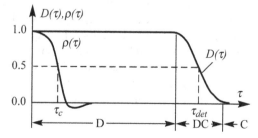

Fig. 2.8 Typical relationship between the degree of determinancy $D(\tau)$ and the autocorrelation function

time τ_c of $x(t)$ characterising the speed of its autocorrelation function decay. The latter can be estimated as $\tau_c \approx 1/\Delta\omega$, where $\Delta\omega$ is the width of the spectrum line. For instance, one gets $\Delta\omega \to \infty$, $\tau_c = 0$ for a white noise, i.e. for process which "forgets" its past at once (Sect. 4.2). An autocorrelation time τ_c can be considered as the least interval of determinacy due to the following consideration. If one has no dynamical equations for a model $z(t)$, then a forecast can be based on previous values of $x(t)$. The simplest principle is "tomorrow is the same as today", i.e. a model $z(t + \tau) = x(t)$. In such a case, one gets $D(\tau) = \rho(\tau)$ and $\tau_{det} = \tau_c$. In general, it can be that $\tau_{det} > \tau_c$ (Fig. 2.8). The same phenomenon can be close to a deterministic one from a viewpoint of one model and fully non-deterministic from a viewpoint of another model.

2.4 Lyapunov Exponents and Limits of Predictability

2.4.1 Practical Prediction Time Estimator

Forecast is a widespread and most intriguing scientific problem. A predictability time for many processes is seemingly limited in principle and even not large from a practical viewpoint. If a process under investigation is chaotic, i.e. close orbits diverge exponentially, it is natural to expect its predictability time to be related to the speed of close orbit divergence. The latter is determined by the value of the *largest Lyapunov exponent* Λ_1 (Sect. 2.1.4). For a dynamical model, it is reasonable to take an interval over which a small perturbation (determined both by model errors and different noise sources in a system) rises up to a characteristic scale of an observed oscillations as an estimator of predictability time. A *predictability time* can be roughly estimated via the following formula (Kravtsov, 1989):

$$\tau_{pred} = \frac{1}{2\Lambda_1} \ln \frac{\sigma_x^2}{\sigma_v^2 + \sigma_\mu^2 + \sigma_{\Delta M}^2}, \qquad (2.34)$$

where σ_μ^2 is the dynamical noise variance, σ_v^2 is the measurement noise variance, $\sigma_{\Delta M}^2$ is the model error (an "ignorance noise" variance), σ_x^2 is the observable quantity variance, and the largest Lyapunov exponent Λ_1 is positive. The formula can be derived from the following qualitative considerations. Let equations of an original system be known exactly and initial conditions only to an error ε (measurement noise). Then, if those "incorrect values" are taken as initial conditions for a model, one gets a prediction error rising in time as $\varepsilon \cdot e^{\Lambda_1 t}$ on average. If a predictability time is defined as a time interval over which a prediction error reaches the value of σ_x, one gets

$$\tau_{pred} = \frac{1}{\Lambda_1} \ln \frac{\sigma_x}{\varepsilon}.$$

The quantity $\tau_\Lambda = 1/\Lambda_1$ is called *Lyapunov time*. Now, let us consider external random influences and model errors along with the measurement noise. If all those factors are regarded as approximately independent, an overall perturbation variance is equal to the sum of variances of the components. Replacing ε in the last formula for τ_{pred} by a square root of an overall perturbation variance, one gets the expression (2.34).

If noises and model errors are small as compared with the signal level, a time interval (2.34) can significantly exceed the autocorrelation time of a process which can be roughly estimated as $\tau_c \sim 1/\Lambda_1$ in many cases. Thus, if the signal level is 1000 times as big as the noise level in terms of root-mean-squared deviations, then a predictability time (2.34) is approximately seven times as big as the autocorrelation time.

The formula (2.34) is not always applicable to estimate a predictability time. The point is that after a certain time interval, any finite perturbation in a chaotic regime reaches a scale where the linearised system (2.10) is no longer appropriate. Further evolution is, strictly speaking, not connected with Lyapunov exponents. Thus, if one is interested in a forecast with a practically acceptable accuracy rather than with a very high accuracy, the Lyapunov exponent is not relevant and cannot impose restrictions on a predictability time. Yet, if the Lyapunov exponent characterises a speed of the perturbation rise at large scales correctly (which is often the case), then one can use it to assess a predictability time even for finite perturbations and errors.

However, under stricter considerations it appears that even in the limit of infinitesimal perturbations the Lyapunov time is not always related to a predictability time. Let us consider this interesting fact in more detail.

2.4.2 Predictability and Lyapunov Exponent: The Case of Infinitesimal Perturbations

The quantity (2.34) can be called a *predictability time* by definition. However, other approaches are also possible. One of the reasonable ideas consists of the following (Smith, 1997). Let us consider how a perturbation of a given initial condition x_0 evolves. According to the definition of the local Lyapunov exponents (2.7), one gets $\|\varepsilon(t_0 + \Delta t)\| = \|\varepsilon_0\| e^{\lambda_1(x_0, \Delta t) \cdot \Delta t}$ in the worst case, i.e. as largest increase in a perturbation. Let us define a predictability time via time intervals over which an initial small perturbation gets q times greater:

$$\tau_q(\mathbf{x}_0) = \frac{\ln q}{\lambda_1(\mathbf{x}_0, \Delta t)}.$$

Such a time interval depends on $\mathbf{x}_0$. To get an overall characteristic of predictability, one can average $\tau_q(\mathbf{x}_0)$ over an invariant measure $p(\mathbf{x}_0)$, i.e. over probability distribution on an attractor:

$$\tau_q \equiv \int p(\mathbf{x}_0)\tau_q(\mathbf{x}_0)d\mathbf{x}_0. \tag{2.35}$$

This definition of a predictability time differs essentially from equation (2.34). Thus, if a time interval over which an error gets q times greater were defined via the largest Lyapunov exponent, then one would get

$$\tau_{q,\Lambda} \equiv \frac{\ln q}{\Lambda_1} = \frac{\ln q}{\int p(\mathbf{x}_0)\lambda_1(\mathbf{x}_0)d\mathbf{x}_0} = \frac{1}{\int p(\mathbf{x}_0)\frac{1}{\tau_q(\mathbf{x}_0)}d\mathbf{x}_0}. \tag{2.36}$$

Here, the Lyapunov exponent (the quantity in the denominator) is expressed as an average over a natural measure[26] which is equivalent to temporal averaging for an ergodic system.

Hence, the situation is analogous to the following one. There are values of a random quantity $x_1, x_2, \ldots, x_N$ and one needs to estimate its expectation $E[x]$. The simplest way is to calculate a sample mean which is a "good" estimator: $\langle x \rangle = (x_1 + \ldots + x_N)/N$. This is an analogue to the formula (2.35) for a mean predictability time. However, one can imagine many other formulas for an estimator. For instance, one may calculate inverse values $1/x_1, 1/x_2, \ldots, 1/x_N$, estimate a quantity $1/E[x]$ as their sample mean and take its inverse. The resulting estimator $\langle x' \rangle = N/(1/x_1 + 1/x_2 + \ldots + 1/x_N)$ is an analogue to equation (2.34). However, a mean value of the inverse quantities is generally a biased estimator of $1/E[x]$. Therefore, $\langle x' \rangle$ is also a "bad" estimator of $E[x]$. The quantities $\langle x \rangle$ and $\langle x' \rangle$ coincide only when $x_1 = x_2 = \ldots = x_N$. In our case it means that the Lyapunov time coincides with τ_q (up to a multiplier $\ln q$) only if the local Lyapunov exponent does not depend on $\mathbf{x}_0$, i.e. orbits diverge at the same speed at any phase space area. This is a condition of applicability of the formula (2.34) even in the linear case.

Thus, a predictability time can be defined without appealing to the Lyapunov exponent which seems even more reasonable. As shown below, the Lyapunov exponent may not relate to a predictability time τ_q, i.e. a system with a greater value of the Lyapunov exponent (a more chaotic system) can have a greater value of τ_q (to be more predictable) compared to a less chaotic system. Besides, systems with the same values of the Lyapunov exponent can have very different predictability times τ_q. Let us discuss an analytic example from Smith (1997). For the sake of definiteness, we speak of the doubling time τ_2.

An example where the Lyapunov time and τ_2 coincide (up to a multiplier $\ln 2$) is a two-dimensional non-linear map which is one of basic models in non-linear dynamics – a baker's map

---

[26] Roughly, this is a probability density p of the visitations of a representative point to different areas of an attractor (see, e.g., Kuznetsov, 2001).

$$x_{n+1} = \begin{cases} \frac{1}{\alpha}x_n, & 0 \le x_n < \alpha, \\ \beta(x_n - \alpha), & \alpha \le x_n < 1, \end{cases}$$

$$y_{n+1} = \begin{cases} \alpha y_n, & 0 \le x_n < \alpha, \\ \alpha + \frac{1}{\beta}y_n, & \alpha \le x_n < 1, \end{cases} \qquad (2.37)$$

with $\alpha = 1/\beta = 1/2$. This map is area preserving (conservative). It maps the square $[0, 1) \times [0, 1)$ on itself. An invariant measure satisfies a condition $p(x, y) = 1$ so that fractal dimension of any kind described in Sect. 2.1.4 is equal to 2. It is called a baker's map since its action on a unit square reminds operations of a baker with a piece of pastry. Firstly, pastry is compressed twice along the y-axis and stretched twice along the x-axis. Secondly, it is cut in half and the right piece is located over the left one via a parallel shift. A single iteration of the map involves all those manipulations (Fig. 2.9). For almost any initial condition on the plane, two nearby points differing only in their x-coordinate are mapped to two points separated by a distance twice as big as the initial one. Similarly, a distance along the y-axis becomes twice as small in a single iteration. Thus, for any point within the square, the direction of the x-axis corresponds to the largest local Lyapunov exponent. The latter does not depend on the interval Δt and equals just to the largest Lyapunov exponent. This is a system with a uniform speed of nearby orbit divergence. Since $\Lambda_1 = \ln 2$, the Lyapunov time is equal to $\tau_\Lambda = 1/\ln 2$. The time $\tau_2(\mathbf{x_0})$ equals 1 for any initial condition, i.e. a perturbation is doubled in a single iteration. Accordingly, an average doubling time is $\tau_2 = 1$.

Let us now consider a modification of the system (2.37) called a baker's apprentice map:

$$x_{n+1} = \begin{cases} \frac{1}{\alpha}x_n, 0 \le x_n < \alpha, \\ (\beta(x_n - \alpha)) \bmod 1, \alpha \le x_n < 1, \end{cases}$$

$$y_{n+1} = \begin{cases} \alpha y_n, 0 \le x_n < \alpha, \\ \alpha + \frac{1}{\beta}([\beta(x_n - \alpha)] + y_n), \alpha \le x_n < 1, \end{cases} \qquad (2.38)$$

where square brackets denote the greatest integer not exceeding the number in the brackets, $\alpha = (2^N - 1)/2^N$ and $\beta = 2^{2^N}$. The action of this map is as follows. A greater piece of pastry $[0, \alpha) \times [0, 1)$ is compressed very weakly along the y-axis and stretched along the x-axis turning into the piece $[0, 1) \times [0, \alpha)$. The right narrow band is compressed very strongly β times along the y-axis. Thereby, one gets a

Fig. 2.9 A single iteration of a baker's map. The square is coloured with *black* and *white* to show where the points from different areas are mapped to

Fig. 2.10 A single iteration of the baker's apprentice map for $N = 1$ (the most "skilful" apprentice)

narrow belt of an integer length which is divided into belts of unit length. The latter ones are located over the big piece $[0, 1) \times [0, \alpha)$ as a pile, the left belts being below the right ones (Fig. 2.10).

This system also preserves an area and has an invariant measure $p(x, y) = 1$ (hence, fractal dimension of any kind equals 2 as for the baker's map). Its directions corresponding to a greater and a smaller local Lyapunov exponents also coincide with the directions of the coordinate axes. It can be shown that its largest Lyapunov exponent is equal to $\Lambda_1 = \alpha \ln 1/\alpha + (1-\alpha) \ln \beta$ and at the above particular values, one gets $\Lambda_1 = \mathbf{ln\,2} - \alpha \ln \alpha > \mathbf{ln\,2}$. Thus, system (2.38) is more chaotic than system (2.37) in the sense of the greater Lyapunov exponent. Its local Lyapunov exponents depend strongly on initial conditions. They are very small for the area of smaller x (a predictability time is big), while they are very big for the area of bigger x (this area is very narrow). An example of operations of both maps with a set of points is shown in Fig. 2.11. A result of four iterates of the baker's map (2.37) is shown in the middle panel. The picture is completely distorted, i.e. predictability is bad. A result of the four iterates of the map (2.38) with $N = 4$ is shown in the right panel. A significant part of the picture is well preserved being just weakly deformed: predictability in this area is good.

The most interesting in this example is the following circumstance. Not only local predictability times $\tau_2(\mathbf{x}_0)$ in some areas are greater for the map (2.38) than for the map (2.37), an average time τ_2 for the map (2.38) is also greater though it has a greater Lyapunov exponent! The value of τ_2 can be found analytically as

$$\tau_2 = \frac{1 - \alpha^j}{1 - \alpha},$$

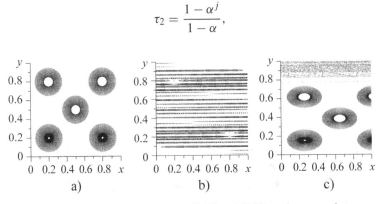

Fig. 2.11 Illustration of dynamics of the maps (2.37) and (2.38) analogous to that presented in Smith (1997): (**a**) an initial set of points; (**b**) an image under the fourth iterate of the baker's map (2.37); (**c**) an image under the fourth iterate of the baker's apprentice map (2.38) with $N = 4$

Table 2.1 Characteristics of the map (2.38) for different N: a maximal local predictability time, a mean predictability time, the largest Lyapunov exponent (Smith, 1997)

N	$\tau_{2,\max}$	τ_2	Λ_1	N	$\tau_{2,\max}$	τ_2	Λ_1
1	1	1.00	$1.5 \cdot \ln 2$	5	22	16.09	$1.04 \cdot \ln 2$
2	3	2.31	$1.31 \cdot \ln 2$	6	45	32.49	$1.02 \cdot \ln 2$
3	6	4.41	$1.17 \cdot \ln 2$	7	89	64.32	$1.01 \cdot \ln 2$
4	11	8.13	$1.09 \cdot \ln 2$				

where $j = \left[-\frac{\ln 2}{\ln \alpha} \right]^*$ and $[\cdot]^*$ denote the smallest integer greater than or equal to the number in the brackets. It can be shown that a predictability time $\tau_2 \approx 2^{N-1} \to \infty$ and $\Lambda_1 \to \ln 2$ for $N \to \infty$. The results of analytic manipulations for some N are brought together in Table 2.1: the predictability time can be arbitrarily high for systems as chaotic as (2.37) and even with a bit greater Lyapunov exponent!

Thus, Lyapunov exponents do not exhaust a question about predictability. Still, they carry certain information and become relevant characteristics of predictability if the speed of the divergence of orbits is uniform over a phase space.

2.5 Scale of Consideration Influences Classification of a Process (Complex Deterministic Dynamics Versus Randomness)

In practice, data are measured at finite accuracy, i.e. arbitrarily small scales of consideration are unavailable. At that, it is often difficult to decide whether an observed irregular behaviour is deterministically chaotic or stochastic (random). Strictly speaking, one can answer such a question only if data are generated with a computer and, therefore, it is known what law they obey. For a real-world process, one should ask which of the two representations are more adequate. A constructive approach is suggested in Cencini et al. (2000), where the answer depends on the *consideration scale*.

To characterise quantitatively an evolution of a perturbation with a size ε in a DS (2.9), it is suggested to use a *finite-size* Lyapunov exponent (FSLE) denoted as $\lambda(\varepsilon)$. It indicates how quickly orbits initially separated by a distance ε diverge. In general, finite perturbations may no longer be described with the linearised equation (2.10). To compute a FSLE, one needs first to introduce a norm (length) of state vectors. In contrast to the case of infinitesimal perturbations, a numerical value of $\lambda(\varepsilon)$ depends on the norm used. For the sake of definiteness, let us speak of the Euclidean norm and denote the norm of an initial perturbation as $\|\boldsymbol{\varepsilon}(0)\| = \varepsilon_0$. The value of a perturbation reaches threshold values $\varepsilon_1, \varepsilon_2, \ldots, \varepsilon_p$ at certain time instants. For instance, let us specify the thresholds as $\varepsilon_n = 2\varepsilon_{n-1}$, $n = 1, \ldots, P - 1$ and speak of a perturbation doubling time for different scales $\tau_2(\varepsilon_n)$. Let us perform N experiments by "launching" neighbouring orbits separated by a distance ε_0 from different initial conditions. We get an individual doubling time of $\tau_2^{(j)}(\varepsilon_n)$, $j = 1, \ldots, N$ for each pair of orbits. A mean doubling time is defined as

$$\tau_2(\varepsilon_n) = \left(1/N\right) \sum_{j=1}^{N} \tau_2^{(j)}(\varepsilon_n)$$

and an FSLE is defined as $\lambda(\varepsilon_n) = \ln 2/\tau_2(\varepsilon_n)$.

If a process is deterministically chaotic and a speed of phase orbit divergence is constant over an entire phase space, then $\lim_{\varepsilon \to 0} \lambda(\varepsilon) = \Lambda_1$ (Sect. 2.4.2).[27] It is important that for a deterministic process, $\lambda(\varepsilon)$ does not depend on ε at small scales: $\lambda(\varepsilon) = \text{const}$. For a stochastic process, $\lambda(\varepsilon) \to \infty$ for $\varepsilon \to 0$. The law of the rise in $\lambda(\varepsilon)$ with decreasing ε may be different, e.g. $\lambda(\varepsilon) \propto \varepsilon^{-2}$ for a Brownian motion (Wiener's process, Sect. 4.2). The authors of Cencini et al. (2000) suggest the following approach to the distinction between deterministically chaotic signals and noisy (random) ones. If one gets for a real-world process that $\lambda(\varepsilon) = \text{const}$ within a certain range of scales, then it is reasonable to describe the process as deterministic in that range of scales. If $\lambda(\varepsilon)$ rises with decreasing ε within a certain range of scales, then the process should be regarded as noisy within that range.

A simple example is a deterministic map exhibiting a "random walk" (*diffusion*) at large scales:

$$x_{n+1} = [x_n] + F(x_n - [x_n]), \qquad (2.39)$$

where

$$F(y) = \begin{cases} (2+\delta)y, & 0 \le y < 0.5, \\ (2+\delta)y - (1+\delta), & 0.5 \le y < 1.0, \end{cases}$$

and square brackets denote an integer part. The function F is plotted in Fig. 2.12 for $\delta = 0.4$. The Lyapunov exponent equals to $\Lambda_1 = \ln|F'| = \ln|2 + \delta|$. The process behaves like Wiener's process (Sect. 4.2) at $\varepsilon > 1$. For instance, $\varepsilon = 1$ means that one traces only an integer part of x. A change of an integer part by ± 1 results from the deterministically chaotic dynamics of a fractional part of x. Since the latter is ignored in consideration at large scales, the former looks like random walk. Figure 2.13 shows that $\lambda(\varepsilon) \approx 0.9$ and the process is classified as deterministic within the range of scales $\varepsilon < 1$. One gets $\lambda(\varepsilon) \propto \varepsilon^{-2}$ for $\varepsilon > 1$ and considers the process as random.

Let us modify the map (2.39) by introducing noise ξ_n uniformly distributed on a segment $[-1, 1]$ and replacing F with its approximation G (10,000 linear pieces with a slope 0.9 instead of the two pieces with a slope 2.4):

$$x_{n+1} = [x_n] + G(x_n - [x_n]) + \sigma \xi_n, \qquad (2.40)$$

---

[27] FSLE defined via doubling times is equal to zero for a process with $\Lambda_1 < 0$.

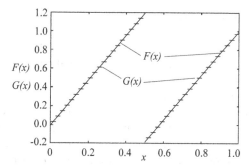

Fig. 2.12 Function $F(x)$ from equation (2.39). Horizontal lines are its approximation $G(x)$ from equation (2.40) consisting of 40 segments with zero slope (Cencini et al., 2000)

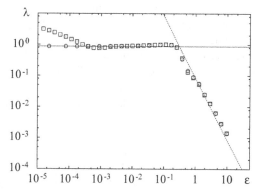

Fig. 2.13 FSLE versus a scale. Circles are shown for the system (2.39), squares for the system (2.40) with $G(x)$ consisting of 10,000 segments with a slope of 0.9 (Cencini et al., 2000)

where the quantity $\sigma = 10^{-4}$ determines the noise standard deviation. The processes (2.39) and (2.40) do not differ for $\varepsilon > 1$ and look like a random walk (Fig. 2.13). They look deterministic with the same Lyapunov exponent in the interval $10^{-4} < \varepsilon < 1$ despite different slopes of their linear pieces: 2.4 in equation (2.39) and 0.9 in equation (2.40). This is the result of averaging of the local linear dynamics of equation (2.40) over the scales $\varepsilon > 10^{-4}$. The processes differ for $\varepsilon < 10^{-4}$ where the process (2.40) behaves again as random from the viewpoint of $\lambda(\varepsilon)$ due to the presence of the noise ξ. Thus, dynamical properties may differ at different scales. It is important to take it into account in describing complex real-world processes.

Based on the described approach, the authors have suggested witty terms to characterise some irregular processes: "*noisy chaos*" and "*chaotic noise*". The first one relates to a process which looks deterministic (chaos) at large scales and random (noise) at small ones, i.e. a macroscopic chaos induced by a micro-level noise. Analogously, the second term describes a process which is random at large scales and deterministic at small ones.

2.6 "Coin Flip" Example

Most likely, everybody used to put a coin on bent fingers (Fig. 2.14a), offer "head" or "tail", flip it and ...relieve him/herself of responsibility for some decision. A small disk falling with rotation is popular as a symbol of candour, an embodiment of chance for different peoples at different times. We use it below to illustrate the discussion of determinancy, randomness, and different approaches to modelling.

We start with a conceptual model. In a typical case, a hand imparts to a coin both a progressive motion with an initial velocity v_0 and a rotation with an initial angular velocity ω_0. Further, the disk flies interacting with the earth and an air until it falls on a surface. If the latter is solid, then it would jump up several times and finally settle down on one of its sides. Without a special practice, one can hardly repeat a flip several times so as to reproduce the same result, e.g. a head. It gets impossible for a strong flip when a coin has enough time to perform many revolutions before landing. The main cause of irreproducibility is a significant scattering of initial velocities and coordinate. In part, one can reach reproducibility if a special device is used, e.g. a steel ruler with a gadget to adjust a deformation[28] (Fig. 2.14b). However, such a device is not a panacea: one can confidently predict a result only for weak flips when a coin performs half a revolution, a single revolution or at most two revolutions (Fig. 2.14c). The longer is the way before landing, the more is an uncertainty in a

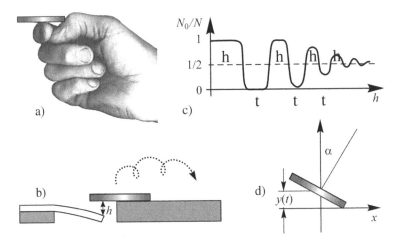

Fig. 2.14 Exercises with a coin: (**a**) a standard situation; (**b**) a physical model with a controllable "strength of a kick" (h is a ruler bend); (**c**) qualitative outlook of an experimental dependency "frequency of a head versus a kick strength" (N experiments were performed at a fixed h, N_0 is the number of resulting heads); (**d**) illustration to a conceptual model

---

[28] A persistent student providing us with experimental data flipped a coin 100 times per experiment with a ruler. He controlled a bend of the ruler by changing the number of pages in a book serving as a ruler support.

final state. Frequencies of a head and a tail equalise despite conditions of successive experiments seem the same.

For a dynamical modelling, let us characterise a coin state with a coordinate y and a velocity v of its centre of mass along with an angle of rotation α about the z-axis perpendicular to x and y (Fig. 2.14d) and an angular velocity ω. Let us single out three qualitatively different stages in the system evolution and introduce special approximation at each of them.

Start. Initial conditions: a coin starts to move having a head as its upper side with a linear velocity v_0 directed vertically; rotation occurs clockwise with an angular velocity ω_0 (Fig. 2.14d). If $2y < d \sin \alpha$ for the starting conditions (where d is the diameter of the coin), then an edge of a coin touches a support after the start of motion (rotation leaves take-off behind) and we regard an outcome as a head. For $2v_0/\omega_0 > d$, the coin flies away without touching the plane of $y = 0$.

Flight. Let us neglect interaction of a coin with air. Let it interact only with the earth. Then, an angular velocity remains constant and is equal to ω_0, while the centre of mass moves with a constant acceleration g.

Finish. Touching a table happens at a time instant t_f, $2y(t_f) = d \sin \alpha(t_f)$, and rotation stops immediately. A coin falls on one of its sides depending on the value of a rotation angle. One gets a head for $0 < (\alpha(t_f) \mathrm{mod} 2\pi) < \pi/2$ or $3\pi/2 < (\alpha(t_f) \mathrm{mod} 2\pi) < 2\pi$ and a tail for $\pi/2 < (\alpha(t_f) \mathrm{mod} 2\pi) < 3\pi/2$.

It is too difficult to specify a single evolution operator for all stages of motion. Therefore, we confine ourselves only with the stage of flight and qualitative considerations for the first and the last stages. Thus, it is obvious that there are many attractors in the phase space of the system: equilibrium points with coordinates $y = v = \omega = 0$, $\alpha = n\pi$, $n = 0, 1, 2, \ldots$, corresponding to final states of a coin lying on one of its sides (Fig. 2.15a shows "tail" points by filled circles and "head" points by open ones). Different attractors correspond to different numbers of coin revolutions before landing. According to the conceptual model, strong dissipation takes place in shaded phase space areas, corresponding to the final stage and to a motion with a small initial velocity v_0, and a representative point reaches one of the two attractors. Boundaries of their basins can be determined from a model of flight. Let us derive them in analogy to Keller (1986) asymptotically from a set of Newton's differential equations $\mathbf{F} = m \cdot \mathbf{a}$ and $\mathbf{M} = I \cdot \boldsymbol{\beta}$, where $\mathbf{F}$ and $\mathbf{M}$ are resultants of forces and their moment of rotation, respectively, $\mathbf{a}$ and $\boldsymbol{\beta}$ are linear and angular accelerations, m and I are the coin mass and moment of inertia. In our case, a model takes the form

$$dy/dt = v, \, dv/dt = -g, \, d\alpha/dt = \omega_0, \, d\omega/dt = 0. \qquad (2.41)$$

Given initial conditions, a solution to equation (2.41) is an orbit

$$y(t) = v_0 t - gt^2 / 2, \, v(t) = v_0 - gt, \, \alpha(t) = \omega_0 t. \qquad (2.42)$$

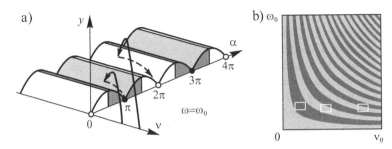

Fig. 2.15 Illustrations to the dynamics of the coin flip model: (**a**) a three-dimensional section of phase space of the dynamical system (2.41) at $\omega = \omega_0$. Wave-shaped surface bounds an area of the final stage where strong dissipation takes place. The considered model of "flight" does not describe the start from the shaded area. Curves with arrows are examples of phase orbits; (**b**) a section of the phase space with a plane ($y = 0$, $\alpha = 0$). Basin of attractors corresponding to the final state "tail" is shaded. *White rectangles* illustrate an accuracy of the initial condition setting (a noise level); their area is $\Delta v \times \Delta \omega$

From here, one gets a dependency $\alpha(t_f) = f(v_0, \omega_0)$ and expressions for the basin boundaries on the plane ω_0, v_0 (a section of phase space by a plane $\alpha = 0, \ y = 0$, Fig. 2.15b)

$$\alpha(t_f) = 2\omega_0 v_0 / g = \pi / 2 + \pi n. \qquad (2.43)$$

Given exact initial conditions, which are obligatory under the dynamical approach, a coin reaches a definite final state. According to this approach, one can predict a final state of a coin, which is illustrated in Fig. 2.16a, where the frequency of "a head" outcome takes only the values of 0 and 1 depending on v_0. It corresponds to reality only for small v_0 (Fig. 2.16b). However, if a flip is sufficiently strong so that a coin performs several revolutions, then such an approach only misleads. Experiments show that by even making efforts to improve accuracy of initial condition setting, one can assure "a head" or "a tail" outcome only for small number of a coin revolution. A significantly more plausible model is obtained if one refuses dynamical description and introduces random quantities into consideration. Let us assume that $v_0 = V_0 + \xi$, where V_0 is a deterministic component, ξ is a random quantity, e.g. distributed uniformly in some interval of Δv with a centre at V_0. Such a stochastic model demonstrates dependency on V_0 qualitatively coinciding with an experiment. Frequencies of both outcomes tend to be 0.5 and vertical and horizontal pieces of the plot are smoothed out for a large number of revolutions.

Given a uniform distribution of ξ, it is convenient to explain observed regularities by selecting a rectangular area $\Delta v \times \Delta \omega$ with a centre at V_0 (Fig. 2.15b). If the entire area is included into a basin of a certain attractor, then an event corresponding to that attractor occurs for sure, i.e. a frequency of one of the outcomes equals unity. If the area intersects both basins (for "a head" and "a tail"), then a frequency of a certain outcome is determined by a portion of the area occupied by the corresponding basin. In general, a frequency of "a head" is defined by an integral taken over the entire

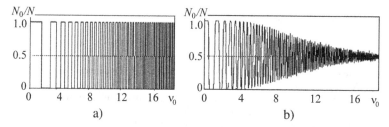

Fig. 2.16 Frequency of "a head" versus an initial velocity v_0 at a fixed value of ω_0: (**a**) exact setting of initial conditions; (**b**) an error in initial conditions

region occupied by its basin of attraction $P\{H\} = \iint\limits_H p(v_0, \omega_0)dv_0\,d\omega_0$, where p is the probability density for observing a "head" in respect of the initial conditions.

Apart from the considered asymptotic and stochastic models, one can suggest a purely empirical probabilistic model. For instance, one can approximate an experimental dependency of the frequency of "a head" on the initial velocity (or on the strength of a flip) shown in Figs. 2.14c and 2.16b with a formula

$$N_0/N = \begin{cases} z, & 0 < z(v) < 1, \\ 1, & z(v) > 1, \\ 0, & z(v) < 0, \end{cases}$$
$$z(v) = 0.5 + a\,e^{-bv}\cos(cv). \tag{2.44}$$

Thus, we have illustrated possibility of the description of a single real-world object with different models, both dynamical and stochastic ones. Each of the models can be useful for certain purposes. It proves again a conventional character of the labels "dynamical system" and "random quantity" in application to real-world situations. In general, even an "international" symbol of randomness, a coin flip, should be considered from the viewpoint of a partial determinancy conception.

Finally, we note that apart from the alternative "deterministic models versus stochastic models", there are other, more complex, interactions between the deterministic and stochastic approaches to modelling. In particular, complicated deterministic small-scale behaviour may be appropriately described by stochastic equations and large-scale averages of a random process may exhibit a good deal of deterministic regularity (Sect. 2.5). Traditional statistical approaches, such as methods of statistical moments or Kalman filtering, are successfully used to estimate parameters in deterministically chaotic systems (see Sects. 8.1.2 and 8.2.2 for concrete examples). Concepts of the theory of probability and the theory of random processes are fruitfully used to describe statistical properties of dynamical chaos, (see, e.g. Anishchenko et al., 2005a, b). Therefore, both approaches discussed throughout this chapter are often used together for the description of complex phenomena in nature.

References

Aivazian, S.A.: Statistical Investigation of Dependencies. Moscow, Metallurgiya (in Russian) (1968)

Alligood, K., Sauer, T., and Yorke, J.A.: Chaos: An Introduction to Dynamical Systems. New York: Springer Verlag, New York (2000)

Andronov, A.A., Vitt, A.A., Khaikin, S.E.: Theory of Oscillators. Nauka, Moscow (1959). Translated into English: Pergamon Press, Oxford (1966)

Andronov, A.A., Leontovich, E.A., Gordon, I.I., Mayer, A.G.: Theory of Bifurcations of Dynamic Systems on a Plane. Nauka, Moscow (1967). Translated into English: Wiley, New York (1973)

Anishchenko, V.S.: Attractors of dynamical systems. Izvestiya VUZ. Applied Nonlinear Dynamics (ISSN 0869-6632). **5**(1), 109–127, (in Russian) (1997)

Anishchenko, V.S., Astakhov, V.V., Neiman, A.B., Vadivasova, T.Ye., Schimansky-Geier, L.: Nonlinear Effects in Chaotic and Stochastic Systems. Tutorial and Modern Development. Springer, Berlin (2002)

Anishchenko, V.S., Okrokvertskhov, G.A., Vadivasova, T.E., Strelkova, G.I.: Mixing and spectral correlation properties of chaotic and stochastic systems: numerical and physical experiments. New J. Phys. **7**, 76–106 (2005a)

Anishchenko, V.S., Vadivasova, T.Ye., Okrokvertskhov, G.A., Strelkova, G.I.: Statistical properties of dynamical chaos. Phys. Uspekhi. **48**, 163–179 (2005b)

Arnold, V.I.: Additional Chapters of the Theory of Ordinary Differential Equations. Nauka, Moscow (in Russian) (1978)

Arnold, V.I.: Ordinary Differential Equations. Nauka, Moscow (1971). Translated into English: MIT Press, Cambridge, MA (1978)

Badii, R., Politi, A.: Complexity: Hierarchical Structures and Scaling in Physics. Cambridge University Press, Cambridge (1997)

Bautin, N.N., Leontovich, E.A.: Methods and Techniques for Qualitative Investigation of Dynamical Systems on a Plane. Nauka, Moscow (in Russian) (1990)

Belykh, V.N.: Elementary introduction into qualitative theory and bifurcation theory in dynamical systems. Soros Educ. J. (1), 115–121 (in Russian) (1997)

Berger, P., Pomeau, Y., Vidal, C.: Order Within Chaos. Hermann, Paris (1984)

Broomhead, D.S., King, G.P.: Extracting qualitative dynamics from experimental data. Physica D. **20**, 217–236 (1986)

Butenin, N.V., Neimark, Yu.I., Fufaev, N.A.: Introduction into the Theory of Nonlinear Oscillations. Nauka, Moscow (in Russian) (1987)

Cencini, M., Falcioni, M., Olbrich, E., et al. Chaos or noise: Difficulties of a distinction. Phys. Rev. E. **62**, 427–437 (2000)

Danilov Yu.A.: Lectures in Nonlinear Dynamics. An Elementary Introduction. Postmarket, Moscow (in Russian) (2001)

Eckmann, J.P., Ruelle, D.: Ergodic theory of chaos and strange attractors. Rev. Mod. Phys. **57**, 617–656 (1985)

Gnedenko, B.V.: A Course in Probability Theory. Gostekhizdat, Moscow (1950). Translated into English: Morikita, Tokyo (1972)

Grassberger, P., Procaccia, I.: Measuring the strangeness of strange attractors. Physica D. **9**, 189–208 (1983)

Guckenheimer, J., Holmes, P.: Nonlinear Oscillations, Dynamical Systems and Bifurcations of Vector-Fields. Springer, Berlin (1983)

Hoel, P.G.: Introduction to Mathematical Statistics, 4th ed. Wiley, New York (1971)

Ibragimov, I.A., Has'minskii, R.Z.: Asymptotic Theory of Estimation. Nauka, Moscow (1979). Translated into English Under the Title Statistical Estimation. Springer, New York (1981)

Katok, A., Hasselblat, B.: Introduction to the modern theory of dynamical systems. Encyclopaedia of Mathematics and its Applications, vol. 54. Cambridge University Press, Cambridge (1995)

Keller, J.B.: The probability of heads. Am. Math. Monthly. **93**(3), 191–197 (1986)

Kendall, M.G., Stuart, A.: The Advanced Theory of Statistics,. vols. 2 and 3. Charles Griffin, London (1979)

Kennel, M.B., Brown, R., Abarbanel, H.D.I.: Determining embedding dimension for phase-space reconstruction using a geometrical construction. Phys. Rev. A. **45**, 3403–3411 (1992)

Kipchatov, A.A.: Estimate of the correlation dimension of attractors, reconstructed from data of finite accuracy and length. Tech. Phys. Lett. **21**(8), 627–629 (1995)

Korn, G., Korn, T.: Handbook of Mathematics for Engineers and Scientists. McGraw-Hill, New York (1961)

Kornfel'd I.P., Sinai Ya.G., Fomin, S.V.: Ergodic Theory. Springer, Berlin (1982)

Kravtsov, Yu.A.: Fundamental and practical limits of predictability. In: Kravtsov, Yu.A. (ed.) Limits of Predictability, pp. 170–200. TsentrCom, Moscow (in Russian) (1997)

Kravtsov, Yu.A.: Randomness, determinancy, predictability. Phys. Uspekhi. **158**(1), 93–115 (in Russian) (1989)

Kuznetsov, S.P.: Complex dynamics of oscillators with delayed feedback (review). Radiophys. Quantum Electr. **25**(12), 1410–1428 (in Russian) (1982)

Kuznetsov, S.P.: Dynamical Chaos. Fizmatlit, Moscow (in Russian) (2001)

Landa, P.S.: Nonlinear Oscillations and Waves in Dynamical Systems. Kluwer Academic Publishers, Dordrecht (1996)

Loskutov, A.Yu., Mikhailov, A.S.: Basics of Complex Systems Theory. Regular and Chaotic Dynamics, Moscow (2007)

Lusternik, L.A., Sobolev, V.I.: Elements of Functional Analysis. Nauka, Moscow (1965). Translated into English: Hindustan Publishing Corpn., Delhi (1974)

Makarenko, N.G.: Fractals, attractors, neural networks and so forth. Procs. IV All-Russian Conf. "Neuroinformatics-2002", Moscow, 2002. M., 2002. Part 2. pp. 121–169 (in Russian)

Malinetsky, G.G.: Chaos. Structures. Numerical Experiment. An Introduction to Nonlinear Dynamics. Editorial URSS, Moscow (in Russian) (2000)

Malinetsky, G.G., Potapov, A.B.: Contemporary Problems of Nonlinear Dynamics. Editorial URSS, Moscow (in Russian) (2000)

Mathematical Encyclopedic Dictionary. 846p. Sov. Encyclopedia, Moscow (in Russian) (1988)

Philosophic Encyclopedic Dictionary. Sov. Encyclopedia, Moscow, 840p. (in Russian) (1983)

Pisarenko, V.F., Sornette, D.: Statistical methods of parameter estimation for deterministically chaotic time series. Phys. Rev. E. **69**, 036122 (2004)

Poincare, H.: Science and Method. Dover, New York (1982)

Press, W.H., Flannery, B.P., Teukolsky, S.A., Vetterling, W.T.: Numerical Recipes in C. Cambridge University Press, Cambridge (1988)

Pugachev, V.S.: Theory of Probabilities and Mathematical Statistics. Nauka, Moscow (in Russian) (1979)

Pugachev, V.S.: Probability Theory and Mathematical Statistics for Engineers. Pergamon Press, Oxford (1984)

Shalizi, C.R.: Methods and techniques of complex systems science: an overview, vol. 3, arXiv:nlin.AO/0307015 (2003). Available at http://www.arxiv.org/abs/nlin.AO/0307015

Shil'nikov L.P., Shil'nikov A.L., Turayev, D.V., Chua, L.: Methods of Qualitative Theory in Nonlinear Dynamics. Parts I and II. World Scientific, Singapore (2001)

Sinai Ya.G.: Contemporary problems of ergodic theory. Fizmatlit, Moscow (in Russian) (1995)

Smith, L.A.: Maintenance of uncertainty. Proc. Int. School of Physics "Enrico Fermi", Course CXXXIII, pp. 177–246. Italian Physical Society, Bologna (1997). Available at http://www.maths.ox.ac.uk/~lenny

Vapnik, V.N.: Estimation of Dependencies Based on Empirical Data. Nauka, Moscow (1979). Translated into English: Springer, New York (1982)

Vapnik, V.N.: The Nature of Statistical Learning Theory. Springer, New York (1995)

von Mises, R.: Mathematical Theory of Probability and Statistics. Academic Press, New York (1964)

Chapter 3
Dynamical (Deterministic) Models of Evolution

3.1 Terminology

3.1.1 Operator, Map, Equation, Evolution Operator

Dynamical modelling requires specification of a D-dimensional state vector $\mathbf{x} = (x_1, x_2, \ldots, x_D)$, where x_i are dynamical variables, and some rule Φ_t allowing unique determination of future states $\mathbf{x}(t)$ based on an initial state $\mathbf{x}(0)$:

$$\mathbf{x}(t) = \Phi_t(\mathbf{x}(0)). \tag{3.1}$$

The rule Φ_t is called an *evolution operator*. "*Operator* is the same as a *mapping*... Mapping is a law according to which an every element x of a given set X is confronted with a uniquely determined element y of another given set Y. At that, X may coincide with Y. The latter situation is called *self-mapping*" (Mathematical dictionary, 1988) (Fig. 3.1a, b). In application to an evolution of a dynamical system state (motion of a representative point in a phase space), one often uses the term "*point map*".

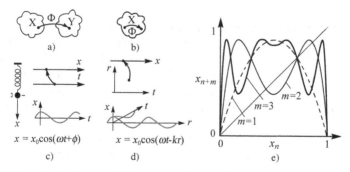

Fig. 3.1 Different kinds of maps: (**a**) from one set into another one; (**b**) self-mapping; (**c**) a function of time describing friction-free oscillations of a pendulum; (**d**) a function of two variables describing a harmonic wave; (**e**) iterates of a quadratic map $x_{n+1} = rx_n(1 - x_n)$ at $r = 3.5$

B.P. Bezruchko, D.A. Smirnov, *Extracting Knowledge From Time Series*, Springer Series in Synergetics, DOI 10.1007/978-3-642-12601-7_3,
© Springer-Verlag Berlin Heidelberg 2010

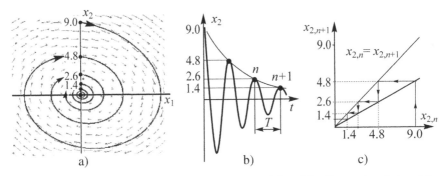

Fig. 3.2 Phase space of the linear dissipative oscillator (3.2) and its discrete description: (**a**) a velocity field specified by equation (3.4), *arrows* denote velocities of a state change; (**b**) a time realisation $x_2(t)$, *filled squares* are the points corresponding to the Poincare section $x_1 = 0$. They are separated by a time interval $\tau = T$. Their relationship is described with a map presented in the panel c; (**c**) a one-dimensional return map for the Poincare section $x_1 = 0$; an evolution can be studied conveniently with Lamerey's diagram (*arrows*)

An evolution operator can be specified directly as a map from a set of initial states $\mathbf{x}(0)$ into a set of future states $\mathbf{x}(t)$. However, it is more often determined indirectly with the aid of equations. "*Equation* is a way to write down a problem of looking for such elements a in a set A which satisfy an equality $F(a) = G(a)$, where F and G are given maps from a set A into a set B"[1] (Mathematical dictionary, 1988). If an equation is given, an evolution operator can be obtained via its solution. Thus, for an ordinary differential equation the theorem about unique existence of a solution assures the existence and one-oneness of a map Φ_t in equation (3.1) under some general conditions. If an exact solution of an equation is impossible, one searches for an approximate solution in the form of a numerical algorithm simulating a representative point motion in a phase space (Fig. 3.2a).

3.1.2 Functions, Continuous and Discrete time

Functions of independent variables (of a single variable $\mathbf{x} = \mathbf{F}(t)$ or of several ones $\mathbf{x} = \mathbf{F}(t, \mathbf{r})$) map a set of the values of the independent variables into a set of the values of the dependent (dynamical) variables. In Fig. 3.1c, d, time t and the vector of spatial coordinates $\mathbf{r}$ are independent variables, while a deviation x from an equilibrium state is a dynamical variable. If a function $\mathbf{F}$ depends explicitly on the initial values of dynamical variables, it can represent an evolution operator, see, e.g., equation (3.3).

A state of an object may be traced either *continuously* in time or *discretely*, i.e. at certain instants t_n separated from each other with a step Δt. In the latter

---

[1] If A and B are number sets, one gets algebraic or transcendental equations. If they are function sets, one gets differential, integral and other equations depending on the kind of the maps.

case, the order number of a time instant $n = 0, 1, 2, 3, \ldots$ is called *discrete time*. If observations are separated by an equal time interval Δt, then the relationship between continuous time t and discrete time n is linear: $t_n = n\Delta t$. For unequal intervals, the dependency can be more complicated. Similarly, one can use discrete versions of spatial coordinates, e.g. a number of steps along a chosen direction, a number of elements in a chain or a lattice.

3.1.3 Discrete Map, Iterate

In "discrete modelling" the values of dynamical variables $\mathbf{x}_n$ at different discrete-time instants n are related to each other via a map from a phase space X into itself $(X \rightarrow X)$: $\mathbf{x}_{n+1} = \mathbf{F}(\mathbf{x}_n, \mathbf{c})$, where $\mathbf{c}$ is a parameter vector. Such a *recurrent formula*[2] for an evolution operator is also called *discrete map*. To study a map, one uses its iterates. *Iterate* (from a Latin word "iteratio", i.e. repetition) is a result of a repeated application of some mathematical operation. Thus, if $\mathbf{F}(\mathbf{x}) \equiv \mathbf{F}^{(1)}(\mathbf{x})$ is a certain function of $\mathbf{x}$ mapping its domain into itself, then functions $\mathbf{F}^{(2)}(\mathbf{x}) \equiv \mathbf{F}[\mathbf{F}(\mathbf{x})]$, $\mathbf{F}^{(3)}(\mathbf{x}) \equiv \mathbf{F}[\mathbf{F}^{(2)}(\mathbf{x})]$, $\ldots$, and $\mathbf{F}^{(m)}(\mathbf{x}) \equiv \mathbf{F}[\mathbf{F}^{(m-1)}(\mathbf{x})]$ are called the second, the third, $\ldots$, and the mth iterates of $\mathbf{F}(\mathbf{x})$, respectively. The index m is the order number of an iterate. For instance, Fig. 3.1e shows three iterates of a quadratic map $x_{n+1} = rx_n(1 - x_n)$, where r is a parameter.

3.1.4 Flows and Cascades, Poincare Section and Poincare Map

In a DS whose evolution operator is specified via differential equations, time is continuous. In a phase space of such a DS, motions starting from close initial points correspond to a beam of phase orbits resembling lines of flow in a fluid (Fig. 3.2a). Such DSs are called *flows* in contrast to *cascades*, i.e. to DSs described with discrete maps, $\mathbf{x}_{n+1} = \mathbf{F}(\mathbf{x}_n, \mathbf{c})$.

The term "Poincare section" denotes a section of a phase space of a flow with a set of dimension $D - 1$, e.g. a section of a three-dimensional space with a surface or a two-dimensional space with a curve. The term "Poincare map" is used for mapping of a set of unidirectional "punctures" of a Poincare section with a phase orbit into itself. It relates a current "puncture" to the next one.

3.1.5 Illustrative Example

Let us illustrate the above-mentioned terms with a model of the oscillations of a load on a spring in a viscous medium. An etalon model of low-amplitude oscillations in a

---

[2] A recurrent formula is the relationship of the form $x_{n+p} = f(x_n, x_{n+1}, \ldots, x_{n+p-1})$ allowing calculation of any element in a sequence if its p starting elements are specified.

viscous fluid under the influence of a restoring force, proportional to the deviation x from an equilibrium state, is an ordinary differential equation of "a linear oscillator":

$$d^2x/dt^2 + 2\delta dx/dt + \omega_0^2 x = 0. \tag{3.2}$$

Similar to system (2.1), one must provide initial conditions $x(0) = x_0$ and $dx(0)/dt = v_0$ to specify an orbit of the two-dimensional system uniquely. An analytic solution to equation (3.2) reads as

$$\begin{aligned}
x(t) &= \left[x_0 \cdot \cos \omega t + \tfrac{v_0 + \delta \cdot x_0}{\omega} \sin \omega t \right] e^{-\delta \cdot t}, \\
v(t) &= \left[v_0 \cdot \cos \omega t - \tfrac{\delta \cdot v_0 + \omega_0^2 x_0}{\omega} \sin \omega t \right] e^{-\delta \cdot t},
\end{aligned} \tag{3.3}$$

where $\omega = \sqrt{\omega_0^2 - \delta^2}$ and $v(t) = dx(t)/dt$. The formula (3.3) determines the relationship between an initial state x_0, v_0 and a future state $x(t)$, $v(t)$. Thus, it gives explicitly an evolution operator of the system (3.2).

Another way to write down evolution equations for the same system is a set of two first-order ordinary differential equations:

$$\begin{aligned}
dx_1/dt &= x_2, \\
dx_2/dt &= -2\delta \cdot x_2 - \omega_0^2 x_1,
\end{aligned} \tag{3.4}$$

where $x_1 = x$, $x_2 = dx/dt$. It is convenient for graphical representations, since it specifies explicitly a velocity field on the phase plane (Fig. 3.2a). Roughly speaking, one can move from an initial state to subsequent ones by doing small steps in the directions of arrows. It is realised in different algorithms for numerical solution of differential equations. To construct a discrete analogue of equation (3.4), one must convert to the discrete time $n = t/\Delta t$. In the simplest case, one can approximately replace the derivatives with finite differences $dx(t)/dt \approx (x_{n+1} - x_n)/\Delta t$ and $dv(t)/dt \approx (v_{n+1} - v_n)/\Delta t$ and get difference equations which can be rewritten in the form of a two-dimensional discrete map

$$\begin{aligned}
x_{n+1} &= x_n + v_n \Delta t, \\
v_{n+1} &= v_n(1 - 2\delta \cdot \Delta t) - \omega_0^2 x_n \Delta t.
\end{aligned} \tag{3.5}$$

At sufficiently small Δt, an orbit of the map approximates well a solution to equation (3.4), i.e. the map (3.5) is a sufficiently *accurate difference scheme*.

In a Poincare section of the phase plane with a straight line $x_1 = 0$ (an ordinate axis), it is possible to establish the relationship between subsequent "punctures" of the axis by a phase orbit (Fig. 3.2). The resulting Poincare map takes the form

$$v_{n+1} = v_n e^{-\delta \cdot T}, \tag{3.6}$$

where $T = 2\pi/\omega$. One can get more detailed information about a modelled object motion from the map (3.5) than from the map (3.6), since the latter describes only the decay of an amplitude. On the other hand, one can use vivid Lamerey's diagram in the one-dimensional case. To construct that diagram on the plane v_n, v_{n+1}, one passes a vertical straight line to the plot of the map, then a horizontal line to the diagonal $v_n = v_{n+1}$, etc., as shown in Fig. 3.2c.

3.2 Systematisation of Model Equations

Mathematicians have developed a rich arsenal of tools for dynamical description of motions. Here, we present their systematisations according to different principles. Firstly, we consider descriptive capabilities in application to objects with various complexity of their spatial structure. Any real-world object is somewhat "spatially extended". Depending on the number and sizes of composing elements, intensity and speed of their interaction, one can model an object as *concentrated* at a single spatial point or at several ones. The latter is the simplest kind of "spatially extended" configuration. A "completely smeared" (continuous) spatial distribution of an object characteristic is also possible. Such an object is also called a "distributed system". Further, we use the term "spatially extended system" more often, since it is more general.

If an object is characterised by a uniform spatial distribution of variables and one can consider only their temporal variations, it is regarded as *concentrated* at a single spatial point. Such a representation is appropriate if a perturbation at a certain spatial point reaches other parts of a system in a time interval much less than time scales of the processes under consideration. In the language of the theory of oscillations and waves, a perturbation wavelength is much greater than the size of an object. Concentrated systems are described with finite-dimensional models such as difference or ordinary differential equations.

If one has to provide a continuous set of values to specify a system state uniquely, then the system is distributed. Classical tools to model such a system are partial differential equations (PDEs), integro-differential equations (IDEs) and delay differential equations (DDEs). For instance, in description of a fluid motion, one refuses consideration of the molecular structure. The properties are regarded uniformly "smeared" within "elementary volumes" which are sufficiently big as compared with a molecule size, but small as compared with macro-scales of a system. This is the so-called *mesoscopic level*.[3] Such "volumes" play a role of elementary particles whose properties vary in space and time according to the Navier–Stokes equations. These famous partial differential equations represent an etalon infinite-dimensional model in hydrodynamics.

---

[3] It is intermediate between a microscopic level, when one studies elements of a system separately (e.g. molecules of a fluid), and a macroscopic one, when an entire system is considered as a whole (e.g. in terms of some averaged characteristics).

Spatially extended systems can be thought of as separated into parts (elements). Each of the parts is a system concentrated at a certain spatial point. Models of such systems are typically multidimensional. One often uses PDEs or a set of coupled maps or ODEs. Depending on the intensity of coupling between elements, a model dimension required for the description of motion and relevant mathematical tools can vary significantly. Thus, if a liquid freezes, one does no longer need PDEs to describe motion of a resulting ice floe and is satisfied with a set of several ODEs for rotational and progressive motions of a solid. If only progressive motions take place, then even a model of a material point suffices.

When a signal with sufficiently broad power spectrum (Sect. 6.4.2), e.g. a short pulse, propagates in a system, variations in its power spectrum and phase shifts at some frequencies may induce a time delay and smearing of the signal. Smearing occurs if a system bandwidth is insufficient to pass all components of a signal, e.g. due to sluggishness. Thus, if one uses a δ-function input, sluggishness of a system leads to a finite width of a response signal waveform. The stronger the sluggishness, the wider the response waveform. A shift of the time instant when a response signal appears relative to the time instant of an input pulse is an estimate of the delay time (Fig. 6.4c). Both sluggishness and delay are often modelled with finite-dimensional models, but the phenomenon of time delay is more naturally described with a DDE. The latter is an infinite-dimensional system, since it requires an *initial curve* over a time-delay interval as an initial state, i.e. a continuous set of values of a dynamical variable.

In Fig. 3.3, mathematical tools for modelling of temporal evolution are systematised according to their *level of generality*, their capability to describe more diverse objects and kinds of motion. As a rule, model equations of greater generality require greater computational efforts for their investigation.

The simplest kind of models is *explicit functions of time* $\mathbf{x} = \mathbf{F}(t)$. In linear problems or special cases, such models can be obtained as *analytic* solutions to evolution equations. Despite an enormous number of functions used in practice (Sect. 3.3), their capabilities for the description of complex (especially, chaotic) time realisations are quite restricted. A somewhat more general case is represented by algebraic or transcendental equations

$$\mathbf{F}(\mathbf{x}, t) = 0. \tag{3.7}$$

If equation (3.7) has no analytic solution, then one says that it defines a dependency $\mathbf{x}(t)$ implicitly.

A "left column" of the scheme consists of various differential equations (DEs). These are equations involving derivatives of dynamical variables in respect of independent variables (time t and spatial coordinates $\mathbf{r}$). For instance, a general first-order DE reads as

$$\mathbf{F}(\mathbf{x}(t, \mathbf{r}), \partial\mathbf{x}(t, \mathbf{r})/\partial t, \ \partial\mathbf{x}(t, \mathbf{r})/\partial\mathbf{r}, t, \mathbf{r}, \mathbf{c}) = 0, \tag{3.8}$$

where $\mathbf{x}$ is a vector of dynamical variables. ODEs were the first differential equations used in scientific practice

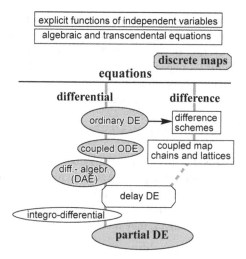

Fig. 3.3 A conventional scheme of dynamical model kinds extending the scheme given in Horbelt (2001). Descriptive capabilities and computational efforts required for investigation increase from top to bottom

$$\mathbf{F}(\mathbf{x}(t), d\mathbf{x}(t)/dt, \ldots, d^n\mathbf{x}(t)/dt^n, t, \mathbf{c}) = 0. \qquad (3.9)$$

ODEs of the form $d\mathbf{x}/dt = \mathbf{F}(\mathbf{x},\mathbf{c})$ allow a clear geometric interpretation. They specify *velocity field*: a direction and an absolute value of a state change velocity $\mathbf{v} = d\mathbf{x}/dt$ at each point of a finite-dimensional phase space. A non-zero vector $\mathbf{v}$ is tangent to a phase orbit at any point. Specification of the velocity field provides a unique prediction of a phase orbit starting from any initial state, i.e. description of all possible motions in the phase space (Fig. 3.2a).

Derivatives of dynamical variables are used in equations of several kinds which differ essentially in the properties of their solutions and the methods of getting the solutions. They are united with a wide vertical line in Fig. 3.3 as branches with a tree stem. ODEs located at the top of a "stem" describe dynamics of concentrated (finite-dimensional) systems, where one does not need to take into account *continuous spatial distribution* of object properties. PDEs also involve spatial coordinates as independent variables and are located at the very bottom of the scheme. They are the most general tool, since they also describe infinite-dimensional motions of spatially distributed systems. However, solving PDEs requires much greater computational efforts compared to solving ODEs. Besides, PDEs loose a vivid geometric interpretation peculiar to ODEs.

Differential algebraic equations (DAEs) are just a union of ODEs and algebraic equations:

$$\begin{aligned}
\mathbf{F}(d\mathbf{x}(t)/dt, \ \mathbf{x}(t), \mathbf{y}(t), t, \mathbf{c}) &= 0, \\
\mathbf{G}(\mathbf{x}(t), \mathbf{y}(t), t, \mathbf{c}) &= 0,
\end{aligned} \qquad (3.10)$$

where $\mathbf{x}$ is a D-dimensional state vector, $\mathbf{y}$ is a K-dimensional vector which does not add new degrees of freedom, $\mathbf{F}$ is a vector-valued function of dimension D, $\mathbf{G}$ is a vector-valued function of dimension K. The second equation is algebraic and determines (implicitly) a dependence of $\mathbf{y}(t)$ on $\mathbf{x}(t)$. Methods to solve such equations are very similar to those for ODEs.

Delay differential equations can, for instance, read as

$$\mathbf{F}(\mathbf{x}(t), d\mathbf{x}(t)/dt, \mathbf{x}(t - \tau), \mathbf{c}) = 0. \tag{3.11}$$

Distinction from ODEs consists in that the values of dynamical variables at a separated past time instant enter the equations along with their current values. ODEs can be regarded as a particular case of DDEs for a zero time delay τ.

Integro-differential equations (IDEs) do not, strictly speaking, belong to the class of DEs. Along with derivatives, they involve integrals of dynamical variables, e.g., as

$$\mathbf{F}\left(\mathbf{x}(t), d\mathbf{x}(t)/dt, \ldots, d^n\mathbf{x}(t)/dt^n, \int_{-\infty}^{\infty} k(t, t')\mathbf{x}(t')dt', t, \mathbf{c}\right) = 0, \tag{3.12}$$

where $k(t, t')$ is a kernel of the linear integral transform. If no derivatives enter an IDE, it is called just an integral equation.

DDEs and IDEs also provide an infinite-dimensional description. DDEs can often be considered as a particular case of IDEs. For instance, an IDE $d\mathbf{x}(t)/dt = \mathbf{F}(\mathbf{x}(t)) + \int_{-\infty}^{\infty} k(t, t')\mathbf{x}(t')dt'$ in the case of $k(t, t') = \delta(t - t' - \tau)$ turns into a DDE $d\mathbf{x}(t)/dt = \mathbf{F}(\mathbf{x}(t)) + \mathbf{x}(t - \tau)$.

To construct a discrete analogue of equation (3.4), one turns to the discrete time $n = t/\Delta t$ and finite differences. At sufficiently small Δt, the difference equation (3.5) have a solution close to that of equation (3.4). With increase in Δt, a difference equation (a discrete map) stops to reflect properties of the original ODEs properly. However, one can construct discrete models exhibiting good correspondence to the original system for large time steps as well. In the example of an oscillator considered above (Fig. 3.2), subsequent values v_n corresponding to the marked points in Fig. 3.2a, b are related at $\Delta t = T$ strictly via the one-dimensional map (3.6) (Fig. 3.2c). The latter map has a dimension smaller than the dimension of the original system and reflects only the monotonous decay of an oscillation amplitude and the transition to an equilibrium state. Here, the loss of information about a system behaviour between observation instants is a payment for a model simplicity.

Both discrete and continuous systems are valuable by themselves so that one could avoid speaking of any priorities. However, modelling practice and recognition of specialists are historically in favour of DEs. It is due to the fact that until the middle of twentieth century, physics was a "scientific prime" and relied mainly on DEs, in particular, on PDEs. To study them, physicists used various analytic techniques. Computers and digital methods, which can now efficiently cope with difference equations, were not yet widely available. Therefore, an arsenal of maps used in modelling was much poorer at that time, than a collection of exemplary flows. However, contemporary tendencies of wider usage of non-linear equations and development of numerical techniques for investigation of multidimensional systems with complex spatial and temporal behaviours seem favourable to the progress of discrete approaches. Currently, popular tools are discrete *ensemble models* called *coupled maps lattices*, which combine a large number of maps with non-trivial temporal dynamics (Sect. 3.7). As models of spatially extended systems, they yield to PDEs in generality but are much simpler for numerical investigation. A specific kind of multidimensional maps or ODEs is represented by artificial neural networks which have recently become a widespread tool, in particular, in the field of function approximation (Sect. 3.8).

3.3 Explicit Functional Dependencies

Dynamical models of evolution in the form of *explicit functions* of time $\mathbf{x} = \mathbf{F}(t)$ can be specified analytically, graphically or as tables and can be obtained in any of the ways described in Sect. 1.5, e.g. by solving a DE or approximating experimental data (Sect. 7.2.1). It is impossible to list all explicit functions used by mathematicians. Yet, it is possible to distinguish some classes of functions. A practically important class of *elementary functions* includes algebraic polynomials, power, rational, exponential, trigonometric and inverse trigonometric functions. As well, it includes functions obtained via a finite number of arithmetical operations and compositions[4] of the listed ones. Let us consider several elementary functions and DEs, whose solutions they represent.

(1) Linear function $x(t) = x_0 + v_0 t$ is a solution to an equation

$$dx/dt = v_0, \tag{3.13}$$

which describes a progressive motion with a constant velocity v_0 and an initial condition $x(0) = x_0$. Its plot is a straight line (Fig. 3.4a).
(2) Algebraic polynomial of an order K reads as

$$x(t) = c_0 + c_1 t + c_2 t^2 + \ldots + c_K t^K, \tag{3.14}$$

---

[4] "Superposition (composition) of functions is arranging a composite function (function of function) from two functions" (Mathematical dictionary, 1988). Here, the terms "superposition" and "composition" are synonyms. However, physicists often call superposition of functions f_1 and f_2 their linear combination $af_1 + bf_2$, where a and b are constants. Then, the meanings of the terms "superposition" and "composition" become different. To avoid misunderstanding, we use only the term "composition" in application to composite functions.

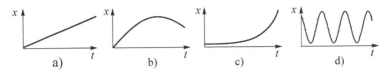

Fig. 3.4 Plots of some elementary functions: (**a**) linear function; (**b**) power function; (**c**) exponential function with $\alpha > 0$; (**d**) sinusoid

where c_i are constant coefficients. It is a solution to an equation $d^K x/dt^K =$ const. A linear function is a particular case of equation (3.14) for $K = 1$. In the case of uniformly accelerated motion of a body thrown up from a height h with an initial velocity v_0, an equation of motion obtained from the Newton's second law and the law of gravity takes the form $d^2 x/dt^2 = -g/m$, where an x-axis is directed upward, m is the mass of a body, g is the gravitational acceleration. The solution is $x(t) = h + v_0 t - gt^2/2$ (Fig. 3.4b). It is valid in a friction-free case and until a body falls down on a land.

(3) Fractional rational function is a ratio of two algebraic polynomials $x(t) = P(t)/Q(t)$. Its particular case for $Q(t) = $ const is an algebraic polynomial.

(4) Power function $x(t) = t^{\alpha}$, where α is an arbitrary real number. If α is a non-integer, only the domain $t > 0$ is considered. For an integer α, it is a particular case of an algebraic polynomial or a fractional rational function.

(5) Exponential function $x(t) = x_0 e^{\alpha t}$ (Fig. 3.4c) is famous due to the property that the speed of its change at a given point t is proportional to its value at the same point. It is the solution to the equation $dx/dt = \alpha x$ with an initial condition $x(0) = x_0$, which describes, for instance, dynamics of a biological population, where α is a constant parameter meaning birth rate.[5]

(6) A harmonic function $x(t) = x_0 \cos(\omega t + \phi_0)$ is one of the trigonometric functions (Fig. 3.4d). It is a solution to an equation of the harmonic oscillator $d^2 x/dt^2 + \omega^2 x = 0$, which is an *exemplary model* of friction-free oscillations of a material point under the influence of a restoring force, proportional to a deviation x from an equilibrium. Its constant parameters are an amplitude of oscillations x_0, angular frequency ω and an initial phase ϕ_0. A bivariate harmonic function $x(t, r) = x_0 \cos(\omega t - kr + \phi_0)$ describes a monochromatic wave of length $\lambda = 2\pi/k$ travelling along the r-axis, which is a solution to the simple wave equation $\partial x/\partial t + V \partial x/\partial r = 0$.

Wide usage of trigonometric functions is to a significant extent due to the fact that according to *Weierstrass' theorem*, any continuous periodic function $x(t)$ can be arbitrarily accurately approximated with a *trigonometric polynomial*

---

[5] Exponential rise of a population observed at $\alpha > 0$ is called the Malthusian rise, since a catholic monk Malthus in the sixteenth century was the first who got this result. It is valid until population gets too large so that there is no longer enough food for everybody.

$$x(t) = \sum_{i=0}^{K} c_k \, \cos(2\pi k/T + \phi_k),\tag{3.15}$$

where K is a polynomial order. A non-periodic function can be approximated with such a polynomial over a finite interval.

An analogous theorem was proved by Weierstrass for the approximation of functions with an algebraic polynomial (3.14). Algebraic and trigonometric polynomials are often used for approximation of dependencies. This is the subject of the theory of approximation (constructive theory of functions), see Sect. 7.2. In the recent decades, artificial neural networks (Sects. 3.8 and 10.2.1), radial basis functions (Sect. 10.2.1) and wavelets compete with polynomials in the approximation practice. Wavelets have become quite popular and are considered in more detail in Sect. 6.4.2. Here, we just note that they are well-localised functions with zero mean, e.g. $x(t) = e^{-t^2/2} - (1/2)e^{-t^2/8}$.

We will consider non-elementary functions and extensions to the class of elementary functions in Sect. 3.5.

3.4 Linearity and Non-linearity

"Nonlinearity is omnipresent, many-sided and inexhaustibly diverse. It is everywhere, in large and small, in phenomena fleeting and lasting for epochs... Nonlinearity is a capacious concept with many tinges and gradations. Nonlinearity of an effect or a phenomenon means one thing, while nonlinearity of a theory means something different" (Danilov, 1982).

3.4.1 Linearity and Non-linearity of Functions and Equations

The word "linear" at a sensory level is close to "rectilinear". It is associated with a straight line, proportional variations of a cause and an effect, a permanent course, as in Fig. 3.4a. However, according to the terminology used in mathematics and non-linear dynamics, all the dynamical systems mentioned in Sect. 3.3 are linear though the plots of their solutions are by no means straight lines (Fig. 3.4b–d). Evolution operators of those dynamical systems (i.e. differential or difference equations and discrete maps) are linear *rather than their solutions* (i.e. functions of time representing time realisations).

What is common in all the evolution equations presented in Sect. 3.3? All of them obey the *superposition principle*: If functions $x_1(t)$ and $x_2(t)$ of an independent variable t are solutions to an equation, then their linear combination $ax_1(t) + bx_2(t)$ is also a solution, i.e. being substituted instead of $x(t)$, it turns an equation into identity. Only the first powers of a dynamical variable and its derivatives (x, dx/dt, ..., $d^n x/dt^n$) may enter a linear DE. No higher powers and products of the derivatives may be present. Accordingly, linear difference equations may include only the first powers of finite differences or a dynamical variable values at discrete-time instants. Equations of any kind are linear if their right-hand and left-hand sides are linear functions of a dynamical variable and its derivatives. Violation of this property means non-linearity of an equation. For instance, equations (3.2), (3.4), (3.5), (3.6) and (3.13) are linear, while equation (2.1) is non-linear. However, linear non-autonomous (involving an explicit time dependence) equations may include non-linear functions of an independent variable (time), e.g. a non-autonomous linear oscillator reads as $\mathrm{d}^2x/\mathrm{d}t^2 + 2\delta\mathrm{d}x/\mathrm{d}t + \omega_0^2 x = A \cos \omega t$.

A linear function "behaves" quite simply: it monotonously decreases or increases with an argument or remains constant. But linearity of a dynamical system does not mean that its motion is inevitably primitive, which can be seen even from several examples presented in Fig. 3.4. Taking into account the superposition principle, one may find a solution for a multidimensional linear equation as a combination of power, exponential and trigonometric functions (each of them being a solution) which demonstrates quite a complicated temporal dependence, indistinguishable in its outward appearance from an irregular, chaotic behaviour over a finite time interval. However, linear systems "cannot afford many things": changes in a waveform determined by the arousal of higher harmonics, dynamical chaos (irregular solutions with exponential sensitivity to small perturbations), multistability (coexistence of several kinds of established motions), etc.

Systems, processes, effects, phenomena are classified as *linear* or *non-linear* depending on whether they are adequately described with linear *equations* or non-linear ones. The world of non-linear operators is far richer than that of linear ones. Further, there are much more kinds of behaviour in non-linear dynamical systems. The place of "linear things" in a predominantly non-linear environment, "particularly" of linear representations, follows already from the fact that non-linear systems can be linearised (i.e. reduced to linear ones) only for low oscillation amplitudes. For that, one replaces dynamical variables x_k in the equations by the sums of their stationary and variable parts as $x_k = x_{0,k} + \tilde{x}_k (x_{0,k} >> \tilde{x}_k)$ and neglects small terms (higher powers of $\tilde{x}_k$, their products, etc.).

Historically, linear equations in a precomputer epoch had incontestable advantage over non-linear ones in scientific practice, since the former could be investigated rigorously and solved analytically. For a long time, one confidently thought that linear phenomena are more important and widespread in nature and linear approaches are all-sufficient (see discussion in Danilov, 1982). Development of computers, numerical techniques for solving non-linear equations and tools for their graphical representation along with the emergence of contemporary mathematical conceptions, including dynamical chaos theory, shifted an attitude of scientists more in favour of non-linear methods and ideas. At that, the linear viewpoint is regarded an important but a special case.

3.4.2 The Nature of Non-linearity

Non-linearity is natural and organically inherent in the world where we live. Its origin can be different and determined by specific properties of objects. One should speak of conditions for linearity to be observed, rather than for non-linearity. However, according to existing traditions one often explains appearance of non-linearity by a competition between eigenmodes of a linearised system or by a parameter dependence on a dynamical variable. The latter dependence is often vitally necessary and can be realised via a feedback. If sensitivity of vision or hearing organs were constant, independent of an influence level (luminosity or sound volume), one might not successfully orient oneself in gloom and avoid becoming blind during a sunny day, hear a rustle of a creeping snake and avoid getting deaf from thunder. "Biological systems which could not capture enormous range of vitally important influences from environment have just died out loosing the struggle for existence. One could write down on their graves: They were too linear for this world" (Danilov, 1982).

Thus, if a coefficient of reproduction k for a population were constant, independent of the number of individuals x_n (n is discrete time), then at $k > 1$ one would observe its unbounded rise with time according to a linear evolution law:

$$x_{n+1} = kx_n. \tag{3.16}$$

In such a case, overpopulation would be inevitable, while at $k < 1$ a total disappearance of a population would come. A more realistic is a dependence of the parameter k on the variable x_n, e.g. $k = r(1 - x_n)$ leading to non-linearity of an evolution operator $x_{n+1} = rx_n(1 - x_n)$. Non-trivial properties of this exemplary one-dimensional dynamical system called the logistic map, including its chaotic behaviour, are well studied (see also Sect. 3.6.2).

3.4.3 Illustration with Pendulums

Widely accepted exemplary objects for illustrations of linear and non-linear oscillatory phenomena are *pendulums*, i.e. systems oscillating near a stable equilibrium state. Their simplest mechanical representatives are a massive load suspended with a thread or a rod (Fig. 3.5a), a load on a spring (Fig. 3.5c), a ball rolling in a pit, a bottle swimming in water, liquid in a U-shaped vessel, and many others. An electric pendulum is the name for a circuit consisting of a capacitor and inductance, an oscillatory circuit (Fig. 3.5b). One speaks of a chemical pendulum (mixture of chemicals reacting in an oscillatory manner) and an ecological pendulum (two interacting populations of predators and preys) (Trubetskov, 1997).

A free real-world pendulum reaches finally a stable equilibrium state (free motions, Fig. 3.5). Depending on initial conditions (a deviation from a stable equilibrium state x and a velocity dx/dt) and object properties, different motions may precede it. In Fig. 3.5 we illustrate two areas of qualitatively different motions: the left one corresponds to relatively large x, when non-linearity is essential, while the right one corresponds to small, "linear" ones. Time realisations of low-amplitude oscillations are identical for all the pendulums considered. The oscillations are *isochronous*, i.e. their quasi-period T_1 does not depend on a current state. They represent a decaying sinusoid which is a solution to the linear

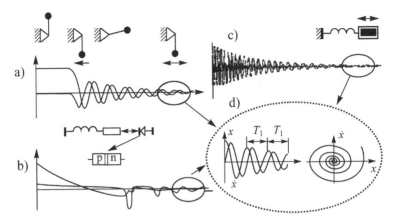

Fig. 3.5 Qualitative outlook of time realisations of x and dx/dt for different pendulums: (**a**) a load on a rod; (**b**) an oscillatory circuit with a diode; (**c**) a load on a spring; (**d**) a linear stage of oscillations (the same for all the examples) is magnified and the corresponding phase portrait on the plane $(x, dx/dt)$ is shown. A representative point moves along an intertwining spiral to an attractor, i.e. to a stable equilibrium point at the origin. Phase portraits of the pendulums are more complicated and diverse in a region of large values of coordinates and velocities

equation (3.2) describing low-amplitude oscillations of all the systems considered up to the coefficients of proportionality. This circumstance is a legal reason to call those oscillations linear. Monotonous decay of oscillations can be modelled with the linear one-dimensional map $x_{n+1} = ax_n$, where $a = \exp(-\delta T_1) < 1$ (Fig. 3.2c).

As compared with a common "standard" for linear oscillations (3.2), types of non-linear behaviour are quite diverse and determined by the properties of each concrete pendulum. Thus, a character of non-linear behaviour differs essentially for the three examples in Fig. 3.5, while their linear stages are identical (Fig. 3.5d). This is related to the peculiarities of each pendulum and to the kinds of their non-linearity (dependency of the parameters on the dynamical variables). For instance, a load on a rod (Fig. 3.5a) exhibits non-linearity due to sinusoidal dependence of a gravitational force moment about a rotation axis on a rotation angle. In an electric pendulum with a semiconductor capacitor (a varactor diode, Fig. 3.5b), non-linearity is related to the properties of a $p - n$ junction, injection and finiteness of charge carrier lifetime. Non-linearity of a spring pendulum (Fig. 3.5c) is determined by the dependence of an elastic force on a spring deformation. For instance, spring coils close up under compression so that an elastic force rises abruptly as compared with the force expected from Hooke's law, i.e. the spring "gets harder". At that, a period of oscillations decreases with their amplitude. In analogy, non-linearity of any oscillator leading to decrease (increase) in a period with an amplitude is called *hard* (*soft*) *spring* non-linearity.

3.5 Models in the form of Ordinary Differential Equations

3.5.1 Kinds of Solutions

Emergence of ordinary differential equations and their history is related to the names of Newton and Leibniz (seventeenth to eighteenth centuries). Afterwards, general procedures to obtain model equations and to find their solutions were developed within analytic mechanics and the theory of differential equations. Here, we describe possible kinds of solutions following the review of Rapp et al. (1999).

3.5.1.1 Elementary Solutions

A solution to a differential equation in the form of an elementary function is called an elementary solution. We confine ourselves with examples from Sect. 3.3. In all of them functions–solutions give exhaustive information about a model dynamics. Interestingly, understanding the behaviour of a dynamical system at Newton's time was tantamount to writing down a formula for a solution $x = F(t)$. That approach even got the name of a *Newtonian paradigm* (Rapp et al., 1999). One spoke of a finite (preferably short) expression consisting of radicals (nth roots), fractional rational, exponential, logarithmic and trigonometric functions. All the solutions considered in Sect. 3.3 are of such a form.

The class of elementary functions (and elementary solutions) is often extended with *algebraic functions*, i.e. solutions to the algebraic equations

$$a_n(t)x^n(t) + a_{n-1}(t)x^{n-1}(t) + \ldots + a_1(t)x(t) + a_0(t) = 0, \qquad (3.17)$$

where n is an integer, $a_i(t)$ are algebraic polynomials. All fractional rational functions and radicals are algebraic functions. The reverse is not true: algebraic functions can be defined by equation (3.17) implicitly.

3.5.1.2 Closed-Form Solutions

Not all differential equations have elementary solutions. There are elementary functions whose integrals are not elementary functions. One of the simplest examples is an elliptic integral

$$\int_0^t \frac{d\tau}{\sqrt{1 + \tau^4}}.$$

The integral exists but is not an elementary function. However, even if an integral of an elementary function is not elementary, one can efficiently cope with it by evaluating it approximately with the aid of available numerical methods.

Expression of a solution via formulas containing integrals of elementary functions is also regarded a complete solution to an equation, so-called integration in finite terms. Thus, a solution to the equation

$$dx/dt + e^{t^2}x = 0 \tag{3.18}$$

given $x(0) = x_0$, reads

$$x(t) = x_0 \exp\left(-\int_0^t e^{\tau^2} d\tau\right).$$

Such a result is called a *closed-form solution*. An elementary solution is its particular case.

Liouville showed that some DEs have no closed-form solutions. For instance, an equation

$$dx/dt + x^2 = t \tag{3.19}$$

which at the first glance seems very simple, cannot be solved in finite terms. A solution exists but cannot be expressed in a closed form. There is no general procedure to get closed-form solutions, though there are many special techniques. In practice, it is often very difficult or even impossible to obtain a closed-form solution.

3.5.1.3 Analytic Solutions

When a closed-form solution is lacking, one can further complicate a technique and try to find a solution in the form of an infinite power series. For instance, let us search for a solution to an equation

$$d^2x/dt^2 - 2t\,dx/dt - 2x = 0 \tag{3.20}$$

in the form

$$x(t) = a_0 + a_1 t + a_2 t^2 + \ldots = \sum_{i=0}^{\infty} a_i t^i. \tag{3.21}$$

Let us substitute the latter formula into the original equation and combine the terms with the same powers of t. Each such combination must be equal to zero. Finally, one gets the following recurrent relationship for the coefficients: $a_{n+2} = 2a_n/(n+2)$. The coefficients a_0 and a_1 are determined by initial conditions. Thus, for $a_0 = 1$ and $a_1 = 0$, one gets

$$x(t) = 1 + t^2 + t^4/2! + t^6/3! + \ldots . \tag{3.22}$$

In this particular case, one gets a Taylor expansion for the function $x(t) = e^{t^2}$ as the answer. If an obtained power series converges, which is not always the case, and one has derived a formula for its coefficients, then such a solution is called an *analytic solution* or a series solution. It is the second-best to the closed-form solution. If an obtained series converges slowly, then its practical application is unfeasible. In particular, such a situation takes place in a famous three-body problem which has a practically useless analytic solution in the form of a very slowly converging series (Wang, 1991).

3.5.1.4 Numerical solutions

Above considered equations with explicit time dependence and elementary non-linearities are relatively simple. In a general case, when a problem cannot be reduced to a linear one or to a certain specific class of equations, one searches for an approximate solution with numerical techniques, given initial and/or boundary conditions. The oldest and simplest one is the Euler technique. However, more accurate and complicated modern methods rely to a significant extent on the same idea. In particular, Runge–Kutta techniques are very popular. Adams integrator and Bulirsch and Stoer technique have their own advantages and shortcomings, they are often superior to Runge–Kutta techniques in terms of both computation time and accuracy (Kalitkin, 1978; Press et al., 1988; Samarsky, 1982).

According to the above-mentioned Newton's paradigm, a numerical solution was by no means satisfactory since it did not allow understanding qualitative features of dynamics and could be useful only for the prediction of future behaviour. The viewpoint changed since efficient computers and rich computer graphical tools arose, which currently allows one both to get qualitative ideas about a model behaviour and to compute a sufficiently accurate approximate solution. Since one can now investigate numerically a very broad class of non-linear equations, researchers pay more attention to the problem of how to get a model DE.

The use of any of the four ways mentioned in Sect. 1.5 is possible for that. Still, the most popular method is a way *from general to particular* since majority of known physical laws take the form of DEs. Besides, the entire apparatus of DEs was created to describe basic mechanical phenomena. Most of the models considered by physicists are asymptotic ones; they are obtained via restrictions imposed on universal formulas by a specific problem. Sometimes, one says that a model is obtained from "the *first principles*" implying some general relationships for a considered range of phenomena, from which one deduces concrete models (yet, such a use of the term "first principles" is criticised from a philosophical viewpoint). These are conservation laws and Newton's laws in mechanics, continuity equations and Navier–Stokes equations in hydrodynamics, Maxwell's equations in electrodynamics, derived special rules like Kirchhoff's laws in the theory of electric circuits, etc. Many non-standard examples of an asymptotic modelling of important physical and biological objects are given by a mathematician Yu.I. Neimark (1994–1997).

Modelling *from simple to complex*, e.g. creation of ensembles, is also typical when DEs are used. It is widely exploited in the description of spatially extended systems. The greater the number of elements included into an ensemble, the wider the class of phenomena covered by the model. Classical models are ensembles of coupled oscillators which represent an approved way of sequential complication of phenomena under consideration in tutorials on the theory of oscillations. An *empirical approach* to obtaining model DEs (reconstruction from time series) is considered in Part II.

3.5.2 Oscillators, a Popular Class of Model Equations

To illustrate possibilities of DE-based models, we select again the class of oscilla-tors. Why is our choice from an "ocean" of models so monotonous? The point is that any number of diverse examples cannot capture all specific descriptive capabilities of DEs. Thus, any example would give just a fragment of a general picture while really general things would be lacking. Therefore, it is reasonable to consider an example whose prehistory and some basic properties are known to a wide audience. Many people have met oscillators for the first time already at the lessons of school physics.

One calls "oscillators" both *objects* capable of oscillating about an equilibrium state and *equations* modelling such motions. Motion of an oscillator occurs within some potential profile either with friction or without it. An etalon *oscillator equation* is a second-order DE

$$d^2x/dt^2 + \gamma(x, dx/dt)dx/dt + f(x) = F(t), \qquad (3.23)$$

where the second term on the left-hand side corresponds to dissipation (fric-tion forces), the third term is determined by a potential U (a restoring force is $-\partial U/\partial x = -f(x)$) and the right-hand side represents an external force. A num-ber of research papers, reviews and dissertations are devoted to different kinds of oscillators (Scheffczyk et al., 1991; http://sgtnd.narod.ru/eng/index.htm).

Linear oscillators correspond to the case of $\gamma = $ const and $f(x) = \omega_0^2 x$. The latter means the quadratic potential $U(x) \sim x^2$. An autonomous oscillator ($F = 0$) is a two-dimensional ($D = 2$) dynamical system. It demonstrates either decaying ($\gamma > 0$, a dissipative oscillator) or diverging ($\gamma < 0$) oscillations. The autonomous dissipative oscillator has a stable fixed point as an attractor. Dimension of this attrac-tor is zero and both Lyapunov exponents are negative. This is one of the simplest dynamical systems in terms of possible kinds of behaviour.

Under a periodic external driving, the dissipative oscillator can be rewritten as a three-dimensional ($D = 3$) autonomous dynamical system (Sect. 3.5.3). It exhibits periodic oscillations with a period of the external force, i.e. has a limit cycle as an attractor in a three-dimensional phase space, and demonstrates a phenomenon of resonance. The dimension of the attractor is one and the largest Lyapunov expo-nent is equal to zero. Thus, the driven linear dissipative oscillator represents a more complex dynamical system compared to the autonomous one.

If a dissipative term involves non-linearity, e.g. like in *van der Pol equation*

$$d^2x/dt^2 - \alpha(1 - \beta x^2)dx/dt + \omega_0^2 x = 0, \tag{3.24}$$

then the oscillator becomes non-linear and is capable of demonstrating its own periodic oscillations (a regime of self-sustained oscillations). The system is two dimensional and its attractor is a limit cycle in a two-dimensional phase space for $\beta > 0$. The dimension of the attractor is then equal to one and its largest Lyapunov exponent is zero. In terms of the complexity of the dynamics (Lyapunov exponents and dimension of an attractor), the system (3.24) is more complex than the autonomous linear dissipative oscillator. In terms of the above dynamical characteristics, it exhibits at fixed values of α, β approximately the same complexity as the harmonically driven linear oscillator. However, the shape of time realisations can be more diverse for the system (3.24), depending on α, β. For instance, it exhibits almost sinusoidal waveform like the driven linear oscillator for small values of α, β and periodic "relaxation" oscillations, where the plot $x(t)$ resembles a saw, for big values of α, β.

In the non-autonomous case, the oscillator (3.24) exhibits much richer dynamics. Even harmonic driving may induce such kinds of behaviour as quasi-periodic oscillations, synchronisation of self-sustained oscillations by an external signal or even chaotic sets in the phase space if the driving amplitude is large enough. Thus, harmonically driven van der Pol oscillator is a considerably more complex system than the above linear oscillators and the autonomous van der Pol oscillator.

A non-quadratic potential profile $U(x)$ also means non-linearity of equation (3.23). Accordingly, its possible solutions get essentially more diverse. Even under a harmonic influence, a non-linear oscillator (which can be rewritten as a three-dimensional autonomous dynamical system) may exhibit a hierarchy of oscillatory regimes and non-linear phenomena including transition to chaos, multistability and hysteresis in a region of resonance. Thus, it can exhibit attractors with fractal dimensions greater than one and a positive largest Lyapunov exponent. Therefore, it is a more complex system than the linear oscillators or the autonomous van der Pol oscillator.

It is not straightforward to decide whether the driven van der Pol oscillator or the driven non-linear dissipative oscillator is more complex. The latter exhibits more diverse dynamical regimes than does the former due to different possible forms of $U(x)$ as described below, but in the case of strong dissipation, it cannot exhibit stable quasi-periodic regimes (where an attractor is a torus, the dimension is equal to two and two largest Lyapunov exponents are equal to zero) which are typical of the driven van der Pol oscillator.

The general non-autonomous oscillator (3.23) with arbitrary non-linear dissipation and arbitrary potential profile includes both the case of the driven non-linear dissipative oscillator and the driven van der Pol oscillator and, hence, may exhibit all the dynamical regimes mentioned above. Specific properties of different non-linear oscillators are determined by the concrete functions entering equation (3.23).

To select general features, one systematises oscillators in respect of (i) the dependency of a period of oscillations on their amplitude ("hard spring" and "soft spring") (Neimark and Landa, 1987; Scheffczyk et al., 1991; Thompson and Stewart, 2002), (ii) an order of the polynomial specifying the potential profile like in the theory of catastrophes (Kuznetsov and Potapova, 2000; Kuznetsova et al., 2004), etc. Below, we characterise complexity of the dynamics of several oscillators on the basis of their parameter space configurations.

When dissipation is too strong so that self-sustained oscillations are impossible, a parameter space of an oscillator is typically characterised by bifurcation sets called "cross-road area" (Carcasses et al., 1991; Mira and Carcasses, 1991) and "spring area" (Fig. 3.6). Cross-road area is a situation where domains of two cycles intersect, their boundaries representing period-doubling lines are stretched along boundaries of a "tongue" formed by saddle-node bifurcation lines and bistability takes place inside the tongue. Spring area is the case where a period-doubling line stretched along the above-mentioned tongue makes a characteristic turn around a "vertex" of the tongue, a point of the "cusp" catastrophe. Those universal configurations fill a parameter space in a self-similar manner (Parlitz, 1991; Schreiber, 1997). A fragment of a typical picture is seen already in Fig. 3.6: the right structure of the "spring area" (born on the basis of a double-period cycle) is built into an analogous upper structure (born on the basis of a "mother" cycle whose period is doubled when one moves to bottom along a parameter plane). A chart of regimes in Fig. 3.11 gives additional illustrations.

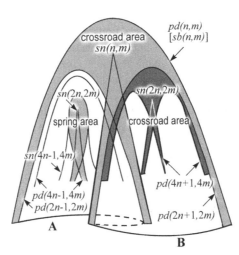

Fig. 3.6 A typical configuration of bifurcation lines "cross-road area" and "spring area" on a parameter plane. Domains of cycle stability are shown in greyscale. Lines of a saddle-node bifurcation are denoted as *sn*, lines of a period-doubling bifurcation are *pd*, lines of a symmetry-breaking bifurcation are *sb*. In parentheses, we show the number of an external force periods *n* and the number of the own periods *m* contained in a single period of a cycle loosing stability on a given line. A and B are conventional sheets used to illustrate bistability

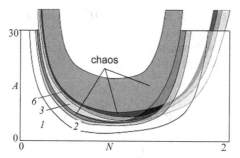

Fig. 3.7 Parameter plane A–N for the Toda oscillator (3.25): A is the driving amplitude, N is the normalised driving frequency. Greyscales show domains of existence of different regimes. Numbers indicate periods of oscillations in units of the driving period

Those self-similar typical configurations do not exhaust a diversity of possible bifurcation structures in parameter spaces of oscillators. For instance, oscillators with strong dissipation and a potential profile essentially different from the quadratic one exhibit a specific configuration of a domain where an arbitrary oscillatory regime exists and evolves to chaos. The domain has the form of a narrow bent strip resembling an "ear" (Fig. 3.7). An equation of the *Toda oscillator* driven by a sinusoidal external force demonstrating the described structure of the parameter space reads as

$$\mathrm{d}^2x/\mathrm{d}t^2 + \gamma\,\mathrm{d}x/\mathrm{d}t + \mathrm{e}^x - 1 = A\sin\omega t. \tag{3.25}$$

Let us denote the normalised frequency of driving $N = \omega/\omega_0$, where ω_0 is the frequency of low-amplitude free oscillations, $\omega_0 = 1$ for the system (3.25).

One more universal configuration of bifurcation sets for non-autonomous oscillators is presented in Sect. 3.6, where Fig. 3.9 illustrates a parameter plane of a circle map. It corresponds to a periodic driving applied to a system capable of exhibiting self-sustained oscillations. A universal configuration on the plane of driving parameters represents a hierarchy of so-called *Arnold's tongues*, i.e. domains where synchronisation takes place. Bifurcation lines inside a tongue exhibit the "cross-road area" structure.

As a solid residual from the current subsection, we stress (i) diversity of evolutionary phenomena which can be modelled with equations of oscillators and with DE-based models in general; (ii) complexity of observed pictures which can be systematised and interpreted in different ways, in particular, on the basis of typical "charts of dynamical regimes" and scaling properties (Kuznetsov and Kuznetsov, 1991, 1993b; Kuznetsov and Potapova, 2000; Neimark and Landa, 1987; Parlitz, 1991; Scheffczyk et al., 1991; http://sgtnd.narod.ru/eng/index.htm); (iii) an opportunity to observe different bifurcation sets and other specific features for different kinds of non-linearity.

3.5.3 "Standard form" of Ordinary Differential Equations

Despite multitude of ODE forms, the following one is the most popular among them and allows a clear geometrical interpretation:

$$
\begin{aligned}
dx_1/dt &= F_1(x_1, x_2, \ldots, x_n), \\
dx_2/dt &= F_2(x_1, x_2, \ldots, x_n), \\
&\cdots, \\
dx_n/dt &= F_n(x_1, x_2, \ldots, x_n).
\end{aligned}
\tag{3.26}
$$

Any set of autonomous ODEs can be reduced to such a form, solved in respect of the highest derivatives. A system (3.26) via a change of variables (probably, at the expense of the dimension increase, i.e. $D > n$) can be rewritten in the form

$$
\begin{aligned}
dy_1/dt &= y_2, \\
dy_2/dt &= y_3, \\
&\cdots, \\
dy_D/dt &= F(y_1, y_2, \ldots, y_D),
\end{aligned}
\tag{3.27}
$$

where y_1 is an arbitrary smooth function of the vector $\mathbf{x}$: $y_1 = h(x_1, x_2, \ldots, x_n)$, e.g. $y_1 = x_1$. Equation (3.27) is sometimes called *standard* (Gouesbet and Letellier, 1994; Gouesbet et al., 2003b). It is widely used in empirical modelling when a model state vector is reconstructed from a scalar observable via sequential differentiation (Sect. 10.2.2). However, it is not always possible to derive the function F in equation (3.27) explicitly. Possibility of reduction of any set of ODEs to the form (3.27) was proven by Dutch mathematician Floris Takens. Formulations of the theorems and some comments are given in Sect. 10.2.1 below.

A simple example is an equation of a dissipative oscillator under an additive harmonic driving:

$$
d^2x/dt^2 + \gamma dx/dt + f(x) = A \cos(\omega t)
\tag{3.28}
$$

with $\gamma = \text{const}$ which can be rewritten as a non-autonomous set of two equations:

$$
\begin{aligned}
dx_1/dt &= x_2, \\
dx_2/dt &= -\gamma x_2 - f(x_1) + A \cos(\omega t),
\end{aligned}
\tag{3.29}
$$

where $x_1 = x$, or as a three-dimensional autonomous system

$$
\begin{aligned}
dx_1/dt &= x_2, \\
dx_2/dt &= -\gamma x_2 - f(x_1) + A \cdot \cos x_3, \\
dx_3/dt &= \omega,
\end{aligned}
\tag{3.30}
$$

where $x_1 = x$ and $x_3 = \omega t$, or as a four-dimensional "standard" system

$$dx_1/dt = x_2,$$
$$dx_2/dt = x_3,$$
$$dx_3/dt = x_4,$$
$$dx_4/dt = -\gamma x_4 - \left(\frac{df(x_1)}{dx_1} + \omega_2\right) x_3 - \frac{d^2 f(x_1)}{dx_1^2} x_2^2 - \omega^2 \gamma x_2 - \omega^2 f(x_1),$$

$$(3.31)$$

where $x_1 = x$. To derive the formula (3.31), one differentiates twice equation (3.28) in respect of time and substitutes the left-hand side of (3.28) instead of $A \cos (\omega t)$ into an obtained equation. Apart from increasing number of dynamical variables (four ones instead of the two), the conversion leads to complication of the right-hand side of the last equation in (3.31) as compared with the original form (3.28). However, all the dynamical variables are related only to the variable x (they are its derivatives) that gives an advantage in the construction of such a model from a time realisation of x.

3.6 Models in the Form of Discrete Maps

3.6.1 Introduction

Similar to DE-based models, *discrete maps* represent a whole "stratum" of mathematical culture with its own history and specific features (Neimark, 1972). This section is a specialised introduction oriented to applications of discrete maps to empirical modelling.

Quite a widespread approach to obtain a model map is to approximate experimental data. In asymptotic modelling, maps are most often derived through a conversion from a DE to a finite-difference scheme or a Poincare section (Sect. 3.1). Creation of an ensemble of maps is a popular way to model a spatially extended system. Usually, such models take the form of coupled map chains and lattices with different coupling architectures: local coupling (only between "neighbours"), global coupling (all-to-all connections), random connections, networks with complex topology (Sect. 3.7.3), etc.

Simplicity of numerical investigation, diversity of dynamical regimes ranging from an equilibrium to chaos exhibited even by one-dimensional maps and the ease of constructing ensembles from simple basic elements have made discrete maps a dominating mathematical tool in non-linear dynamics. Currently, they are a full-fledged "competitor" of flows. Let us discuss several examples.

3.6.2 Exemplary Non-linear Maps

3.6.2.1 Piecewise Linear Maps

Seemingly, piecewise linear maps have the second simplest form after the linear map (3.16) capable of demonstrating simple rise or decay of a variable. Different kinds of such maps were studied, in particular:

(i) The "saw tooth" is the map $x_{n+1} = \{2x_n\}$, where braces denote fractional part of a number. Its plot is shown in Fig. 3.8a. The map is remarkable since it allows strictly and clearly prove an existence of dynamical chaos in simple non-linear systems. In a binary system, the map in a single iteration shifts a binary point one position to the right (Bernoulli's shift) and throws away an integer part of the resulting number. To illustrate irregularity and high sensitivity to small perturbations inherent in chaotic motions, let us specify an irrational number as an initial condition and write it down as an infinite non-periodic binary fraction $x_0 = 0.01001010100010100100010011010...$. Then, a sequence x_n generated by the map is also non-periodic: whether x_n belongs to the interval $(0,0.5)$ or $(0.5,1)$ is determined by the first figure after the decimal point which behaves according to the sequence of "0" and "1" in the binary fraction x_0. Moreover, a variation in any figure in the fraction x_0, even arbitrarily far from the decimal point (i.e. arbitrarily small), leads to a change in x_n of the order of 1 in a finite number of steps.

Thus, the saw tooth is an example of a one-dimensional system. Its chaotic "attractor" contains all irrational numbers; therefore, this is a set of full measure. Thus, its fractal dimension is equal to one. Its only Lyapunov exponent is positive: it equals ln 2 as can be readily shown. Hence, in terms of Lyapunov exponents, complexity of the dynamics is greater than that for the above-mentioned two- and three-dimensional continuous-time systems like the autonomous and harmonically driven linear oscillators and the autonomous van der Pol oscillator (Sect. 3.5) which cannot have attractors with positive Lyapunov exponents. In this respect, the saw tooth is as complex as driven non-linear oscillators in chaotic regimes. In terms of the attractor geometry, the saw tooth is simpler since its "attractor" does not exhibit any fractal structure.

(ii) Models of neurons. Modelling a neuron dynamics is a problem topical both in biophysics and non-linear dynamics (see, e.g., Izhikevich, 2000; Kazantsev, 2004; Kazantsev and Nekorkin, 2003; 2005; Kazantsev et al., 2005; Nekorkin et al., 2005) where one considers mainly ODE-based models. However, discrete map models are also developed in the last years, since investigation of their dynamics requires less computational efforts and extends possibilities of modelling large ensembles of coupled neurons. Simple model maps capable of generating "spikes" and "bursts" (i.e. short pulses and "packets" of pulses) have been suggested. A pioneering work considering a two-dimensional piecewise smooth map is Rulkov (2001). The piecewise linear map illustrated in

Fig. 3.8b (Andreev and Krasichkov, 2003) can also exhibit those dynamical features. Since it is two-dimensional, the plotted dependence x_{n+1} versus x_n is non-unique. The choice of a branch is governed by the second dynamical variable y (we do not present the entire cumbersome equations). Complexity of this model is greater than that of the saw tooth since it exhibits different periodic and chaotic regimes depending on the parameter values. The two-dimensional phase space of the neuron model map allows richer possibilities of dynamics than does the one-dimensional phase space of the saw tooth.

(iii) Maps for information storage and processing (Fig. 3.8c) illustrate a practical application to information recording with the use of the multitude of generated cycles (Andreev and Dmitriev, 1994; Dmitriev, 1991). Their authors have created a special software allowing to store and selectively process amounts of information compared to the contents of big libraries with such maps (http://www.cplire.ru/win/InformChaosLab/index.html).

3.6.2.2 One-Dimensional Quadratic Map

Non-linearity which seems the most natural and widespread in real-world systems is the quadratic non-linearity. Its properties are reflected by the class of one-

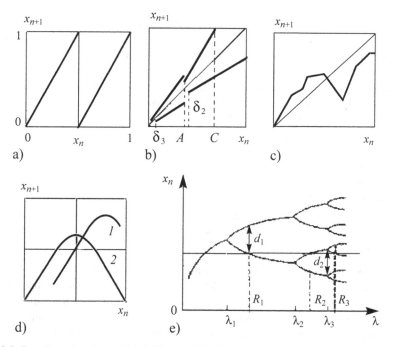

Fig. 3.8 One-dimensional maps: (**a**) the "saw tooth"; (**b**) a neuron model; (**c**) a map for information recording; (**d**) quadratic maps with different locations of the maximum; (**e**) Feigenbaum's "tree" for a quadratic map

dimensional maps $x_{n+1} = f(x_n)$, where the function f exhibits a quadratic maximum. The most eminent representative of this class is the *logistic map* (Fig. 3.8d, curve 1):

$$x_{n+1} = rx_n(1 - x_n). \tag{3.32}$$

The parameter r plays a role of the birth rate in population dynamics. As well, savings in bank account with "floating" bank interest would rise according to the same rule if the interest were introduced so as to restrict infinite enrichment of depositors (Schuster, 1984). Namely, if x_n is a value in account at an nth year and interest per annum is $\varepsilon = $ const, then a simple interest gives a sum of $x_{n+1} = (1+\varepsilon)x_n$ at the next year and the sum rises infinitely. At that, a small deposit does not promise essential change in a depositor's prosperity within nearest years as compared with good prospects of a person having a big initial sum of money. If one introduced a floating interest from the "considerations of justice", then one would get a map $x_{n+1} = \varepsilon_0(1 - x_n/x_{max})x_n$ which is reduced to the logistic map with a parameter $r = x_{max}(1+\varepsilon_0)^2/\varepsilon_0$ via the change of variable $z_n = x_n\varepsilon_0/x_{max}(1+\varepsilon_0)$. It is possible to list more examples from diverse fields. Any map $x_{n+1} = f(x_n)$ with the second-order polynomial f can be rewritten in the form (3.32) or in another often used form $x_{n+1} = \lambda - x_n^2$ (Fig. 3.8d, curve 2). Among "services" of the quadratic map, the following ones can be distinguished:

(1) M. Feigenbaum detected transition to chaos via a period-doubling sequence and described its universal regularities at the chaos boundary using this map as an example (Feigenbaum, 1980; Kuznetsov and Kuznetsov, 1993a). Figure 3.8e shows famous Feigenbaum's "tree", "established" values of the dynamical variable x_n versus the parameter λ. Universal quantities are, for instance, the ratios of the parameter bifurcation values near a point of transition to chaos λ_∞: $(\lambda_\infty - \lambda_n)/(\lambda_\infty - \lambda_{n+1}) = $ const $= \delta$ or, in another form, $\lambda_n = \lambda_\infty - $const$\cdot\delta^{-n}$, where $\delta = 4,6692016091\ldots$ and $n >> 1$.

(2) It is a basic element for the construction of non-linear models in the form of chains and lattices (Kuznetsov and Kuznetsov, 1991) and for illustration of non-linear phenomena under periodic and quasi-periodic external driving (Bezruchko et al., 1997b).

(3) It was used to demonstrate the phenomena of hysteresis and symmetry breaking under fast change of a parameter value across a bifurcation point (Butkovsky et al., 1998).

In terms of Lyapunov exponents and fractal dimensions, complexity of the logistic map (3.32) is greater than that of the saw tooth. At $r = 4$, its attractor is a full measure set similar to the saw tooth dynamics. However, the logistic map exhibits dynamics with different fractal dimensions less than one at different parameter values. Thus, it has richer dynamical properties compared to the saw tooth.

Circle map. This is a one-dimensional map

$$\theta_{n+1} = \theta_n + \Delta + (k/2\pi)\sin\theta_n \pmod{2\pi}, \tag{3.33}$$

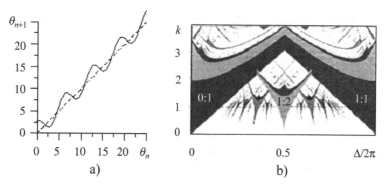

Fig. 3.9 The circle map (3.33): (**a**) its plot without taking modulo 2π; (**b**) its parameter plane (k, Δ) where domains of periodic regimes are shown in greyscale, while domains of quasi-periodic dynamics and chaos are shown in white

whose plot is shown in Fig. 3.9a. It can be interpreted from a physical viewpoint. Under certain assumptions, one can reduce model DEs for a self-sustained oscillator driven by a periodic sequence of pulses to such a map. An attractor of an original system can be a torus, while the map (3.33) can be considered as a Poincare map in a plane cross section of the torus (Kuznetsov, 2001).

In a cross section of a torus, a representative point under subsequent "punctures" draws a closed curve whose points can be described with an angular coordinate θ_n, where n is the order number of a puncture. The parameter Δ is determined by the ratio of periods of rotation along "big" and "small" circumferences, i.e. the ratio of frequencies of autonomous self-sustained oscillations and driving. The parameter k characterises the driving amplitude. Structure of the parameter plane for the system (3.33) is shown in Fig. 3.9b. Different greyscale tones mark domains of stable periodic regimes. Periodic regimes corresponding to the synchronisation of self-sustained oscillations by an external signal exist in domains resembling beaks. These domains are called *Arnold's tongues* by the name of a soviet mathematician V.I. Arnold. At that, an orbit on a torus becomes a closed curve in the cross section. Different tongues correspond to different values of the rotation number, i.e. the number of revolutions of a representative point along a small circumference during a single revolution along a big circumference. The dynamics of the circle map has been studied in detail, in particular, a characteristic dependence of the total width of synchronisation intervals versus k is described, regularities of chaos domain location are established, etc.

In terms of the Lyapunov exponents and fractal dimensions, the circle map complexity is similar to that of the logistic map. Both systems can exhibit periodic and chaotic regimes at different parameter values. However, the circle map can also exhibit quasi-periodic regimes with zero Lyapunov exponents which are not observed in the logistic map. Accordingly, it exhibits additional bifurcation mechanisms and the corresponding structures on the parameter plane. Thus, the circle map is, in some sense, a more complex object than the logistic map.

3.6.2.3 A Model Map for a *Non-isochronous Non-linear Oscillator* Under Dissipative Pulse Driving

A plot of a one-dimensional multi-parametric map

$$x_{n+1} = x_n e^{-d/N} \cos(2\pi/(N(1 + \beta x_n))) + A. \tag{3.34}$$

is shown in Fig. 3.10a (Bezruchko et al., 1995).

The idea behind this map and meaning of its four parameters are illustrated in Fig. 3.14c with a time realisation of a dissipative oscillator, e.g. a mathematical pendulum, driven periodically in a specific manner. Namely, a load is taken aside by the same value A along the x-axis. After that, it starts to oscillate with the same initial phase. For instance, one can take a load by hand and leave it with zero initial velocity. In the case of an electric pendulum, an RL diode circuit shown in Fig. 3.5b, such driving is realised via pulses of current with direct polarity for the diode. At that, big active conductance of a diode quickly cancels free oscillations so that an initial phase of free oscillations does not vary between pulses (Fig. 3.14a). If a quasi-period T during exponentially decaying free oscillations $x(t) = x_n e^{-\delta \cdot t} \cos(2\pi t/T)$ is regarded constant between two pulses and non-isochronism is taken into account in a simplified manner as a dependence of T on an initial amplitude $T = T_0(1 + \beta x_n)$, where x_n is a starting value in the nth train of free oscillations (i.e. after the nth pulse), then a model map takes the form (3.34). Here, A is the amplitude of a driving pulse, $N = T_0/T$ is a normalised driving frequency, $d = \delta \cdot T_0$ is a damping coefficient, β is a coefficient of non-linearity, which is positive for a "soft spring" and negative for a "hard" one.

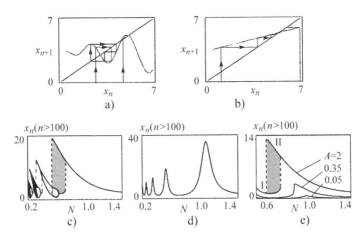

Fig. 3.10 Investigations of the map (3.34): (**a, b**) its plots and Lamerey's diagrams at $A = 3.4$, $N = 0.1$, $d = 0.1$ and $\beta = 0.05$ (**a**) or $\beta = 0$ (**b**); (**c, d**) bifurcation diagrams $x_n(N)$ for $\beta = 0.05$ (**c**) or $\beta = 0$ (**d**) which can be interpreted as resonance curves. Intervals of single valuedness correspond to period-1 oscillations, divarication of a curve means period doubling, "smeared" intervals show chaos. Resonance curves at different values of A exhibit a transition from a linear resonance to a non-linear one (**e**). Domains of bistability and chaos are shaded

Despite being one-dimensional and relatively simple in its mathematical form, the map exhibits practically all basic non-linear phenomena inherent in low-dimensional non-linear dynamical systems: a multitude of oscillatory regimes on the basis of different kinds of oscillations (modes), linear and non-linear resonance, bi- and multistability, complexity and fractal character of basins of attraction, hysteresis and dynamical chaos (Prokhorov and Smirnov, 1996). Thus, its dynamical complexity is similar to that of the circle map and logistic map and is even greater in some respects (Fig. 3.11).

Higher dimensional maps are capable of demonstrating even more diverse dynamics and, hence, greater complexity in terms of the number of positive Lyapunov exponents and big value of fractal dimension. However, the phenomena illustrated above with one-dimensional examples can already convince a reader that discrete maps represent a very fruitful and efficient research tool.

3.6.3 Role of Discrete Models

We consider the role of discrete models with a concrete example. In 1981, Lindsay reported an observation of dynamical chaos in quite an accessible (cheap) and popular system, a circuit with an inductance coil and a varactor diode driven by a harmonic electromotive force (Linsay, 1981). Since then a piece of wire convoluted in a coil and a piece of semiconductor supplied with contacts are actively used for experimental demonstrations of non-linear phenomena. A paper in "Scientific American" even recommended to have such systems "on a windowsill in each house".[6] Below, we demonstrate capabilities of discrete models of this object.

3.6.3.1 "Ancestors" of the Object

A circuit consisting of an inductance coil and a capacitor (an oscillatory circuit) is an electric analogue of a mechanical pendulum. Similar to how mechanical pendulum properties are determined by its shape and parameters, processes in a circuit depend on the construction of its elements. For the simplest case when plates of an air capacitor are connected with a wire coils (Fig. 3.12a), a conceptual model (an equivalent scheme) takes the form shown in Fig. 3.12b. Given the parameters L, C, R of the scheme,[7] one can readily derive a model of the circuit in the form of the linear dissipative oscillator (3.2) from Kirchhoff's laws, where x is a dynamical variable

---

[6] This is a diode with a $p - n$ junction whose capacity depends on voltage, i.e. an electrically controlled capacitor. Circuits with such diodes are used in radioengineering for more than half a century. They were even suggested as memory elements for computers. Different kinds of such circuits are widely presented in contemporary radio sets and TV sets.

[7] When a charge is accumulated on the capacitor plates and a current flows in the wires, electric and magnetic forces appear and tend to compress or stretch the wires. Therefore, if substances of the coil and capacitor are not hard enough, their size (and, hence, C and L) can depend on the current and voltage (dynamical variables) implying emergence of nonlinearity.

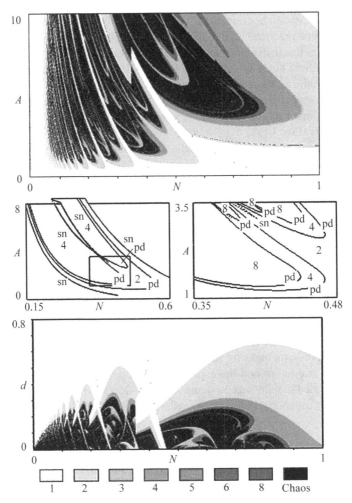

Fig. 3.11 Chart of the dynamical regimes for the map (3.34) on the parameter planes A–N and d–N. Greyscale tones show domains of oscillations whose periods are reported at the bottom. The same tone may correspond to different motions with the same period or to chaotic regimes developed on the basis of different cycles. Bifurcations occur at the boundaries of the domains. The fragment in the middle shows domains of bi- and multistability. A domain of existence and evolution of a certain cycle is shown with a separate sheet: *sn* are boundaries of the sheets, *pd* are period-doubling curves. Overlap of two sheets corresponds to bistability and hysteresis

(charge) and $\delta = R/2L$ is a damping coefficient. Free oscillations of the dissipative oscillator decay, while oscillations driven by a periodic signal are periodic with a period of driving T. The only oscillatory effect demonstrated by the system is a *resonance*, an increase in the driven oscillation amplitude when natural and driving frequencies get closer.

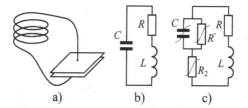

Fig. 3.12 Electric pendulums: (**a**) the simplest oscillatory circuit; (**b**) its equivalent scheme; (**c**) an equivalent scheme where the role of a capacitor is played by a diode represented by a combination of non-linear capacitor and resistors

3.6.3.2 Consequences of a Capacitor Replacement

Inclusion of a diode whose equivalent parameters R and C depend on the current and voltage into the circuit (Fig. 3.12c) leads to a striking extension of the range of observed oscillatory phenomena. Even under the simplest harmonic driving, the "electric pendulum" demonstrates a hierarchy of driven motions of various complexity: harmonic, more complex periodic and chaotic ones. Similar picture is observed under a pulse driving. Bifurcation sets (surfaces in three-dimensional spaces, curves on two-dimensional cross sections) bound domains of existence of different oscillatory regimes in a parameter space forming the structures presented in Fig. 3.13. The driving amplitude V and the normalised frequency $N = \omega/\omega_0$ are shown along the horizontal axes of the three-dimensional picture and the linear resistance R along the vertical axis. The structure can be understood better by considering different plane cross sections of the parameter space. Oscillation type within a domain can be illustrated with a respective time realisation.

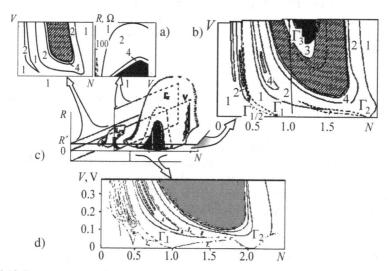

Fig. 3.13 Parameter space of an RL-diode circuit under harmonic external driving obtained from experimental investigations. *Dashes* show curves of hysteresis (hard) transitions. Chaos domains are shaded. Numbers denote period of oscillations in the respective domains

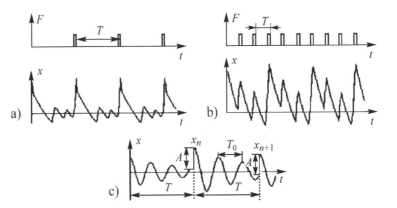

Fig. 3.14 Time realisation of a current in an RL-diode circuit under periodic pulse driving of direct polarity $F(t)$: (**a**) a cycle belonging to the class of subharmonic oscillations, a driving period T is three times as big as a quasi-period of free oscillations ($\Gamma_{1/3}$); (**b**) a cycle belonging to the class of "period adding sequence", a driving period T is three times as small as a quasi-period of free oscillations ($\Gamma_{3/1} = \Gamma_3$); (**c**) a time realisation-based model of subharmonic oscillations, where a quasi-period of decaying oscillation within a train is constant and depends only on an initial deviation

Figure 3.14 shows typical time realisations of a current in the case of pulse driving emf whose polarity is "direct" for the diode. Pulses come with a period T, $\omega = 2\pi/T$. Despite small duration of pulses, free oscillations quickly decay during a pulse since an equivalent capacity of a diode for a direct current (Fig. 3.12c) is shunted by its low active resistance. When a pulse ends, oscillations start almost with the same initial phase (Fig. 3.14a, b), while a time realisation between two pulses represents decaying free oscillations. Depending on the driving amplitude and period, the damping rate, the kind of non-linearity and initial conditions, different repeated motions (*cycles*) can be established in the system. *Periods of the cycles* are equal to the driving period or divisible by it, i.e. kT, where k is an integer. Possible variants of periodic motions are diverse but can be systematised as follows. All cycles can be conventionally divided into two groups based on the similarity property. Each of the groups preserves some peculiarities of the waveform of time realisations and the shape of limit cycles in the phase space.

The first group is formed by cycles whose period is equal to the driving period $1T$ and exists in the low-frequency domain $N < 1$. Such cycles are usually called *subharmonic cycles*. Since the driving period is big as compared with the time scale of free motions, there are generally several maxima in a time realisation within a train (Fig. 3.14a). The second group consists of cycles with periods kT, where $k = 2, 3, \ldots$, which are observed for bigger driving frequencies $0.5 < N < 2$. Examples of such cycles are shown in Fig. 3.14b. Since a change of such a regime under the increase in amplitude is accompanied by subsequent increase in k by 1, they are called cycles of "period adding sequence". A conventional notation of the cycles is $\Gamma_{m/k}$. Here, k corresponds to the ratio of the driving period to the quasi-period of free oscillations. It can be estimated as the number of maxima within an interval T in an oscillogram. The value of m is a period of cycle measured in units of the driving period.

3.6.3.3 Mathematical Models

Processes in semiconductor diodes whose properties determine non-linearity of a system are analysed most strictly with the use of partial differential equations. However, for sufficiently slow motions, a diode can be considered as a bipole with some equivalent properties reflecting relationship between the voltage on its contacts and the current in connecting wires so that one can use ODEs. Even simpler models can be obtained in the form of maps, if one restricts the consideration only with a part of possible motions. Further, we consider models capable of describing fragments of the above-mentioned (Fig. 3.13) complex picture.

A Continuous-Time Model

Let us represent a semiconductor diode as a non-linear capacitor, whose capacity C depends on the voltage as $C = C_0/(1 - U/\varphi)$, where C_0 is the initial diode capacity, U is the voltage on the diode, φ is the contact potential. Then, a model equation for the circuit derived from Kirchhoff's laws takes the form of Toda oscillator (3.25):

$$d^2x/d\tau^2 + \gamma dx/d\tau + e^x - 1 = A \sin N\tau,$$

where x is the dimensionless charge on the capacitor plates, γ is the damping coefficient, A is the dimensionless driving amplitude, $N = \omega/\omega_0$ is the normalised driving frequency, $\tau = \omega_0 t$ is dimensionless time. Results of numerical simulations presented in Fig. 3.7 demonstrate good qualitative description of an object in the *entire parameter space*.

Discrete-Time Models

(i) One can successfully use one-dimensional multimodal map (3.34) as a discrete-time model for subharmonic oscillations, i.e. in the low-frequency domain $N = \omega/\omega_0 \leq 1$. A model is adequate to the real-world system in the parameter space domains where motions on the basis of the cycles $\Gamma_{1/2}$, $\Gamma_{1/3}$ and so on take place (Fig. 3.11). Those domains have qualitatively the same structure. They are similar to each other and self-similar. Self-similarity means a configuration like in "matreshka" (a set of nesting dolls): a basic constructive element is reproduced at smaller and smaller scales. However, in contrast to matreshka, the domains of existence of various oscillation kinds on the parameter plane at sufficiently low levels of dissipation can overlap forming domains of multistability (Fig. 3.11, bottom panel)

(ii) A *two-dimensional map* modelling driven dynamics of the circuit in a higher-frequency domain $0.8 \leq N \leq 2$ is suggested in Bezruchko et al. (1997a) based on the characteristic waveform of time realisation of the cycles belonging to

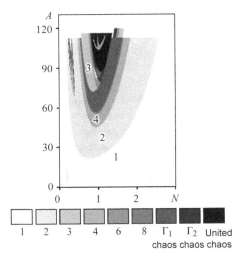

Fig. 3.15 Chart of the dynamical regimes of the map suggested in Bezruchko et al. (1997a), which describes the cycles of "period adding sequence"

the "period adding sequence" (Fig. 3.14b). It is more complicated than map (3.34) and reproduces well the structure of the parameter plane of an original circuit (Kipchatov, 1990) for large driving frequencies and amplitudes, where the cycles of the "period adding sequence" exist (cf. Figs. 3.13b and 3.15). At that, it does not reflect diversity of the basis cycles and other peculiarities of the circuit dynamics described above.

(iii) In the domains where any of the basis cycles demonstrate period-doubling sequence under a parameter change, a good model of the circuit is the one-dimensional quadratic map $x_{n+1} = \lambda - x_n^2$.

(iv) In a domain of negative resistance where an object demonstrates self-sustained oscillations, its dynamics is modelled well with the circle map (3.33) which exhibits the phenomena of synchronisation by a weak periodic driving and of suppression of the oscillations by a strong periodic driving.

Thus, the Toda oscillator equation (3.25) describes the circuit dynamics in the most complete way among all the considered models. It reflects all families of the characteristic cycles of the RL-diode circuit and peculiarities of its parameter space structure. The discrete-time model (3.34) and the two-dimensional model map describe only one of the two existing families of cycles, either "subharmonic" or "period adding" one. In particular, map (3.34) reflects such phenomena as linear and non-linear resonance, multistability and hysteresis. The quadratic map is universal but does not capture specificity of the object. The same holds true for the circle map. Is it possible to create a model map which could compete with the differential equation of the Toda oscillator? Currently, we could not say how complex a formula for such a map might be.

3.7 Models of Spatially Extended Systems

To model spatially extended objects, one often uses ensembles of coupled ODEs or coupled maps (e.g. Afraimovich et al., 1989; Nekorkin and Velarde, 2002; Shalfeev and Matrosov, 2005). Spatial properties of such systems manifest themselves in solutions with different spatial profiles of characterising variables. For instance, oscillations of two coupled linear oscillators can be represented as a superposition of two basic sinusoidal regimes with different frequencies. One of them corresponds to in-phase oscillations, when the elements move in a completely identical manner, while the other one reflects anti-phase oscillation, when there is a constant phase shift between the oscillators by π. This peculiarity of a spatially extended system, consisting of concentrated elements, can be considered as an analogue of spatial modes in a bounded continuously distributed system (Fig. 3.16).

A property of *multistability* resembling multitude of spatial modes is ubiquitous in ensembles of oscillatory systems. Such a principal multimodality and the corresponding sensitivity to weak parameter variations (when possible kinds of motions are numerous and their basins of attraction form complicated and even fractal structures) is a typical property of spatially extended non-linear systems. Capabilities of relatively simple discrete models to describe this basic phenomenon are illustrated in Sect. 3.7.1, while more complicated tools are briefly considered after that.

3.7.1 Coupled Map Lattices

In chains and lattices, identical basis maps $x_{n+1} = f(x_n)$ are usually coupled to each other in a certain manner: locally (only nearest neighbours), globally (all to all) or within groups. Complexity of these models rises with the number of coupled maps, i.e. with the dimension of a model. In general, the greater is the model dimension, the greater can be the number of coexisting attractors, their fractal dimension and the number of positive Lyapunov exponents.

3.7.1.1 Two Dissipatively Coupled Quadratic Maps

A symmetric coupling, when elements influence each other in the same way, is shown by a rectangle in Fig. 3.17. A triangle marks a unidirectional coupling, when

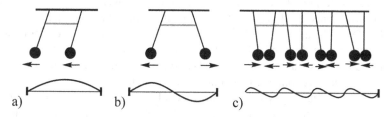

Fig. 3.16 Oscillatory modes in an ensemble of two (**a, b**) and several (**c**) pendulums. *Top panels* illustrate the systems, bottom ones illustrate their spatial modes

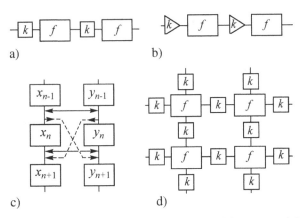

Fig. 3.17 Coupled map lattices: (**a, b**) one-dimensional lattices; (**c**) a space-and-time diagram for interacting populations to illustrate symmetric coupling kinds; (**d**) a two-dimensional lattice

only one element affects another one. An intermediate case of asymmetric coupling is also possible. A systematisation of the *coupling kinds* is given in Kuznetsov (1986), where symmetric couplings between maps are reduced to the following types: *dissipative* coupling

$$
\begin{aligned}
x_{n+1} &= f(x_n) + k(f(y_n) - f(x_n)), \\
y_{n+1} &= f(y_n) + k(f(x_n) - f(y_n)),
\end{aligned}
\tag{3.35}
$$

inertial coupling

$$
\begin{aligned}
x_{n+1} &= f(x_n) + k(y_n - x_n), \\
y_{n+1} &= f(y_n) + k(x_n - y_n),
\end{aligned}
\tag{3.36}
$$

or their combination. Here, x and y are dynamical variables, k is the coupling coefficient, f is the non-linear function of the basis map.

The systematisation allows an interesting interpretation in the language of the population biology. One can assume that individuals first breed in their population and then get an opportunity to migrate to another population. "First breed, then creep away". The cycle is repeated next year. Solid lines on a space – time diagram (Fig. 3.17c) correspond to such a case. Such coupling tends to make simultaneous states of subsystems equal to each other so that it can be naturally called *dissipative coupling*. Dashed lines in Fig. 3.17c correspond to a situation when individuals may migrate before the cycle of breeding and death within their population. Such coupling can be reasonably called *inertial coupling* since it promotes memorising a previous-step state. A combined coupling is also possible.

The choice of coupling type in practical modelling is non-trivial. In particular, it is illustrated by experimental investigations of a set of coupled non-linear electric circuits (Sect. 3.6.3) in the domain of parameter space, where each system transits to chaos via period doublings. It appears that coupling via a resistor (a dissipative element) is adequately described as dissipative, while coupling via a capacitor (a purely reactive element) as combined one, rather than purely inertial (Astakhov et al., 1991b).

The choice of the basis map and of the kind of coupling introduce specific features into a model behaviour, but the phenomenon of *multistability* in ensembles of coupled maps is always determinative. It is illustrated by the simplest set of two quadratic maps $f(x_n) = \lambda - x_n^2$ with the dissipative coupling (3.36):

$$
\begin{aligned}
x_{n+1} &= \lambda - x_n^2 + k\left(x_n^2 - y_n^2\right), \\
y_{n+1} &= \lambda - y_n^2 + k\left(y_n^2 - x_n^2\right).
\end{aligned}
\tag{3.37}
$$

For the same value of λ in both subsystems, we introduce the following systematisation of oscillatory modes. In the limit of zero coupling ($k = 0$), each regime of a period N can be realised in N ways differing by the shifts between the subsystems oscillations in time by $m = 0, 1, 2, \ldots, N-1$ steps as shown in Fig. 3.18 for $N = 2$ and 4. We call those N ways the *oscillation kinds* and use them to describe a hierarchy of the oscillatory regimes in the presence of coupling when interaction leads to different variants of mutual synchronisation. We denote periodic regimes as N_m. Despite the lack of repeatability, the same classification principle can be maintained for chaotic regimes N^m if one interprets N as the number of the attractor strips and m as a time shift between maximal values of x_n and y_n. Regimes with $m = 0$ are called in-phase.

By showing the domains of existence and evolution of each oscillation kind on a separate sheet, one can get a vivid multi-sheet scheme of the domains of existence (stability) of oscillatory regimes on the plane (k, λ). Figure 3.19a shows the domains for all oscillatory regimes of the periods 1, 2, 4 and 8 at $k < 0.5$. Figure 3.19b represents a cross section of the multi-sheet picture shown in Fig. 3.19a with a plane $k = 0.05$ and qualitatively illustrates an evolution of motion in system (3.37) under the variation of the parameter λ at a fixed weak coupling. Solid lines correspond to stable regimes and dashed ones to unstable regimes. Points indicate bifurcation transitions. The letters A, B, C and D mark branches combining certain groups of regimes[8]: they start with periodic regimes whose number rises with λ and end with chaotic ones.

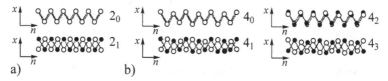

a)   b)

Fig. 3.18 Coupled map dynamics: time realisations (**a**) for the period $N = 2$; (**b**) for the period $N = 4$. Dynamical variables of the first and second subsystems are shown by filled and open circles, respectively. Notations of the oscillation kinds N_m are shown to the right

---

[8] The branch A corresponds to the evolution of in-phase regimes ($m = 0$), $B - D$ to the others.

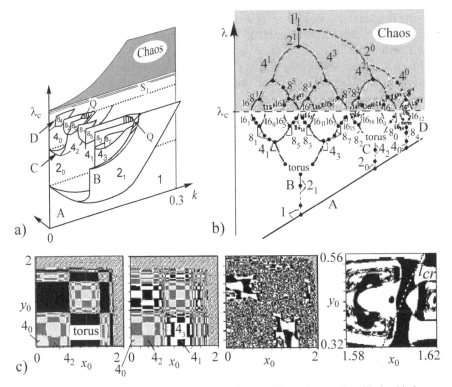

Fig. 3.19 Dynamics of the systems (3.37): (**a**) a scheme of the evolution of oscillation kinds on a parameter plane; (**b**) its section for $k = 0.05$; (**c**) phase space division into basins of attractors in cases of multistability (they change each other when one moves along the parameter plane in the panel **b**)

Domains of chaotic regimes are shaded. A critical value of the non-linearity parameter at which a transition to chaos occurs (an accumulation point of a period-doubling sequence) is denoted λ_c. Domains denoted by a letter Q or a word *torus* correspond to quasi-periodic oscillations and transition to chaos via their breaking. Different oscillation kinds can divide phase space into *basins of attraction* with a fractal structure (Fig. 3.19c). Increase in the dissipative coupling strength k is accompanied by reduction in the number of coexisting states so that only in-phase motion is stable at large k, i.e. the system becomes in effect one-dimensional.[9]

3.7.1.2 Complex Dynamics of a Chain: Consequences of the Increase in the Number of Elements

It is not surprising that the increase in the number of elements in an ensemble leads to even more complicated oscillatory picture. Indeed, the longer the chain,

---

[9] Complex dynamics of this non-linear system is illustrated by a computer program available at http://www.nonlinmod.sgu.ru and in research papers Astakhov et al. (1989, 1991a).

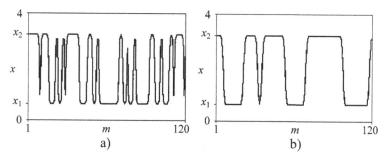

Fig. 3.20 A period-1 regime with a non-uniform spatial distribution in the chain (3.38) of 120 elements, $A = 0.965$, $N = 0.64$, $d = 0.2$, $\beta = 0.2$: (**a**) $k = 0.1$; (**b**) $k = 0.35$. The quantities x_1 and x_2 are equilibrium states of a bistable elementary cell

the more the kinds of motions with different temporal and spatial profiles possible. Figure 3.20 illustrates it with numerical results for the chain of dissipatively coupled pendulum maps

$$x_{n+1}^m = (1-k)f\left(x_n^m\right) + (k/2)\left(f\left(x_n^{m+1}\right) + f\left(x_n^{m-1}\right)\right), \qquad (3.38)$$

where n is the discrete time, m is the number of a chain element, k is the coupling coefficient, f is the multimodal map (3.34). A number of an element is shown along the horizontal axis and its instantaneous state along the vertical one. There are flat intervals in the pattern (*domains*) and fast transitions between them (*kinks*).[10]

The structures evolve under the parameter changes: temporal and spatial periods double; periodic, quasi-periodic and chaotic configurations arise. Domains widen with the coupling coefficient rise, while kinks get flatter. Finally, a very large k provides a spatially uniform regime for any initial conditions, an analogue to the emergence of an ice floe in the above-mentioned example with cooled water (Sect. 3.2). Details on the dynamical properties of the chain are given in Bezruchko and Prokhorov (1999).

3.7.1.3 Two-Dimensional Map Lattice

Further complication of the model (3.38) in the sense of its spatial development can be performed both via the increase in the number of elements and via changes in the coupling architecture. In the next example, the same multimodal maps constitute a lattice, where each element interacts with its four nearest neighbours. Coupling is local and dissipative:

---

[10] Map (3.34) is taken for simplicity. Thus, multistability in a set of quadratic maps is formed on the basis of a period-doubled cycle, while in a set of maps (3.34) it is observed already for the period-1 cycles. When an isolated map (3.34) has two period-1 states, there are four period-1 oscillation kinds in a set of two maps (Bezruchko and Prokhorov, 1999).

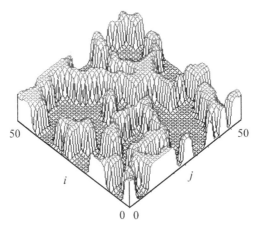

Fig. 3.21 A dynamical regime with a non-uniform spatial distribution in the two-dimensional lattice (3.39) consisting of 50×50 period-1 elements, $A = 0.965$, $N = 0.64$, $d = 0.2$, $\beta = 0.2$, $k = 0.2$. Boundary conditions are periodic

$$x_{n+1}^{i,j} = (1 - k)f\left(x_n^{i,j}\right) + (k/4)\left(f\left(x_n^{i+1,j}\right) + f\left(x_n^{i-1,j}\right) + f\left(x_n^{i,j+1}\right) + f\left(x_n^{i,j-1}\right)\right),$$
(3.39)

where i and j determine spatial location of a lattice element. An example of instantaneous snapshot of such a lattice having the same numbers of elements along each direction is shown in Fig. 3.21. This is a stationary structure achieved from random initial conditions. At weak couplings, one can get almost any required stationary distribution by specifying different initial conditions. Under the increase in the coupling coefficient, the number of possible structures reduces. Above some threshold value of k, the only attractor is a spatially uniform distribution.

3.7.2 Cellular Automata

A *cellular automaton* is a discrete dynamical system representing a set of identical cells coupled with each other in the same way. All the cells form a cellular automaton lattice. Lattices may be of various types differing both in dimension and shape of the cells (Minsky, 1967). Cellular automata were suggested in the work of von Neumann (1966) and became a universal model of parallel computations like Turing's machine for sequential computations. Any cell computes its new state at each step from the states of its nearest neighbours. Thus, the laws in a system are local and everywhere the same. "Local" means that it is sufficient to look at the neighbourhood state to learn what will happen at a future instant; no long-range interactions are allowed. "Sameness" means that one can distinguish one place from another one by a landscape,[11] not by any difference in the laws (Margolus and Tof-foli, 1990). Based on this description, one can single out the following characteristic properties of cellular automata:

(i)   A lattice is uniform and evolution law for the cells is everywhere the same.
(ii) Changes in the states of all cells occur simultaneously, after calculation of a new state of each cell in a lattice.
(iii) Interactions are local, only neighbouring cells can affect a given cell.
(iv) A set of cell states is finite.

Usually, one illustrates a cellular automaton with an example of a model called the *game "Life"* created by D. Conway, a mathematician from Cambridge University, in 1970. It is widely presented in the Internet (see, e.g., http://www.famlife.narod.ru). Rules of functioning of that automaton somewhat mimic real-world processes observed in birth, development and death of a colony of living organisms. One considers an infinite flat lattice of square cells (Fig. 3.22). Living cells are shown by dark colour. Time is discrete ($n = 0, 1, 2, \ldots$) and a situation at the next time step $n + 1$ is determined by the presence of living neighbours for each living cell. Neighbouring cells are those having common edges. Evolution is governed by the following laws:

(i)   *Survival*. Each cell having two or three neighbouring living cells survives and transits to the next generation.
(ii) *Death*. Each cell with more than three neighbours dies due to overpopulation. Each cell with less than two neighbours dies due to solitude.
(iii) *Birth*. If the number of living cells neighbouring to a certain empty cell is equal exactly to three, then a new organism is born in that empty cell.

Thus, if an initial distribution of living cells (a landscape) has the form shown in Fig. 3.22a, then a configuration shown in Fig. 3.22b appears in a single time step.

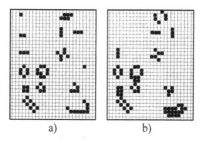

a)   b)

Fig. 3.22 Examples of landscapes for the cellular automaton "Life": (**a**) an initial profile, $n = 0$; (**b**) a situation after the first step $n = 1$

---

[11] A distribution of the values of a characterising quantity over an automaton workspace.

Furthermore, some initial structures may die out, while the others survive and get stationary, or are repeated periodically, or move in space, and so on. Basic properties of this automaton are as follows: structures separated by two empty cells do not affect each other at once; a configuration at a time instant n completely determines the future (states at time steps $n + 1$, $n + 2$, etc.); one cannot restore the past of the system from its present (the dynamics is non-invertible); stable forms are, as a rule, symmetric, etc. The larger the area occupied by a population, the more complicated its behaviour.

Currently, the game "Life" has got further development. Thus, in modern versions the automaton is three dimensional and capable of modelling several populations like interacting "herbivores" and "predators". However, even more sophisticated versions of this simple system do not represent a limit complexity level for the problems which can be solved with cellular automata.

A cellular automaton can be equivalently described with a set of coupled maps with discrete states. Its peculiarity is the simplicity of construction and convenience of computer investigation. Cellular automata are used to model hydrodynamic and gas-dynamic flows, electric circuits, heat propagation, movement of a crowd, etc. (Loskutov and Mikhailov, 2007; Malinetsky and Stepantsev, 1997; Shalizi, 2003). They are applied to create genetic algorithms, to find a shortest way on a graph and so forth (see, e.g., Margolus and Toffoli, 1990; http://is.ifmo.ru).

3.7.3 Networks with Complex Topology

The above coupled map lattices and cellular automata describe spatially extended systems with local interactions between the elements. More complicated coupling architectures involve various non-local interactions. In such a case, one speaks of a *network of coupled maps* rather than a coupled map lattice. Coupling architecture is often called *topology* of a network. Topology is said to be regular if it is described by a simple regular law, e.g. the above locally coupled maps (where only the nearest neighbours interact) or globally coupled maps (where every element is connected to every other element, all-to-all coupling).

It is easy to imagine other topologies which are not described as simply as the above regular topologies. For example, one can artificially generate a completely random coupling architecture. Typically, one speaks of a complex topology if it looks rather complicated and irregular and, simultaneously, exhibits some non-trivial statistical properties different from completely random networks. During the last years *networks with complex topology* are actively studied in different fields as reflected by the reviews (Albert and Barabasi, 2002; Arenas et al., 2008; Boccaletti et al., 2006; Dorogovtsev and Mendes, 2003; Kurths et al., 2009; Osipov et al., 2007; Strogatz, 2001; Watts, 1999). A special attention is paid to the so-called "small-world" and "scale-free" properties of a network. To define them, one must introduce the concepts of node, link, degree and path. Each element of a network is called a *node*. If two nodes are coupled (interact with each other), then one says that there exists a *link* between them. These two nodes are called *vertices* of the link. If a node is a vertex of M links ($0 \leq M \leq N - 1$, where N is the size of the network), then the number M is called a *degree* of that node. A *path* connecting two nodes A and B is the sequence of vertices which one has to pass by (via the existing links) to reach the node B from the node A. The shortest path between two nodes is a path consisting of the smallest number of vertices.

The *small-world property* means that any two elements of a network are connected by a sufficiently short path. To give a quantitative formulation, one can notice that a mean (over all pairs of nodes) shortest path length in a regular hypercubic lattice with d dimensions grows with the lattice size as $N^{1/d}$. Small-world property is defined as follows: a mean shortest path length grows at most logarithmically with N. This notion was first introduced in Watts and Strogatz (1998). The small-world property has been observed in a variety of real-world networks, including biological and technological ones (see Sects. 2.2.1 and 2.2.3 in Boccaletti et al., 2006 and references therein).

The *scale-free property* concerns heterogeneity of couplings. Homogeneity in coupling structure means that all nodes are topologically equivalent, e.g. regular lattices or random networks. In regular lattices, each node has the same degree except for the edges. In random networks, each of the $N(N - 1)/2$ possible links is present with equal probability. Therefore, a degree distribution is binomial or Poisson in the limit of large network size. However, it was found that many real complex networks exhibit a degree distribution $p(M) \sim M^{-\gamma}$, where $2 < \gamma < 3$. Such networks were introduced in Barabasi and Albert (1999) and called scale-free, since the power law has the same functional form at all scales. There are many examples of technical, biological and social networks characterised as scale free (see Sects. 2.2.2 and 2.2.3 in Boccaletti et al., 2006 and references therein).

As for the dynamical aspect, many studies have been devoted to studying synchronisation in complex networks of coupled dynamical systems (see, e.g., Arenas et al., 2008; Osipov et al., 2007). It was found that the small-world property often enhances synchronisation as compared with regular lattices. Under some conditions, the scale-free property may lead to similar enhancement, see e.g. Motter et al. (2005).

Finally, we note that a network with complex topology consisting of N coupled one-dimensional maps is an N-dimensional dynamical system, similar to a coupled map lattice consisting of N one-dimensional maps. Thus, both models are equivalent in terms of their state vector dimension. However, a network with complex topology is a much more complicated object in terms of coupling structure. Thus, a network with complex topology can be considered as a more complex model of spatially extended systems.

3.7.4 Delay Differential Equations

Delay differential equations are typically used to model systems whose behaviour at present is determined not only by a present state but also by the values of dynamical variables at previous time instants. Such objects are widely presented in nature. They are studied in physics, biology, physiology and chemistry. Causes of a time delay can be different. Thus, in a population dynamics, a delay is connected with the fact that individuals participate in a reproduction process only after becoming adult. In spatially extended radio-physical systems, a delay is determined by a finite speed of signal propagation. A delay time τ is related to the time necessary for a signal to overpass a distance between elements. In a sufficiently general case, a time-delay system is described with the equation

$$\varepsilon_n \frac{d^n x(t)}{dt^n} + \varepsilon_{n-1} \frac{d^{n-1} x(t)}{dt^{n-1}} + \ldots + \varepsilon_1 \frac{dx(t)}{dt} = F(x(t), x(t - \tau_1), \ldots, x(t - \tau_k)), \tag{3.40}$$

where $\tau_1, \ldots, \tau_k$ stands for several possible time delays caused by different factors. Particular cases of equation (3.40) are as follows: *Ikeda equation* $\dot{x}(t) = -x(t) + \mu \cdot \sin(x(t - \tau) - x_0)$ describing the dynamics of a passive optical resonator; *Mackey – Glass equation* $\dot{x}(t) = -b \cdot x(t) + a \cdot x(t - \tau)/(1 + x^c(t - \tau))$ describing the process of red corpuscle generation in living organisms; the delayed feedback generator[12] $\varepsilon \cdot \dot{x}(t) = -x(t) + f(x(t - \tau))$, which is a very popular model in radio-physics.

Despite the only scalar dynamical variable, all the listed dynamical systems are infinite-dimensional, since one must specify a distribution of a dynamical variable over the interval $[0, \tau]$ as an initial condition. Even a first-order non-linear DDE can exhibit complex motions corresponding to attractors of very high dimensionality, chaos, multistability and other non-linear phenomena. In general, infinite dimensionality of the phase space leads to the possibility of observing attractors of arbitrary high dimension and with arbitrarily many positive Lyapunov exponents. In this sense, DDEs are more complex systems than previously described finite-dimensional model maps and ODEs. For more detailed information about DDE-based models, we refer to the review on complex dynamics of the feedback generator (Kuznetsov, 1982), the monograph Dmitriev and Kislov (1989), the research paper Kislov et al. (1979) and the website http://www.cplire.ru/win/InformChaosLab/index.html.

3.7.5 Partial Differential Equations

This is probably the most extensively studied mathematical tool developed specially for modelling of spatially extended systems. PDEs are used in very different scientific disciplines ranging from physics, chemistry and biology to ecology and economics. It is sufficient to recall famous Maxwell's equations in electrodynamics, Schrödinger's equation in quantum mechanics, reaction – diffusion equations in

---

[12] It is a ring consisting of a non-linear amplifier (characterised by a function f), an inertial element (a filter with a response time determined by ε) and a delay line (with a delay time τ).

chemistry and biology, and Ginzburg – Landau equation everywhere. Many classical models of the wave theory take the form of PDEs:

Simple wave equation $\partial x/\partial t + v(x)\partial x/\partial z = 0$, where x is the characterising quantity, v is the velocity of a perturbation propagation (depending on the perturbation value, in general); z is the spatial coordinate. The model can describe steepening and turnover of a wave profile.

Corteveg – de Vries equation $\partial x/\partial t + v(x)\partial x/\partial z + \beta\partial^3 x/\partial z^3 = 0$ is the simplest model exhibiting soliton-like solutions. Roughly, the latter ones are localised perturbations propagating with a constant waveform and velocity and preserving these characteristics after collision with each other.

Burgers' equation $\partial x/\partial t + v(x)\partial x/\partial z - \alpha\partial^2 x/\partial z^2 = 0$ is the simplest model describing waves in a medium with dissipation, in particular, shock waves (i.e. movements of a region of fast change in the value of x).

A dynamical system described with a PDE is infinite-dimensional even for a single spatial coordinate. To specify its state, one must provide an initial function $x(0, z)$. If a system without spatial boundaries is considered (such an idealisation is convenient if a system is very lengthy so that any phenomena at its boundaries do not significantly affect the dynamics under study and are not of interest for a researcher), then an initial function must be defined over an entire axis $-\infty < z < \infty$. If a system is bounded, then an initial function must be defined only over a corresponding interval $0 < z < L$, while boundary conditions are specified at its edges (e.g. fixed values $x(t, 0) = x(t, L) = 0$). In the latter case, one speaks of a boundary problem.

PDEs can exhibit both such attractors as fixed points, limit cycles, other kinds of low-dimensional behaviour and a very high-dimensional dynamics. This mathematical tool is even richer with properties and more complex for investigation compared to all the above-mentioned model equations. Of basic interest is the question about conditions of existence and uniqueness of a solution to a PDE. In part due to it, recently researchers have paid much attention to *regimes with sharpening* (when a solution exists only over a finite time interval) which are quite typical (Malinetsky and Potapov, 2000, pp. 148–170).

A huge body of literature is devoted to PDEs, (e.g. Loskutov and Mikhailov, 2007; Mikhailov and Loskutov, 1989; Sveshnikov et al., 1993; Tikhonov and Samarsky, 1972; Vladimirov, 1976).

3.8 Artificial Neural Networks

Artificial neural network (ANN) is a kind of mathematical model whose construction mimics some principles of organisation and functioning of networks of brain nerve cells (neurons). The idea is that each neuron can be modelled with a sufficiently simple automaton (an artificial neuron), while the entire brain complexity, flexibility and other important properties are determined by the couplings between neurons. The term "neural networks" was established in the middle of the 1940s (McCulloc and Pitts, 1943). Very active investigations in this field were carried out until the 1970s. After that, a significant decrease in the attention of researchers took place. In the 1980s, the interest reappeared due to problems of associative memory and neurocomputers so that the number of international conferences on ANNs and neurocomputers has reached a hundred by the end of twentieth century.

If an artificial neuron represents a function relating input and output values and a signal can propagate in a network only in one direction (no feedbacks), then an ANN is also just a function transforming an input signal into an output value. Below, we briefly consider mainly such a simple version. If feedbacks are present and/or a neuron is a system with its own dynamics (namely, a discrete map), then an ANN is a multidimensional map, i.e. a set of coupled maps with specific properties of the elements and couplings (see, e.g., Ivanchenko et al., 2004). Analogously, if a neuron is described with ordinary differential equations, then the respective ANN is a set of coupled ODEs (see, e.g., Kazantsev, 2004; Kazantsev and Nekorkin, 2003; 2005). Thus, complexity of an ANN dynamics depends on the kind of basic elements, the number of basic elements and couplings between them.

3.8.1 Standard Formal Neuron

Such an artificial neuron consists of an adaptive integrator and a non-linear converter (Fig. 3.23a). A vector of values $\{x_i\}$ is fed to its inputs. Each input x_i is supplied with a certain weight w_i. The integrator performs weighted (adaptive) summation of inputs

$$S = \sum_{i=1}^{n} w_i x_i. \tag{3.41}$$

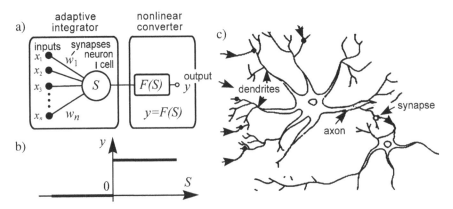

Fig. 3.23 Formal and biological neurons: (**a**) a scheme of an artificial neuron; (**b**) a plot of the unit step function; (**c**) a biological neuron (*filled circles* mark input synapses, *open ones* mark an output, triangles mark direction of excitation propagation)

The non-linear converter forms an output signal of a neuron as

$$y = F(S). \tag{3.42}$$

The choice of the neuron activation function F is determined by (i) specificity of a problem; (ii) convenience of realisation with a computer, an electric scheme or another tool; (iii) a "learning" algorithm (some learning algorithms impose constraints on the activation function properties, Sect. 3.8.4). Most often, the kind of non-linearity does not principally affect a problem solution. However, a successful choice can reduce duration of learning several times. Initially, one used the "unit step" as a function F:

$$F(S) = \begin{cases} 0, & S < 0, \\ 1, & S \geq 0, \end{cases} \tag{3.43}$$

whose plot is shown in Fig. 3.23b. Currently, a list of possible activation functions would occupy a huge space (e.g. Gorban' et al., 1998; http://www.neuropower.de). In particular, a widespread version is a non-linear function with saturation, the so-called logistic function or classical sigmoid:

$$F(S) = \frac{1}{1 + e^{-\alpha S}}. \tag{3.44}$$

With decrease in α, sigmoid gets flatter tending to a horizontal straight line at the level of 0.5 in the limit of $\alpha \to 0$. With increase in α, sigmoid tends to the unit step function (3.43).

Input values of the variable x can be likened to excitations of a real-world (biological) neuron (Fig. 3.23c) coming from dendrites of surrounding neurons via synapses (connection places). A real-world neuron can have the number of dendrites ranging from units to dozens of thousands. They provide information about the states of surrounding cells coupled to the neuron. Coupling strengths in a model are reflected by the weight coefficients w_i. Carrier of information in nerve cells is a jump of a membrane potential (a neural pulse, a *spike*). It is formed in a cell after a joint action of dendrites exceeds some critical value that is modelled in a formal neuron by the summation and the non-linear function. A spike propagates via an axon as a wave of membrane polarisation. Coming to a synapse, such a wave induces secretion of substances (neurotransmitters) which diffuse into dendrites of the neurons coupled to a given axon and are converted by receptors into an electric excitation pulse.[13] After generation of a pulse, a cell turns out unreceptive to external influences for a certain time interval. Such a state is called *refractory*. In other words, one deals with an excitable system which can be in the *resting phase* (before generation), *excitation phase* (during conduction of a pulse) and *refractory*

---

[13] As well, there are purely electric mechanisms of neuron coupling.

phase (during a certain interval after a pulse). The refractory period determines a limit possible frequency of pulse generation (less than 200 Hz).

3.8.2 Architecture and Classification of Neural Networks

To construct an ANN, one usually selects one of several standard architectures and removes superfluous elements or adds (more rarely) new ones. Two architectures are regarded as basic ones: fully connected and (multi)layered networks. In *fully connected neural networks* each neuron sends its output signal to all the neurons including itself. All input signals are sent to all neurons. Output signals of a network can be defined as all or some of neuron output signals after several steps of network functioning.

In *multi-layer neural networks*, neurons are combined in layers (Fig. 3.24). A layer contains neurons with the same input signals. The number of neurons in a layer is arbitrary and does not depend on the number of neurons in other layers. In general, a network consists of several layers which are enumerated from left to right in Fig. 3.24. External input signals are fed to inputs of neurons of an input layer (it is often enumerated as 0th), while the network outputs are output signals of the last layer. Apart from an input and an output layer, a multi-layered network may contain one or several *hidden layers*. Depending on whether the next layers send their signals to previous ones, one distinguishes between feed-forward networks (without feedbacks) and recurrent networks (with feedbacks). We note that after introduction of feedbacks a network is no longer a simple mapping from a set of input vectors to a set of output vectors. It becomes a dynamical system of high dimensionality and the question about its stability arises. Besides, neural networks can be divided into the following:

(i) Homogeneous and heterogeneous (i.e. with the same activation function for all neurons or with different activation functions);
(ii) Binary (operate with binary signals consisting of 0s and 1s) and analogous (operate with real-valued numbers);
(iii) Synchronous and asynchronous.

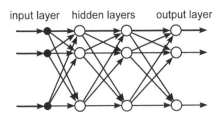

Fig. 3.24 A three-layer feed-forward network (a perceptron)

As well, ANNs differ in the number of layers. Theoretically, the number of layers and the number of neurons in each layer can be arbitrary. However, they are bounded in fact by computational resources realising a neural network. The more complex a network is, the more complicated tasks it can solve.

3.8.3 Basic Properties and Problems

Despite primitivism in comparison with biological systems, even multi-layer feedforward ANNs possess a number of useful properties and are capable of solving quite important tasks. Those properties are as follows.

(i) *Learning ability*. After selection of an ANN architecture and neuron properties, one can "train" an ANN to solve some problem with the aid of a certain learning algorithm. There are no guarantees that it is always possible but in many cases learning appears successful.

(ii) *Generalisation capability*. After the learning stage, a network becomes insensitive to small variations in an input signal (noise) and gives a correct result at its output.

(iii) *Abstraction capability*. If several distorted variants of an input image are presented to a network, the latter can itself create at its output an ideal image, which has never been met by it previously.

Among the tasks solved with ANNs, we note pattern (e.g. visual or auditory images) recognition, associative memory[14] realisation, clustering (division of an investigated set of objects into groups of similar ones), approximation of functions, time series prediction (Sect. 10.2.1), automatic control, decision making, diagnostics.

Many of the listed tasks are reduced to the following mathematical formulation. It is necessary to construct a map $X \to Y$ such that a correct output signal Y is formed in response to each possible input X. The map is specified by a finite number of pairs (an input, a correct output). The number of those pairs (learning examples) is significantly less than the total number of possible input signals. A set of all learning examples is called a *learning sample*. For instance, in image recognition, an input X is some representation of an image (a figure, a vector), an output Y is the number of a class to which an input image belongs. In automatic control, X is a set of values of the control parameters of an object, Y is a code determining an action appropriate for the current values of control parameters. In forecast, an input signal is a set of

---

[14] In von Neumann's model of computations (realised in a usual computer), memory access is possible only via an address, which does not depend on the memory contents. Associative memory is accessible based on the current contents. Memory contents can be called even by partial or distorted contents.

values of an observable quantity until a current time instant and an output is a set of the next values of an observable.

All these and many other applied problems can be reduced to a problem of construction of some multivariate function. What are capabilities of ANNs in this respect? As it was illustrated, they compute *univariate* linear and non-linear functions and their compositions obtained due to cascade connection of neurons. What can one get with the use of such operations? What functions can be accurately approximated with ANNs? As a result of long-lasted polemics between Kolmogorov and Arnold, a possibility of exact representation of a continuous multivariate function via a composition of univariate continuous functions and summation was shown (Arnold, 1959; Kolmogorov, 1957). The most complete answer to the question about approximating properties of neural networks is given by Stone's theorem (Stone, 1948) stating universal approximating capabilities of an arbitrary non-linearity: linear operations and cascade connection allow to get a device approximating any continuous multivariate function to any required accuracy on the basis of an arbitrary non-linear element. A popular exposition of the theorems of Kolmogorov and Stone in application to ANNs is given in Gorban' (1998). Thus, neurons in a network may have practically any non-linear activation function, only the fact of its non-linearity is important. In principle, ANNs are capable of doing "very many things". Yet, an open question is: How to teach them to do it?

3.8.4 Learning

During its functioning, a neural network forms an output signal Y corresponding to an input signal X, i.e. realises a certain function $Y = g(X)$. If a network architecture is specified, then the values of g are determined by synaptic weights. The choice of their optimal values is called *network learning*. There are various approaches to learning.

Learning by instruction. Here, one uses a learning sample, i.e. pairs of known input and output values $(X_1, Y_1), \ldots, (X_N, Y_N)$.

Let the values of vectors X and Y be related via $Y = g(X)$, in particular, $Y_i = g(X_i)$, $i = 1, \ldots, N$. A function g is unknown. We denote E as an error function assessing deviation of an arbitrary function f from the function g. Solving a problem with an ANN of a given architecture means to construct a function f by selecting synaptic weights so as to minimise the error function. In the simplest case, learning consists of searching for a function f which minimises E over a learning sample. Given a learning sample and the form of function E, learning of a network turns into a multidimensional non-linear optimisation problem (Dennis and Schnabel, 1983), which is often very complicated in practice (see also Sect. 10.2.1). It requires time-consuming computations and represents an iterative procedure; the number of iterations ranges typically from 10^3 to 10^8.

Since creation of intellectual schemes is based to a significant extent on biological prototypes, researchers still discuss whether the algorithms of learning by instruction can be considered as analogues to natural learning processes or they are completely artificial. It is known that, for instance, neurons of visual cortex learn to react on light pulses only under the influence of the pulses themselves without an external teacher. In particular, we are able to solve such a complicated task as image recognition. However, higher stages of learning (e.g. for children) are impossible without a teacher (their parents). Besides, some brain areas are quite able to play a role of "teacher" for other areas by controlling their activity. Therefore, it is not possible to claim uniquely which type of learning (with a teacher or without it) is more biologically plausible.

Learning without a teacher. In a widespread version it is as follows. There is a set of input vectors. A set of output vectors is absent. Learning a network means selecting its parameter values so that it would classify input vectors in some "optimal" way. An ANN must divide a set of input vectors into groups (classes) so that each class contains vectors close to each other while differences between classes are relatively big. This is done via optimisation of a cost function involving the two mentioned factors. When a new input vector is presented, a learned network attributes it to one of the classes which have been formed by it previously (without a teacher). One of the most well-known examples of such a way to solve classification problems is the learning of Cohonen network (see, e.g., Gorban' et al., 1998).

Currently, there is a huge body of literature on neural networks highlighting very different questions ranging from the choice of the ANN architecture to its learning and practical applications. In particular, there are many works accessible to a wide readership (Gorban', 1998; Gorban' et al., 1998; Loskutov and Mikhailov, 2007; Malinetsky and Potapov, 2000, pp. 171–203; Wasserman, 1989; http://www.neuropower.de). Some additional details and examples of ANN applications to modelling from time series are given in Sect. 10.2.1.

Thus, several representative classes of deterministic models are discussed in Sects. 3.5, 3.6, 3.7 and 3.8. Roughly speaking, we have described them in the order of increasing complexity in terms of the phase space dimension, fractal dimension of possible attractors, the number of positive Lyapunov exponents, the diversity of possible dynamical regimes and configurations of the parameter space. Yet, linear ordering most often appears impossible; therefore, we have presented more specific discussion of the complexity for each example separately. To summarise, the presented deterministic models are capable of describing huge number of phenomena observed in real-world systems ranging from quite simple ones (e.g. an equilibrium state, a limit cycle and a linear resonance) to very complex (e.g. high-dimensional chaotic motions, transition to chaos and diverse bifurcations in multi-parametric non-linear systems).

References

Afraimovich, V.S., Nekorkin, V.I., Osipov, G.V., Shalfeev, V.D.: Stability, structures, and chaos in nonlinear synchronisation networks. Gor'ky Inst. Appl. Phys. RAS (in Russian) (1989)

Albert, R., Barabasi, A.-L.: Statistical mechanics of complex networks. Rev. Modern Phys. **74**, 47–97 (2002)

Andreev, K.V., Krasichkov, L.V.: Electrical activity of a neuron modeled by piecewise continuous maps. Tech. Phys. Lett. **29**(2), 105–108 (2003)

Andreev, Yu.V., Dmitriev, A.S.: Recording and restoration of images in one-dimensional dynamical systems. J. Commun. Technol. Electron. **39**(1), 104–113, (in Russian) (1994)

Arenas, A., Diaz-Guilera, A., Kurths, J., Moreno, Y., Zhou, C.: Synchronization in complex networks. Physics Reports. **469**, 93–153 (2008)

Arnold, V.I.: On the representation of continuous functions of three variables by the superpositions of continuous functions of two variables. Matem. Sbornik. **48**(1), 3–74, (in Russian) (1959)

Astakhov, V.V., Bezruchko, B.P., Gulyaev, Yu.V., Seleznev, Ye.P.: Multistable states of dissipatively coupled Feigenbaum systems. Tech. Phys. Letters. **15**(3), 60–65, (in Russian) (1989)

Astakhov, V.V., Bezruchko, B.P., Ponomarenko, V.I.: Formation of multistability, classification of isomers, and their evolution in coupled Feigenbaum systems. Radiophys. Quantum Electron. **34**(1), 35–39, (in Russian) (1991a)

Astakhov, V.V., Bezruchko, B.P., Ponomarenko, V.I., Seleznev, Ye.P.: Multistability in system of radiotechnical oscillators with capacity coupling. J. Commun. Technol. Electron. **36**(11), 2167–2170, (in Russian) (1991b)

Barabasi, A.L., Albert, R.: Emergence of scaling in Random Networks. Science. **286**, 509–512 (1999)

Bezruchko, B.P., Prokhorov, M.D., Seleznev, Ye.P.: Multiparameter model of a dissipative nonlinear oscillator in the form of one dimensional map. Chaos, Solitons, Fractals. **5**(11), 2095–2107 (1995)

Bezruchko, B.P., Prokhorov, M.D.: Control of spatio-temporal chaos in a chain of bi-stable oscillators. Tech. Phys. Lett. **25**(12), 51–57, (in Russian) (1999)

Bezruchko, B.P., Jalnine, A.Yu., Prokhorov, M.D., Seleznev, Ye.P.: Discrete nonlinear models of a periodically driven RL-diode circuit. Izvestiya VUZ. Appl. Nonlinear Dynamics (ISSN 0869-6632). **5**(2), 48–62, (in Russian) (1997a)

Boccaletti, S., Latora, V., Moreno, Y., Chavez, M., Hwang, D.U.: Complex networks: structure and dynamics. Phys. Rep. **424**, 175–308 (2006)

Butkovsky, O.Ya., Kravtsov Yu.A., Surovyatkina, E.D.: Structure of the attraction zones of the final states in the presence of dynamical period doubling bifurcations. J. Exp. Theor. Phys. **86**(1), 206–212 (1998)

Carcasses, J., Mira, C., Bosh, M., et al. Crossroad area – spring area transition. Parameter plane representation. Int. J. Bif. Chaos. **1**, 183 (1991)

Danilov Yu.A. Nonlinearity. Znaniye – sila. **11**, 34–36, (in Russian) (1982)

Dennis, J., Schnabel, R.: Numerical Methods for Unconstrained Optimization and Nonlinear Equations. Prentice-Hall, Upper Saddle River, (1983)

Dmitriev, A.S.: Recording and retrieval of information in one-dimensional dynamical systems. J. Commun. Technol. Electron. **36**(1), 101–108, (in Russian) (1991)

Dmitriev, A.S., Kislov, V.Ya. Stochastic oscillations in radiophysics and electronics. Nauka, Moscow, (in Russian) (1989)

Dorogovtsev, S.N., Mendes, J.F.F.: Evolution of Networks. Oxford University Press, Oxford (2003)

Feigenbaum, M.J.: Universal behavior in nonlinear systems. Los Alamos Sci. **1**(1), 4–27 (1980)

Gorban' A.N., Dunin-Barkovsky, V.L., Kirdin, A.N., et al. Neuroinformatics. Nauka, Novosibirsk, (in Russian) (1998)

Gorban' A.N.: Multivariable functions and neural networks. Soros Educ. J. **12**, 105–112, (in Russian) (1998)

Gouesbet, G., Letellier, C.: Global vector-field approximation by using a multivariate polynomial L_2 approximation on nets. Phys. Rev. E. **49**, 4955–4972 (1994)

Gouesbet, G., Meunier-Guttin-Cluzel, S., Ménard, O.: Global reconstructions of equations of motion from data series, and validation techniques, a review. In: Gouesbet, G., Meunier-Guttin-Cluzel, S., Ménard, O. (eds.) Chaos and Its Reconstructions, pp. 1–160. Nova Science Publishers, New York, (2003)

Horbelt, W.: Maximum likelihood estimation in dynamical systems: PhD thesis. University of Freiburg, Freiburg, Available at http://webber.physik.uni-freiburg.de/~horbelt/diss (2001)

Ivanchenko, M.V., Osipov, G.V., Schalfeev, V.D., Kurths, J.: Phase synchronization in ensembles of bursting oscillators. Phys. Rev. Lett. **93**, 134101 (2004)

Izhikevich, E.M.: Neural excitability, spiking and bursting. Int. J. Bif. Chaos. **10**, 1171–1266 (2000)

Kalitkin, N.N.: Numerical Methods. Nauka, Moscow, (in Russian) (1978)

Kazantsev, V.B.: Dynamical transformation of pulse signals in neuron systems. Izvestiya VUZ. Appl. Nonlinear Dynamics (ISSN 0869-6632). **12**(6), 118–128, (in Russian) (2004)

Kazantsev, V.B., Nekorkin, V.I.: Dynamics and oscillatory neurons. In: Gaponov-Grekhov, A.V., Nekorkin, V.I. (eds.) Informational Aspects. Nonlinear Waves – 2002, pp. 9–33 Institute of Applied Physics RAS, Nizhny Novgorod, (in Russian) (2003)

Kazantsev, V.B., Nekorkin, V.I.: Phase-controlled oscillations in neurodynamics. In: Gaponov-Grekhov, A.V., Nekorkin, V.I. Nonlinear Waves – 2004, pp. 345–361. Institute of Applied Physics RAS, Nizhny Novgorod, (in Russian) (2005)

Kazantsev, V.B., Nekorkin, V.I., Binczak, S., et al. Spiking dynamics of interacting oscillatory neurons. Chaos. **15**, 023103 (2005)

Kazantsev, V.B., Nekorkin, V.I.: Dynamics and oscillatory neurons. In: Gaponov-Grekhov, A.V., Nekorkin, V.I. (eds.) Informational Aspects. Nonlinear Waves – 2002, pp. 9–33 Institute of Applied Physics RAS, Nizhny Novgorod, (in Russian) (2003)

Kipchatov, A.A.: Peculiarities of nonlinear dynamics of a nonautonomous nonlinear circuit. Radiophys. Quantum Electr. **33**(2), 182–190, (in Russian) (1990)

Kislov, V.Ya., Zalogin, N.N., Myasin, E.A.: Investigation of stochastic self-sustained oscillatory processes in self-sustained generators with time delay. J. Commun. Technol. Electr. **24**(6), 1118–1130, (in Russian) (1979)

Kolmogorov, A.N.: About representation of continuous multivariable functions as superposition of continuous one-variable functions. Doklady Acad. Sci. USSR. **114**(5), 953–956, (in Russian) (1957)

Kurths, J., Maraun, D., Zhou, C.S., Zamora-Lopez, G., Zou, Y.: Dynamics in Complex Systems. Eur. Rev. **17**(2), 357–370 (2009)

Kuznetsov, A.P., Kuznetsov, S.P.: Critical dynamics of coupled map lattices at boundary of chaos (a review). Radiophys. Quantum Electr. **34**(10–12), 1079–1115, (in Russian) (1991)

Kuznetsov, A.P., Kuznetsov, S.P.: Critical dynamics of one-dimensional maps. Part 1. Feigenbaum scenario. Izvestiya VUZ. Appl. Nonlinear Dynamics (ISSN 0869-6632). **1**(1–2), 15–33, (in Russian) (1993a)

Kuznetsov, A.P., Kuznetsov, S.P.: Critical dynamics of one-dimensional maps. Part 2. Two-parametric transition to chaos. Izvestiya VUZ. Appl. Nonlinear Dynamics (ISSN 0869-6632). **1**(3–4), 17–35, (in Russian) (1993b)

Kuznetsov, A.P., Potapova, A.Yu. Peculiarities of complex dynamics of nonlinear non-autonomous oscillators with Thom's catastrophe. Izvestiya VUZ. Appl. Nonlinear Dynamics (ISSN 0869-6632). **8**(6), 94–120, (in Russian) (2000)

Kuznetsov, S.P.: Complex dynamics of oscillators with delayed feedback (review). Radiophys. Quantum Electr. **25**(12), 1410–1428, (in Russian) (1982)

Kuznetsov, S.P.: Dynamical chaos. Fizmatlit, Moscow, (in Russian) (2001)

Kuznetsov, S.P.: Universality and scaling in the behavior of coupled Feigenbaum systems. Radiophys. Quantum Electr. **28**, 681–695 (1986)

Kuznetsova, A.Yu., Kuznetsov, A.P., Knudsen, C., Mosekilde, E.: Catastrophe theoretic classification of nonlinear oscillators. Int. J. Bif. Chaos. **14**, 1241–1266 (2004)

Linsay, P.S.: Period doubling and chaotic behaviour in a driven anharmonic oscillator. Phys. Rev. Lett. **47**, 1349–1352 (1981)

Loskutov, A.Yu., Mikhailov, A.S.: Basics of Complex Systems Theory. Regular and Chaotic Dynamics, Moscow (2007)

Malinetsky, G.G., Potapov, A.B.: Contemporary Problems of Nonlinear Dynamics. Editorial URSS, Moscow, (in Russian) (2000)

Malinetsky, G.G., Stepantsev, M.E.: Modelling of crowd movement with cellular automata. Izvestiya VUZ. Appl. Nonlinear Dynamics (ISSN 0869-6632). **5**(5), 75–79, (in Russian) (1997)

Margolus, N., Toffoli, T.: Cellular automata machines. Addison-Wesley, New York (1990)

Mathematical Encyclopedic Dictionary. Sov. Encyclopedia, Moscow, 846p., (in Russian) (1988)

McCulloc, W.S., Pitts, W. A logical calculus of the ideas immanent in nervous activity. Bull. Math. Biophys. **5**, 115–133 (1943)

Mikhailov, A.S., Loskutov, A.Yu. Foundations of Synergetics. An Introduction. W.H. Freeman, New York (1989)

Minsky, M.: Computation: Finite and Infinite Machines. Prentice-Hall, New York (1967)

Mira, C., Carcasses, J.: On the crossroad area – saddle area and spring area transition. Int. J. Bif. Chaos. **1**, 643 (1991)

Motter, A.E., Zhou, C.S., Kurths, J.: Enhancing complex-network synchronization. Europhys. Lett. **69**, 334–340 (2005)

Neimark, Yu.I., Landa, P.S.: Stochastic and Chaotic Oscillations, 424p. Nauka, Moscow (1987). Translated into English: Kluwer Academic, Dordrecht and Boston (1992)

Neimark Yu.I.: Method of Point Mapping in the Theory of Nonlinear Oscillations. Nauka, Moscow, (in Russian) (1972)

Neimark Yu.I.: Mathematical models in natural science and technology. Nizhny Novgorod Inst. Appl. Phys. RAS. **Part 1**, (1994); **Part 2**, (1996); **Part 3**, (1997). (in Russian)

Nekorkin, V.I., Dmitrichev, A.S., Shchapin, D.S., Kazantsev, V.B.: Dynamics of neuron model with complex threshold excitation. Math. Model. **17**(6), 75–91, (in Russian) (2005)

Nekorkin, V.I., Velarde, M.G.: Synergetic Phenomena in Active Lattices. Springer, Berlin (2002)

Osipov, G.V., Kurths, J., Zhou, C.: Synchronization in Oscillatory Networks. Springer, Berlin (2007)

Parlitz, U.: Common dynamical features of periodically driven strictly dissipative oscillators. Int. J. Bif. Chaos. **3**, 703–715 (1991)

Press, W.H., Flannery, B.P., Teukolsky, S.A., Vetterling, W.T.: Numerical Recipes in C. Cambridge University Press, Cambridge (1988)

Prokhorov, M.D., Smirnov, D.A.: Empirical discrete model of the diode oscillating loop. J. Commun. Technol. Electr. **41**(14), 1245–1248 (1996)

Rapp, P.E., Schmah, T.I., Mees, A.I.: Models of knowing and the investigation of dynamical systems. Phys. D. **132**, 133–149 (1999)

Rulkov, N.F.: Regularization of synchronized chaotic bursts. Phys. Rev. Lett. **86**, 183–186. Modeling of spiking-bursting neural behavior using two-dimensional map. Phys. Rev. E. **65**, 041922 (2001)

Samarsky, A.A.: Introduction to Numerical Methods. Nauka, Moscow, (in Russian) (1982)

Scheffczyk, C., Parlitz, U., Kurz, T., et al. Comparison of bifurcation structures of driven dissipative nonlinear oscillators. Phys. Rev. A. **43**, 6495–6502 (1991)

Schreiber, T.: Detecting and analyzing nonstationarity in a time series using nonlinear cross predictions. Phys. Rev. Lett. **78**, 843–846 (1997)

Schuster, H.G.: Deterministic chaos. Physik Verlag, Weinheim (1984)

Shalfeev, V.D., Matrosov V.V.: Chaotically modulated oscillations in coupled phase systems. In: Gaponov-Grekhov, A.V., Nekorkin , V.I. (eds) Nonlinear Waves – 2004, pp. 77–89. Institute of Applied Physics RAS, Nizhny Novgorod, (In Russian) 2005

Shalizi, C.R.: Methods and Techniques of Complex Systems Science: An Overview, vol. 3, arXiv:nlin.AO/0307015. Available at http://www.arxiv.org/abs/nlin.AO/0307015 (2003)

Stone, M.N.: The generalized Weierstrass approximation theorem. Math. Mag. **21**, 167–183, 237–254 (1948)

Strogatz, S.H.: Exploring complex networks. Nature **410**, 268–276 (2001)

Sveshnikov, A.G., Bogolyubov, A.N., Kravtsov, V.V.: Lectures in mathematical physics. Moscow State University, Moscow (in Russian) (1993)

Thompson, J.M., Stewart, H.B.: Nonlinear Dynamics and Chaos. Wiley, New York (2002)

Tikhonov, A.N., Samarsky, A.A.: Equations of Mathematical Physics Nauka, Moscow (1972). Translated into English: Dover Publications (1990)

Trubetskov, D.I.: Oscillations and Waves for Humanitarians. College, Saratov, (in Russian) (1997)

Vladimirov, V.S.: Equations of Mathematical Physics. Nauka, Moscow (1976). Translated into English: Mir Publ., Moscow (1984)

von Neumann, J.: Theory of Self-Reproducing Automata. University of Illinois Press, Chicago, IL (1966)

Wang, Q.: The global solution of the n-body problem. Celestial Mech. **50**, 73–88 (1991)

Wasserman, P.: Neural Computing. Van Nostrand Reinhold, New York (1989)

Watts, D.J.: Small Worlds: The Dynamics of Networks between Order and Randomness. Princeton University Press, Princeton, NJ (1999)

Watts, D.J., Strogatz, S.H.: Collective dynamics of small-world networks. Nature **393**, 440–442 (1998)

Chapter 4
Stochastic Models of Evolution

To continue the discussion of randomness given in Sect. 2.2.1, we briefly touch on *stochastic models* of temporal evolution (random processes). They can be specified either via explicit definition of their statistical properties (probability density functions, correlation functions, etc., Sects. 4.1, 4.2 and 4.3) or via stochastic difference or differential equations. Some of the most widely known equations, their properties and applications are discussed in Sects. 4.4 and 4.5.

4.1 Elements of the Theory of Random Processes

If, given initial conditions $\mathbf{x}(t_0)$ and fixed parameter values, a process demonstrates the same time realisation in each trial, then its natural description is deterministic (Chap. 3). However, such a situation is often not met in practice: different trials "under the same conditions" give different realisations of a process. One relates such non-uniqueness to influences from multiple uncontrolled factors, which are often present in the real world. Then, it is reasonable to refuse deterministic description and exploit an apparatus of the theory of probability and theory of random processes (see, e.g. Gihman and Skorohod, 1974; Kendall and Stuart, 1979; Malakhov, 1968; Rytov et al., 1978; Stratonovich, 1967; Volkov et al., 2000; Wentzel', 1975).

4.1.1 Concept of Random Process

Random process (random function of time) is a generalisation of the concept of random quantity to describe time-dependent variables. More precisely, its definition is given as follows. Firstly, *random function* is a random quantity depending not only on a random event ω but also on some parameter. If that parameter is time, then the random function is called random process and denoted $\xi(t, \omega)$. A quantity ξ may be both scalar (a scalar random process) and vector (a vector or a multidimensional random process). It may run either a discrete range of values (a process with discrete

B.P. Bezruchko, D.A. Smirnov, *Extracting Knowledge From Time Series*, Springer Series in Synergetics, DOI 10.1007/978-3-642-12601-7_4,

© Springer-Verlag Berlin Heidelberg 2010

states) or a continuous one. For the sake of definiteness, we further speak of the latter case. Studying and development of such models is the subject of the theory of random processes (Gihman and Skorohod, 1974; Stratonovich, 1967; Volkov et al., 2000; Wentzel', 1975). In the case of discrete time $t = 0, 1, 2, \ldots$, a random process is called a *random sequence*.

For a random process, an outcome of a single trial is not a single number (like for a random quantity) but a function $\xi(t, \omega_1)$, where ω_1 is a random event realised in a given trial. The random event can be interpreted as a collection of uncontrolled factors influencing a process during a trial. The function $\xi(t, \omega_1)$ is called a *realisation of a random process*. It is a deterministic (non-random) function of time, because a random event $\omega = \omega_1$ is fixed. In general, one gets different realisations as outcomes of different trials. A set of realisations obtained from various trials (i.e. for different ω) is called an *ensemble* of realisations (Fig. 4.1).

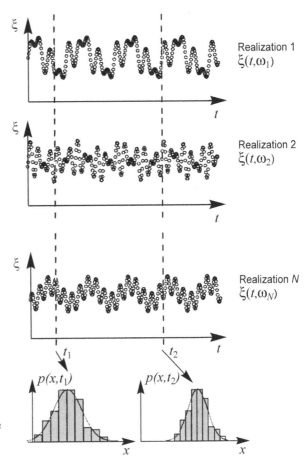

Fig. 4.1 An ensemble of N realisations (three of them are shown) and two sections of a random process

4.1.2 Characteristics of Random Process

At any fixed time instant t, a random process $\xi(t, \omega)$ is a random quantity. The latter is called a *section of a random process* at a time instant t and characterised by a probability density function $p(x, t)$. This distribution law is called *one-dimensional distribution* of the random process. It depends on time and may differ for two different time instants. Knowledge of one-dimensional distribution law $p(x, t)$ allows one to calculate expectation and variance of the process at any time instant t. If the distribution law varies in time, then the expectation

$$m(t) = E\left[\xi(t, \omega)\right] = \int_{-\infty}^{\infty} xp(x, t)dx \tag{4.1}$$

and the variance

$$\sigma_\xi^2(t) = E[\xi(t, \omega) - m(t)]^2 = \int_{-\infty}^{\infty} [x - m(t)]^2 p(x, t)dx \tag{4.2}$$

may vary in time as well. They are deterministic (non-random) functions of time, since dependence on random events is eliminated due to integration.

In general, sections $\xi(t, \omega)$ at different time instants t_1 and t_2 exhibit different probability density functions $p(x, t_1)$ and $p(x, t_2)$, Fig. 4.1. Joint behaviour of the sections is described by two-dimensional probability density function $p_2(x_1, t_1, x_2, t_2)$. One can define n-dimensional distribution laws p_n for any sets t_1, t_2, $\ldots$, t_n in the same way. These laws constitute a collection of finite-dimensional distributions of a random process $\xi(t, \omega)$. Probabilistic properties of a process are fully defined only if the entire collection is given. However, since the latter represents an infinite number of distribution laws, one cannot in general fully describe a random process.

To be realistic, one must confine him/herself with the use of some characteristics, e.g. one- and two-dimensional distributions or low-order moments (expectation, variance, auto-covariance function). Thus, *auto-covariance function* depends on two arguments:

$$K(t_1, t_2) = E\left[(\xi(t_1, \omega) - m(t_1))(\xi(t_2, \omega) - m(t_2))\right] =$$
$$= \iint (x_1 - m(t_1))(x_2 - m(t_2)) \, p_2(x_1, t_1, x_2, t_2)dx_1 \, dx_2. \tag{4.3}$$

For fixed t_1 and t_2, the expression (4.3) defines *covariance* of the random quantities $\xi(t_1, \omega)$ and $\xi(t_2, \omega)$. If it is normalised by root-mean-squared deviations, one gets *autocorrelation function* $\rho(t_1, t_2) = K(t_1, t_2)/(\sigma_\xi(t_1)\sigma_\xi(t_2))$, i.e. correlation coefficient between random quantities $\xi(t_1, \omega)$ and $\xi(t_2, \omega)$.[1] Autocorrelation function takes values ranging from -1 to 1. The value of $|\rho(t_1, t_2)| = 1$ corresponds to a deterministic linear dependence $\xi(t_1, \omega) = \text{const} \cdot \xi(t_2, \omega)$.

To characterise a process, one often uses conditional one-dimensional distribution $p_1(x, t | x_1, t_1)$, i.e. distribution of a section $\xi(t)$ under the condition that at a time instant t_1 the quantity ξ takes a value of $\xi(t_1) = x_1$. The function $p_1(x, t | x_1, t_1)$ is called probability density of the transition from a state x_1 at a time instant t_1 to a state x at a time instant t.

4.1.3 Stationarity *and* Ergodicity *of Random Processes*

An important property of a process is its stationarity or non-stationarity. A process is called *strongly stationary* (*stationary in a narrow sense*) if all its finite-dimensional distributions do not change under a time shift, i.e. $p_n(x_1, t_1, \ldots, x_n, t_n) = p_n(x_1, t_1 + \tau, \ldots, x_n, t_n + \tau)$, $\forall n, t_1, \ldots, t_n, \tau$. In other words, neither characteristic of a process changes under a time shift. A process is called *weakly stationary* (*stationary in a wide sense*) if its expectation, variance and autocorrelation function (i.e. moments up to the second order inclusively) do not change under a time shift. For a stationary process (in any of the two senses), one has $m(t) = \text{const}$, $\sigma_\xi^2(t) = \text{const}$ and $K(t_1, t_2) = k(\tau)$ with $\tau = t_2 - t_1$. The strong stationarity implies the weak one.[2]

In general, "stationarity" means invariance of some property under a time shift. If a property of interest (e.g. nth-order moment of the one-dimensional distribution) does not change in time, then a process is called stationary with regard to that property.

A process is called *ergodic* if all its characteristics can be determined from its single (infinitely long) realisation. For instance, the expectation is then determined as

$$m = \lim_{T \to \infty} \frac{1}{T} \int\limits_0^T \xi(t, \omega_1) dt$$

for almost any ω_1, i.e. temporal averaging and ensemble (state space) averaging give the same result. If one can get *all* characteristics of a process in such a way, the process is called *strictly ergodic*. If only certain characteristics can be restored from a single realisation, the process is called ergodic with regard to those characteristics. Thus, one introduces the concept of the first-order ergodicity, i.e. ergodicity with regard to the first-order moments, and so forth. Ergodic processes constitute an

---

[1] There is also another terminology where equation (4.3) is called an auto-correlation function, and after the normalisation, a normalised auto-correlation function. To avoid misunderstanding, we do not use it in this book.

[2] There is also somewhat different interpretation related to an additional requirement of a *finite* variance for a weakly stationary process. In such a case, the weak stationarity does not follow from the strong one.

important class, since in practice one often has a single realisation rather than a big ensemble of realisations. Only for an ergodic process one can restore its properties from such a data set. Therefore, one often assumes ergodicity of a process under investigation when time series analysis (Chaps. 5, 6, 7, 8, 9, 10, 11, 12 and 13) is performed. An ergodic process is stationary, while the reverse is not compulsorily true.

An example (a real-world analogue) of a random process is provided by almost any physical measurement. It may be measurements of a current in a non-linear circuit, exhibiting self-sustained oscillations. Measured time realisations differ for different trials due to thermal noise, various interference, etc. Moreover, having a realisation over a certain time interval, one cannot uniquely and precisely predict its future behaviour, since the latter is determined by random factors which will affect a process in the future.

A simpler example of a random process is a strongly simplified model representation of photon emission by an excited atom. An emission instant, initial phase, direction and polarisation are unpredictable. However, as soon as a photon is emitted and its starting behaviour gets known, the entire future is uniquely predictable. According to the representation considered, the process is described as sinusoidal function of time with a random initial phase (harmonic noise). Random processes of such type are sometimes called *quasi-deterministic*, since random factors determine only initial conditions, while further behaviour obeys a deterministic law.

4.1.4 Statistical Estimates of Random Process Characteristics

To get statistical estimates of a one-dimensional distribution law $p(x, t)$ and its moments, one can perform many trials and obtain a set of realisations $\xi(t, \omega_1)$, $\xi(t, \omega_2)$, ..., $\xi(t, \omega_n)$. The values of the realisations at a given time instant $t = t^*$ constitute a sample of size n from the distribution of a random quantity $\xi(t^*, \omega)$, Fig. 4.1. One can estimate a distribution law $p(x, t^*)$ and other characteristics based on that sample. It can be done for each time instant.

Multidimensional distribution laws can be estimated from an ensemble of realisations in an analogous way. However, the number of realisations for their reliable estimation must be much greater than that for the estimation of statistical moments or one-dimensional distributions.

A situation when one has only a single realisation is more complex. Only for an ergodic process and a sufficiently long realisation, one can estimate characteristics of interest by replacing ensemble averaging with temporal averaging (Sect. 4.1.3).

4.2 Basic Models of Random Processes

A random process can be specified via explicit description of its finite-dimensional probability distributions. In such a way, one introduces basic models in the theory of random processes. Below, we consider several of them (Gihman and Skorohod, 1974; Volkov et al., 2000; Wentzel', 1975).

(i) One of the most important models in the theory of random processes is the *normal (Gaussian) random process*. This is a process whose finite-dimensional distribution laws are all normal. Namely, an n-dimensional distribution law of this process reads as

$$p_n(x_1, t_1, \ldots, x_n, t_n) = \frac{1}{\sqrt{(2\pi)^n \, |\mathbf{V}_n|}} \exp\left(-\frac{1}{2}(\mathbf{x}_n - \mathbf{m}_n)^{\mathrm{T}} \cdot \mathbf{V}_n^{-1} \cdot (\mathbf{x}_n - \mathbf{m}_n)\right)$$

$$(4.4)$$

for any n, where

$$\mathbf{x}_n = \begin{bmatrix} x_1 \\ x_2 \\ \ldots \\ x_n \end{bmatrix}, \mathbf{m}_n = \begin{bmatrix} m(t_1) \\ m(t_2) \\ \ldots \\ m(t_n) \end{bmatrix}, \mathbf{V}_n = \begin{bmatrix} K(t_1, t_1) & K(t_1, t_2) & \ldots & K(t_1, t_n) \\ K(t_2, t_1) & K(t_2, t_2) & \ldots & K(t_2, t_n) \\ \ldots & \ldots & \ldots & \ldots \\ K(t_n, t_1) & K(t_n, t_2) & \ldots & K(t_n, t_n) \end{bmatrix},$$

$$(4.5)$$

$m(t)$ is the expectation, $K(t_1, t_2)$ is the auto-covariance function (4.3), T stands for transposition and $|\mathbf{V}_n|$ is a determinant of a matrix $\mathbf{V}_n$. Here, all the finite-dimensional distributions are known (a process is fully determined) if the expectation and the auto-covariance function are specified. A normal process remains normal under any linear transform.

(ii) A process *with independent increments*. This is a process for which the quantities $\xi(t_1, \omega)$, $\xi(t_2, \omega) - \xi(t_1, \omega)$, $\ldots$, $\xi(t_n, \omega) - \xi(t_{n-1}, \omega)$ (increments) are statistically independent for any n, $t_1, \ldots, t_n$, such that $n > 1$ and $t_1 < t_2 < \ldots < t_n$.

(iii) *Wiener's process*. This is an N-dimensional random process with independent increments for which a random vector $\xi(t_2, \omega) - \xi(t_1, \omega)$ for any $t_1 < t_2$ is distributed according to the normal law with zero mean and the covariance matrix $(t_2 - t_1)s^2 I_n$, where I_n is the nth-order unit matrix and $s = \text{const}$. This is a non-stationary process. In the one-dimensional case, its variance linearly rises with time as $\sigma^2(t) = \sigma^2(t_0) + s^2 \cdot (t - t_0)$.

Wiener's process describes, for instance, a Brownian motion, i.e. movements of a Brownian particle under random independent shocks from molecules of a surrounding medium.

One can show that Wiener's process is a particular case of the normal process. Wiener's process with $s = 1$ is called standard.

(iv) A (first-order) *Markovian process* is a random process whose conditional probability density function for any $n, t_1, \ldots, t_n$, such that $t_1 < t_2 < \ldots < t_n$, satisfies the property $p_1(x_n, t_n | x_{n-1}, t_{n-1}, \ldots, x_1, t_1) = p_1(x_n, t_n | x_{n-1}, t_{n-1})$. This is also expressed as "the future depends on the past only via the present". Any finite-dimensional distribution law of this process is expressed via its one-dimensional and two-dimensional laws. One can show that Wiener's process is a particular case of a Markovian process.

One more important particular case is a Markovian process with a finite number K of possible states $S_1, \ldots, S_K$. Due to discreteness of the states, it is described in terms of probabilities rather than probability densities. Conditional probability $P\{\xi(t+\Delta t) = S_j | \xi(t) = S_i\}$ is called transition probability, since it describes the transition from the state i to the state j. A quantity $\lambda_{i,j}(t) = \lim_{\Delta t \to +0} P\{\xi(t + \Delta t) = S_j | \xi(t) = S_i\}/\Delta t$ is called transition probability density.

Markovian processes play an especial role in the theory of random processes. Multitude of investigations are devoted to them.

(v) *Poisson process* with a parameter $\lambda > 0$ is a scalar random process with discrete states possessing the following properties: (a) $\xi(0, \omega) = 0$; (b) increments of the process are independent; (c) for any $0 \leq t_1 < t_2$, a quantity $\xi(t_2, \omega) - \xi(t_1, \omega)$ is distributed according to the Poisson law with a parameter $\lambda(t_2 - t_1)$, i.e.

$$
P\{\xi(t_2, \omega) - \xi(t_1, \omega) = k\} = \frac{\lambda(t_2 - t_1)^k}{k!} \exp(-\lambda(t_2 - t_1)),
$$

where k is a non-negative integer. Poisson process is often used in applications, e.g. in the queueing theory.

(vi) *White noise* is a weakly stationary (according to one of the definitions of stationarity, see Sect. 4.1.3) random process whose values at different time instants are uncorrelated, i.e. its auto-covariance function is $k(\tau) = \text{const} \cdot \delta(\tau)$. It is called "white", since its power spectrum is a constant, i.e. all frequencies are equally presented in it. Here, one draws an analogy to the white light, which involves all frequencies (colours) of the visible part of spectrum. White noise variance is infinitely large: $\sigma_\xi^2 = k(0) = \infty$.

A widespread model is Gaussian white noise. This is a stationary process with a Gaussian one-dimensional distribution law and auto-covariance function $k(\tau) = \text{const} \cdot \delta(\tau)$. Strictly speaking, such a combination is contradictory, since white noise has infinite variance, while a normal random process has a finite variance. Yet, somewhat contradictory concept of Gaussian white noise is useful in practice and in investigations of stochastic differential equations (Sect. 4.5). Gaussian white noise can be interpreted as a process with a very large variance, while a time interval over which its auto-covariance function decreases down to zero is very small as compared with the other characteristic timescales of a problem under consideration.

(vii) A discrete-time analogue of white noise is a sequence of *independent identically distributed* random quantities. It is also often called white noise. Most often, one considers normal one-dimensional distribution, even though any other distribution is also possible. In case of discrete time, variance σ_ξ^2 is finite so that the process is weakly stationary, no matter what definition of the weak stationarity is used.

White noise is the "most unpredictable" process, since any interdependence between its successive values is absent. A sequence of independent normally distributed quantities serves as a basic model in the construction of discrete-time stochastic models in the form of stochastic difference equations (Sect. 4.4).

(viii) *Markov chain* is a Markovian process with discrete states and discrete time. This simple model is widely used in practice. Its main characteristics are probabilities of transitions from one state to another one. Graph-theoretic tools are used for the analysis and representation of such models.

4.3 Evolutionary Equations for Probability Distribution Laws

Exemplary random processes derived from intuitive conceptual considerations are listed above. Thus, the normal random process can be obtained from the idea about big number of independent factors, white noise from independence of subsequent values and Poisson process from an assumption of rare events (Gihman and Skorohod 1974; Volkov et al., 2000; Wentzel', 1975). All "essential properties" of these three processes are known: finite-dimensional distribution laws, statistical moments, etc.

As for Markovian processes, they are based on the ideas about relationships between the future states and the previous ones. In general, a Markovian process may be non-stationary. Thus, one can ask how an initial probability distribution changes in time, whether it converges to some stationary one and what such a limit distribution looks like. Answers to those questions are not formulated explicitly in the definition of a Markovian process. However, to get the answers, one can derive evolutionary equations for a probability distribution law based directly on the definition. For a process with a finite number of states, they take the form of a set of ordinary differential equations (*Kolmogorov equations*):

$$
\begin{bmatrix} dp_1/dt \\ dp_2/dt \\ \dots \\ dp_K/dt \end{bmatrix} = \begin{bmatrix} -(\lambda_{1,2} + \dots + \lambda_{1,K}) & \lambda_{2,1} & \dots & \lambda_{K,1} \\ \lambda_{1,2} & -(\lambda_{2,1} + \lambda_{2,3} + \dots + \lambda_{2,K}) & \dots & \lambda_{K,2} \\ \dots & \dots & \dots & \dots \\ \lambda_{1,K} & \lambda_{2,K} & \dots & -(\lambda_{K,1} + \dots + \lambda_{K,K-1}) \end{bmatrix}
$$
$$
\cdot \begin{bmatrix} p_1 \\ p_2 \\ \dots \\ p_K \end{bmatrix}, \tag{4.6}
$$

where $p_i(t)$ is a probability of a state S_i, $\lambda_{i,j}(t)$ are transition probability densities. If the functions $\lambda_{i,j}(t)$ are given, one can trace an evolution of the probability distribution starting from any initial distribution by integrating the Kolmogorov equations. In simple particular cases, e.g. for constant $\lambda_{i,j}$, a solution can be found analytically.

A problem is somewhat simpler in the case of Markov chains (at least, for numerical investigation), since an evolution of a probability vector is described with a Kth-order difference equation. For a vivid representation of Markovian processes with discrete states, one often uses graphs where circles and arrows indicate different states and possible transitions between them.

In the case of a continuous-valued Markovian process, a state is to be described with a probability density function rather than a probability vector. Therefore, one derives partial differential equations for an evolution of the probability distribution law rather than ordinary differential equations (4.6). This is a *generalised Markov equation* (the other names are Kolmogorov – Chapman equation and direct Kolmogorov equation) for a conditional probability density function:

$$\frac{\partial p(x, t|x_0, t_0)}{\partial t} = \sum_{k=1}^{\infty} \frac{(-1)^k}{k!} \frac{\partial^k}{\partial x^k} \left[c_k(x, t) p(x, t|x_0, t_0) \right], \qquad (4.7)$$

where $c_k(x, t) = \lim\limits_{\Delta t \to 0} \frac{1}{\Delta t} \int\limits_{-\infty}^{\infty} (x' - x)^k p(x', t + \Delta t|x, t) dx'$ are coefficients related to "probabilities of change" of a state x and determining "smoothness" of the process realisations.

In an important particular case of a *diffusive Markovian process* (where $c_k = 0$ for any $k > 2$), the equation simplifies and reduces to

$$\frac{\partial p(x, t)}{\partial t} = -\frac{\partial}{\partial x} (c_1(x, t) p(x, t)) + \frac{1}{2} \frac{\partial^2}{\partial x^2} (c_2(x, t) p(x, t)), \qquad (4.8)$$

where c_1 is called the drift coefficient and c_2 is the diffusion coefficient. Equation (4.8) is also called *Fokker – Planck equation* (Wentzel', 1975; Risken, 1989). It is an equation of a parabolic type. Of the same form are diffusion and heat conduction equations in mathematical physics. The names of the coefficients originate from the same field. Relationships between parameters of stochastic differential equation specifying an original process and the drift and diffusion coefficients in the Fokker–Planck equation are considered in Sect. 4.5.

4.4 Autoregression and Moving Average Processes

A random process can be specified via a stochastic equation. Then, it is defined as a solution to a stochastic equation, i.e. its substitution into an equation makes the latter an identity. In particular, discrete-time stochastic equations which define random processes of "autoregression and moving average" (Box and Jenkins, 1970) are considered below. They are very often used in modelling from observed data.

Linear filter. As a basic model for the description of complex real-world processes, one often uses Gaussian white noise $\xi(t)$. Let it have zero mean and the variance σ_ξ^2. Properties of a real-world signal may differ from those of Gaussian white noise, e.g. an estimate of the autocorrelation function $\rho(\tau)$ may significantly differ from zero at non-zero time lags τ. Then, a fruitful approach is to construct a model as Gaussian white noise transformed by a *linear filter*. In general, such a transform in discrete time is defined as

$$x_n = \xi_n + \sum_{i=1}^{\infty} \psi_i \xi_{n-i}. \tag{4.9}$$

For the variance of x_n to be finite (i.e. for x_n to be stationary), the weights ψ_i must satisfy $\sum_{i=1}^{\infty} \psi_i^2 \leq$ const. Linear transform (4.9) preserves normality of a process and introduces non-zero autocorrelations $\rho(\tau)$ at non-zero time lags.

Moving average processes. Using a model with an infinite number of weights is impossible in practice. However, one may reasonably assume that the value of ψ_i decreases quickly with i, i.e. the remote past weakly affects the present, and consider the model (4.9) containing only a finite number of weights q. Thereby, one gets a "moving average" process which is denoted as MA(q) and defined by the difference equation

$$x_n = a_n - \sum_{i=1}^{q} \theta_i \xi_{n-i} \tag{4.10}$$

involving $q + 1$ free parameters: θ_1, θ_2, ..., θ_q, and σ_ξ^2.

Autoregression processes. A general expression (4.9) can be equivalently rewritten in the form

$$x_n = \xi_n + \sum_{i=1}^{\infty} \pi_i x_{n-i}, \tag{4.11}$$

where the weights π_i are uniquely expressed via ψ_i. In more detail, conversion from (4.9) to (4.11) can be realised through subsequent exclusion of the quantities ξ_{n-1}, ξ_{n-2}, etc. from equation (4.9). For that, one expresses ξ_{n-1} via x_{n-1} and previous values of ξ as $\xi_{n-1} = x_{n-1} - \sum_{i=1}^{\infty} \psi_i \xi_{n-1-i}$. Then, one substitutes this expression into equation (4.9), thereby excluding ξ_{n-1} from the latter. Next, one excludes ξ_{n-2} and so on in the same manner. The process (4.11) involves an infinite number of parameters π_i. However, a fast decrease $\pi_i \to 0$ at $i \to \infty$ often takes place, i.e. the remote past weakly affects the present (already in terms of the values of x variable). Then, one may take into account only a finite number of terms in equation (4.11). As a result, one gets an "autoregression" process of an order p which is denoted as AR(p) and defined as

$$x_n = \xi_n + \sum_{i=1}^{p} \phi_i x_{n-i}. \tag{4.12}$$

This model contains $p + 1$ free parameters: ϕ_1, ϕ_2, ..., ϕ_p and σ_ξ^2. The values of weights must satisfy certain relationships (Box and Jenkins, 1970) for a process to be stationary. Thus, in the case of $p = 1$, the variance of a process (4.12) is $\sigma_x^2 = \sigma_\xi^2 / (1 - \phi_1^2)$ so that one needs $|\phi_1| < 1$ to provide the stationarity. The term "autoregression" appears, since the sum in equation (4.12) determines regression of the current value of x on the previous values of the *same* process. The latter circumstance inspires the prefix "auto". The general concept of regression is described in Sect. 7.2.1.

AR processes represent an extremely popular class of models. One of the reasons is the simplicity of their parameter estimation (see Chaps. 7 and 8). Moreover, they are often readily interpretable from the physical viewpoint. In particular, the AR(2) process given by $x_n = \phi_1 x_{n-1} + \phi_2 x_{n-2} + \xi_n$ with appropriate values of the parameters ϕ_1 and ϕ_2 describes a stochastically perturbed linear damped oscillator, i.e. a generalization of the deterministic oscillator (3.2). Its characteristic period T and relaxation time τ are related to the parameters ϕ_1 and ϕ_2 as $\phi_1 = 2 \cos(2\pi/T) \exp(-1/\tau)$ and $\phi_2 = -\exp(-2/\tau)$, see Timmer et al. (1998) for a further discussion and applications of the AR(2) process to empirical modelling of physiological tremor.

The same model equation was first used for the analysis of solar activity in the celebrated work (Yule, 1927), where parameters of an AR(2) process were estimated from the time series of annual sunspot numbers. It was shown that an obtained AR(2) model could reproduce 11-year cycle of solar activity and gave better predictions than a traditional description with explicit periodic functions of time, which had been used before. Since then, linear stochastic models have become a widely used tool in many fields of data analysis. As for the modelling of solar activity, it was considered in many works after 1927. In particular, non-linear improvements of AR models are discussed, e.g., in Judd and Mees (1995; 1998); Kugiumtzis et al. (1998). Additional results on the analysis of solar activity data are presented in Sect. 12.6.

Autoregression and moving average processes. To get a more efficient construction for the description of a wide range of processes, one can combine equations (4.10) and (4.12). Reasons for their combining are as follows. Let us assume that an observed time series is generated by an AR(1) process. If one tries to describe it as an MA process, then an infinite (or at least very large) order q is necessary. Estimation of a large number of parameters is less reliable that leads to an essential reduction of the model quality. Inversely, if a time series is generated with an MA(1) process, then an AR process of a very high order p is necessary for its description. Therefore, it is reasonable to combine equations (4.10) and (4.12) in a single model to describe an observed process most parsimoniously. Thus, one gets an autoregression and moving average process of an order (p, q) which is denoted ARMA (p, q) and defined as

$$x_n = \xi_n + \sum_{i=1}^{p} \phi_i x_{n-i} - \sum_{i=1}^{q} \theta_i \xi_{n-i}. \tag{4.13}$$

It involves $p + q + 1$ free parameters.

Autoregression and integrated moving average processes. A stationary process (4.13) cannot be an adequate model for non-stationary processes with either deterministic trends or stochastic ones. The term "stochastic trend" means *irregular* alternation of intervals, where a process follows almost deterministic law. However, an adequate model in the case of polynomial trends is a process whose finite difference is a stationary ARMA process. A finite difference of an order d is defined as $y_n = \nabla^d x_n$, where $\nabla x_n = x_n - x_{n-1}$ is the first difference (an analogue to differentiation) and ∇^d denotes d sequential applications of the operator ∇. Thus, one gets an autoregression and integrated moving average process of an order (p, d, q) denoted ARIMA(p, d, q) and defined via the set of difference equations

$$y_n = \xi_n + \mu + \sum_{i=1}^{p} \phi_i y_{n-i} - \sum_{i=1}^{q} \theta_i \xi_{n-i},$$

$$\nabla^d x_n = y_n. \tag{4.14}$$

An intercept μ determines a deterministic trend. To express x_n via the values of the ARMA process y_n, one should use summation operator (an analogue to integration), which is inverse to the operator ∇. It explains the word "integrated" in the title of an ARIMA process.

ARMA and ARIMA processes were the main tools to model and predict complex real-world processes for more than half a century (1920–1970s). They were widely used to control technological processes (Box and Jenkins, 1970, vol. 2). Their various modifications were developed, in particular, seasonal ARIMA models defined as ARIMA processes for a seasonal difference $\nabla_s x_n = x_n - x_{n-s}$ of the kind

$$y_n = \xi_n + \sum_{i=1}^{P} \Phi_i y_{n-is} - \sum_{i=1}^{Q} \Theta_i \xi_{n-is},$$

$$\nabla_s^D x_n = y_n, \tag{4.15}$$

where ξ_n is an ARIMA (p, d, q) process. A process (4.15) is called a seasonal ARIMA process of an order $(P, D, Q) \times (p, d, q)$. Such models are relevant to describe processes with seasonal trends (i.e. a characteristic timescale s).

Only during the last two decades due to the development of computers and concepts of non-linear dynamics, ARIMA models more and more "step back" in a competition with non-linear models (Chaps. 8, 9, 10, 11, 12 and 13), though they remain the main tool in many fields of knowledge and practice.

4.5 Stochastic Differential Equations and White Noise

4.5.1 The Concept of Stochastic Differential Equation

To describe continuous-time random processes, one uses *stochastic differential equations* (SDEs). The most well known is the first-order equation (so-called Langevin equation)

$$dx(t)/dt = F(x, t) + G(x, t) \cdot \xi(t), \tag{4.16}$$

where F and G are smooth functions of their arguments, $\xi(t)$ is zero-mean Gaussian white nose with the auto-covariance function $\langle \xi(t)\xi(t+\tau)\rangle = \delta(\tau)$.

Introducing a concept of SDE is by no means trivial since it includes a concept of the random process derivative $dx(t)/dt$. How should one understand such a derivative? The simplest way would be to assume that all the realisations of a process x are continuously differentiable and define the derivative at a point t as a random quantity, whose value is an ordinary derivative of a single realisation of x at t. However, this is possible only for processes $\xi(t)$ with sufficiently smooth realisations so that for each specific realisation of $\xi(t)$, equation (4.16) can be considered and solved as a usual ODE. However, white noise does not belong to such class of processes but reasonably describes multitude of practical situations (a series of independent shocks) and allows simplification of mathematical manipulations. To have an opportunity to analyse equation (4.16) with white noise $\xi(t)$, one generalises the concept of derivative $dx(t)/dt$ of a random process x at a point t. The derivative is defined as a random quantity

$$\frac{dx(t)}{dt} = \lim_{\Delta t \to 0} \frac{x(t+\Delta t) - x(t)}{\Delta t},$$

where the limit is taken in the root-mean-squared sense (see, e.g., Gihman and Skorohod, 1974; Volkov et al., 2000; Wentzel', 1975). However, even such a concept does not help much in practice. The point is that one should somehow integrate equation (4.16) to get a solution. Formally, one can write

$$x(t) - x(t_0) = \int_{t_0}^{t} F(x(t'), t')dt' + \int_{t_0}^{t} G(x(t'), t') \cdot \xi(t')dt' \tag{4.17}$$

and estimate a solution over an interval $[t_0, t]$ via the estimation of the integrals. A stochastic integral is also defined via the limit in the root-mean-squared sense. However, its definition is not unique. There are two most popular forms of the stochastic integral: (i) Ito's integral is defined analogous to the usual Riemann's integral via the left rectangle formula and allows to get many analytic results (Oksendal, 1995); (ii) Stratonovich's integral is defined via the central rectangles formula (a symmetrised form of the stochastic integral) (Stratonovich, 1967); it is more readily interpreted from the physical viewpoint since it is symmetric with respect to time (Mannella, 1997). Moreover, one can define the generalised stochastic integral, whose particular cases are Ito's and Stratonovich's integrals (Gihman and Skorohod, 1974; Mannella, 1997; Volkov et al., 2000; Wentzel', 1975). Thus, the stochastic DE (4.16) gets an exact meaning if one indicates in which sense the stochastic integrals are to be understood.

If $G(x, t) = G_0 = $ const, all the above-mentioned forms of the stochastic integral $\int_{t_0}^{t} G(x(t'), t') \cdot \xi(t')dt'$ coincide. Below, we consider this simple case in more detail. One can show that a process x in equation (4.17) is Markovian. Thus, one can write down the corresponding Fokker – Planck equation where the drift coefficient is $F(x, t)$ and the diffusion coefficient is G_0^2. Let us consider a particular case of $F = 0$:

$$\mathrm{d}x(t)\big/\mathrm{d}t = G_0\xi(t). \tag{4.18}$$

The solution to this equation can be written down formally as

$$x(t) - x(t_0) = G_0 \int_{t_0}^{t} \xi(t')\mathrm{d}t'. \tag{4.19}$$

One can show that the process (4.19) is Wiener's process. Its variance linearly depends on time as $\sigma_x^2(t) = \sigma_x^2(t_0) + G_0^2 \cdot (t - t_0)$. The variance of its increments over an interval Δt is equal to $G_0^2\Delta t$. It agrees well with known observations of Brownian particle motion, where mean square of the deviation from a starting point is also proportional to time.

A geophysical example. Equation (4.18) allows to derive an empirically established *Gutenberg – Richter law* for the repetition time of earthquakes depending on their intensity (Golitzyn, 2003). Let x be a value of a mechanical tension (proportional to deformations) at a given domain of the earth's crust. Let us assume that it is accumulated due to different random factors (various shocks and so forth) described as white noise. On average, its square rises as $G_0^2(t - t_0)$ (4.18) starting from a certain zero time instant when the tension is weak. Earthquakes arise when the system accumulates sufficient elastic energy during a certain time interval and releases it in some way. If the release occurs when a fixed threshold E is reached, then a time interval necessary to accumulate such energy reads as $\tau = E/G_0^2$. From here, it follows that the frequency of occurrence of earthquakes with energy exceeding E is $\sim 1/\tau \sim G_0^2/E$, i.e. the frequency of occurrence is inversely proportional to energy. The Gutenberg – Richter law reduces to the same form under certain assumptions. Analogous laws describe appearance of tsunami, landslides and similar events (Golitzyn, 2003).

An example from molecular physics. Under an assumption that independent shocks abruptly change a velocity of a particle rather than its coordinate, i.e. the white noise represents random forces acting on a particle, one gets the second-order SDE:

$$\mathrm{d}^2x(t)\big/\mathrm{d}t^2 = G_0\xi(t). \tag{4.20}$$

It allows to derive analytically the *Richardson – Obukhov law* stating that mean square of the displacement of a Brownian particle rises with time as $(t - t_0)^3$ under certain conditions. This law holds true for the sea surface within some range of scales (it is known as relative diffusion) (Golitzyn, 2003).

4.5.2 Numerical Integration of Stochastic Differential Equations

The above examples allow an analytic solution but for non-linear F and/or G, one has to use numerical techniques, which differ from those for ODEs. For simplicity, we start again with equation (4.16) with $G(x, t) = G_0 = \text{const}$:

$$\mathrm{d}x/\mathrm{d}t = F(x(t)) + G_0 \xi(t). \tag{4.21}$$

At a given initial condition $x(t)$, an SDE determines an ensemble of possible future realisations rather than a single realisation. The function F uniquely determines only the conditional probability density functions $p(x(t + \Delta t)|x(t))$. If F is non-linear, one cannot derive analytic formulas for the conditional distributions. However, one can get those distributions numerically via the generation of an ensemble of the SDE realisations. For that, the noise term $\xi(t')$ over an interval $[t, t + \Delta t]$ is simulated with the aid of pseudo-random number generator and the SDE is numerically integrated step by step.

The simplest approach is to use the Euler technique with a small integration step h (see, e.g., Mannella, 1997; Nikitin and Razevig, 1978). The respective difference scheme for equation (4.21) reads as

$$x(t + h) - x(t) = F(x(t)) \cdot h + \varepsilon_0(t) \cdot G_0 \sqrt{h}, \tag{4.22}$$

where $\varepsilon_0(t)$, $\varepsilon_0(t + h)$, $\varepsilon_0(t + 2h)$, ... are independent identically distributed Gaussian random quantities with zero mean and unit variance. The second term in the right-hand side of equation (4.22) shows that the noise contribution to the difference scheme scales with the integration step as $\sqrt{h}$. This effect is not observed in ODEs where the contribution of the entire right-hand side is of the order of h or higher. For SDEs, such an effect takes place due to the integration of the white noise $\xi(t)$: the difference scheme includes the random term whose variance is proportional to the integration step. The random term dominates for very small integration steps h. The scheme (4.22) is characterised by an integration error of the order $h^{3/2}$, while for ODEs the Euler technique gives an error of the order of h^2.

Further, at a fixed step h, one can generate an ensemble of noise realisations $\varepsilon_0(t)$, $\varepsilon_0(t + h)$, $\varepsilon_0(t + 2h)$, ... and compute for each realisation the value of $x(t + \Delta t)$ at the end of the time interval of interest via the formula (4.22). From an obtained set of values of $x(t + \Delta t)$, one can construct a histogram, which is an estimate of the conditional probability density $p(x(t + \Delta t)|x(t))$. This estimate varies under the variation of h and tends to a true distribution only in the limit $h \to 0$ like an approximate solution of an ODE tends to a true one for $h \to 0$. In practice, one should specify so small integration step h that an approximate distribution would weakly change under further decrease in h. Typically, to get the same order of accuracy, one must use smaller steps for the integration of SDEs as compared with the corresponding ODEs to get similar convergence. This is due to the above-mentioned lower order of accuracy for the SDEs.

A process x in equation (4.16) or (4.21) may well be vector valued. Then, white noise $\xi(t)$ is also a multidimensional process. All the above considerations remain the same for vector processes. As an example, let us consider integration of stochastic equations of the van der Pol oscillator:

$$
\begin{aligned}
dx_1/dt &= x_2, \\
dx_2/dt &= \mu(1 - x_1^2)x_2 - x_1 + \xi(t),
\end{aligned}
\tag{4.23}
$$

with $\mu = 3$ (Timmer, 2000) and $G_0 = 1$. Estimates of the conditional distribution $p(x_1(t + \Delta t)|x(t))$ are shown in Fig. 4.2. We take the initial conditions $\mathbf{x}(t) = (0, -4)$ lying close to probable states of the system observed in a long numerically obtained orbit, $\Delta t = 0.5$ corresponding approximately to 1/18 of a basic period, and integration steps $h = 0.1, 0.01, 0.001$ and 0.0001. A distribution estimate stabilises at $h = 0.001$. Thus, an integration step should be small enough to give a good approximation to conditional distributions, often not more than about 0.0001 of a basic period. For a reasonable convergence of a numerical technique for the corresponding ODE, i.e. equation (4.23) without noise, a step 0.01 always suffices.

One more reason why dealing with SDEs is more complicated than numerical integration of ODEs is the above-mentioned circumstance that the integral of a random process $\xi(t)$ is an intricate concept. Let us now consider equation (4.16) with a non-constant function G. The corresponding Fokker – Planck equation takes different forms depending on the definition of the stochastic integrals. Namely, the drift coefficient reads as $F(x, t)$ under Ito's definition and as

$$
F(x, t) + \frac{1}{2}\frac{\partial G(x, t)}{\partial x}G(x, t)
$$

under Stratonovich's definition (Nikitin and Razevig, 1978; Mannella, 1997; Risken, 1989) (the diffusion coefficient is $G^2(x, t)$ in both cases). Accordingly, the

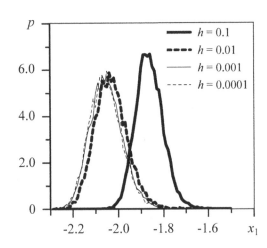

Fig. 4.2 Probability density estimates $p(x_1(t + \Delta t)|\mathbf{x}(t))$ for the system (4.23) at integration step $h = 0.1$, $0.01, 0.001, 0.0001$ and initial conditions $\mathbf{x}(t) = (0, -4)$. Each histogram is constructed from an ensemble of 10,000 time realisations with a bin size of 0.01. Good convergence is observed at $h = 0.001$

Euler scheme, i.e. a scheme accurate up to the terms of the order $h^{3/2}$, depends on the meaning of the stochastic integrals. For Ito's integrals, the Euler scheme is similar to the above case (Nikitin and Razevig, 1978) and reads as

$$x(t + h) - x(t) = F(x(t)) \cdot h + \varepsilon_0(t)G(x(t))\sqrt{h}. \tag{4.24}$$

For Stratonovich's integrals often considered in physics (Risken, 1989; Mannella, 1997; Siegert et al., 1998), the Euler scheme takes the form

$$x(t+h)-x(t) = F(x(t))\cdot h + \frac{1}{2}\frac{\partial G(x, t)}{\partial x}G(x(t))\varepsilon_0^2(t)h + G(x(t))\varepsilon_0(t)\sqrt{h}, \tag{4.25}$$

where an additional term of the order $O(h)$ appears. The latter is necessary to provide the integration error not greater than $O(h^{3/2})$ (Mannella, 1997; Nikitin and Razevig, 1978).

If one needs a higher order of accuracy, then the formula gets even more complicated, especially in the case of Stratonovich's integrals. It leads to several widespread pitfalls. In particular, a seemingly reasonable idea to integrate "deterministic" ($F(x, t)$) and "stochastic" ($G(x, t)\xi(t)$) parts of equation (4.16) separately, representing the "deterministic" term with the usual higher order Runge – Kutta formulas and the "stochastic" term in the form $\varepsilon_0(t) \cdot G(x(t))\sqrt{h}$, is called an "exact propagator". However, it appears to give even worse accuracy than the simple Euler technique (4.25) since the integration of the "deterministic" and "stochastic" parts is unbalanced (Mannella, 1997). An interested reader can find correct formulas for integration of SDEs with higher orders of accuracy in Mannella (1997) and Nikitin and Razevig (1978).

4.5.3 Constructive Role of Noise

Noise (random perturbations of dynamics) is often thought of as an interference, an obstacle, something harmful for the functioning of a communication system, detection of an auditory signal and other tasks. However, it appears that noise in non-linear systems can often play a constructive role leading to enhancement of their performance. The most striking and widely studied phenomena of this type are called "stochastic resonance" and "coherence resonance".

The term *stochastic resonance* was introduced in Benzi et al. (1981) where the authors found an effect, which they tried to use for explanation of the ice age periodicity (Benzi et al., 1982). The same idea was developed independently in Nicolis (1981; 1982). The point is that the evolution of the global ice volume on the Earth during the last million years exhibits a kind of periodicity with an average period of about 10^5 years (a glaciation cycle). The only known similar timescale is observed for the variations in the eccentricity of the Earth's orbit around the Sun determined by influences of other bodies of the solar system. The perturbation in the total amount of solar energy received by the Earth due to this effect is about 0.1%. Then, the question arises whether such a small periodic perturbation can be amplified so strongly to induce such a large-scale phenomenon as alternation of ice ages.

Benzi and co-authors considered an overdamped bistable oscillator driven by a Gaussian white noise ξ and a weak periodic signal $A \cos(\Omega t)$:

$$\mathrm{d}x \big/ \mathrm{d}t = x(a - x^2) + A \cos(\Omega t) + \xi(t) \tag{4.26}$$

with the parameter $a > 0$. The corresponding autonomous system $\mathrm{d}x/\mathrm{d}t = x(a - x^2)$ has an unstable fixed point $x = 0$ and two stable fixed points $x = \pm\sqrt{a}$. In the presence of noise, an orbit spends a long time near one of the two stable states but sometimes jumps to another state due to the noise influence. Switching between the two states is quite irregular: only on average it exhibits a characteristics timescale, the so-called Kramers' rate.

In the presence of the periodic driving $A \cos(\Omega t)$, one can consider the system (4.26) as a transformation of an input signal $A \cos(\Omega t)$ into "output" signal $x(t)$. In other words, the system (4.26) performs signal detection. Performance of the system is better if $x(t)$ is closer to a harmonic one with the frequency Ω. Its closeness to periodicity can be quantified in different ways. In particular, a signal-to-noise ratio (SNR) can be introduced as the ratio of its power spectral density (Sect. 6.4.2) at the frequency Ω to its "noise-floor" spectral density. It appears that the dependence of SNR on the intensity of the noise ξ has a clear maximum at a non-zero noise level. The curve "SNR versus noise intensity" resembles the resonance curve of "output amplitude versus driving frequency". In the particular example of the system (4.26), the phenomenon can be explained by the dependence of Kramers' rate on the noise level so that the resonance takes place when Kramers' rate becomes equal to the driving frequency. Therefore, the phenomenon was called "stochastic *resonance*".

Thus, a weak periodic input may have stronger periodic output for some intermediate (non-zero) noise level. In other words, non-zero noise improves the system performance as compared with the noise-free case. Whether this phenomenon is appropriate to describe the glaciation cycles is still the matter of debate, but the effect was then observed in many non-linear systems of different origin (see the reviews Anishchenko et al., 1999; Bulsara et al., 1993; Ermentrout et al., 2008; Gammaitoni et al., 1998; McDonnell and Abbott, 2009; Moss et al., 2004; Nicolis, 1993; Wiesenfeldt and Moss, 1995). In particular, many works report stochastic resonance in neural systems such as mechanoreceptors of crayfish (Douglass et al., 1993), sensory neurons of paddlefish (Greenwood et al., 2000 ; Russell et al., 1999), other physiological systems (Cordo et al., 1996; Levin and Miller, 1996), different neuron models (Longtin, 1993; Volkov et al., 2003b, c) and so on. There appeared many extensions and reformulations of the concept such as aperiodic stochastic resonance (Collins et al., 1996), stochastic synchronisation (Silchenko and Hu, 2001; Silchenko et al., 1999; Neiman et al., 1998) and stochastic multiresonance (Volkov et al., 2005).

Currently, many researchers speak of a stochastic resonance in the following situation: (i) one can define input and output signals for a non-linear system;

(ii) performance of the system improves at some non-zero noise level as compared to the noise-free setting. Thus, the formulation is no longer restricted to weak and/or periodic input signals and bistable systems. The counterintuitive idea that noise can improve functioning of a system finds the following fundamental explanation (McDonnell and Abbott, 2009): The system is non-linear and its parameter values in a noise-free setting are suboptimal for the performance of a required task. Hypothetically, its performance could be improved by adjusting the parameters. The other way is the noise influence which may improve functioning of the system. Thus, stochastic resonance is a result of an interplay between noise and non-linearity. It cannot be observed in a linear system.

A similar phenomenon introduced in Pikovsky and Kurths (1997) is called *coherence resonance*. It is observed in excitable non-linear systems without input signals. Its essence is that an output signal of a system is most coherent (exhibits the sharpest peak in a power spectrum) at a non-zero noise level and becomes less regular both for weaker and stronger noises. The phenomenon was first observed in FitzHugh – Nagumo system which is sometimes used as a simple neuron model:

$$\varepsilon \, \mathrm{d}x \big/ \mathrm{d}t = x - x^3 \big/ 3 - y,$$
$$\mathrm{d}y \big/ \mathrm{d}t = x + a + \xi(t). \tag{4.27}$$

The parameter $\varepsilon \ll 1$ determines the existence of fast motions (where only x changes) and slow motions (where $y \approx x - x^3/3$). The parameter $|a| > 1$ so that a stable fixed point is the only attractor of the noise-free system. The noise ξ is Gaussian and white. A stable limit cycle appears for $|a| < 1$. Thus, for $|a|$ slightly greater than 1, the system becomes excitable, i.e. a small but finite deviation from the fixed point (induced by the noise) can produce a large pulse (spike) in the realisation $x(t)$. These spikes are generated quite irregularly. The quantitative "degree of regularity" can be defined as the ratio of the mean interspike interval to the standard deviation of the interspike intervals. This quantity depends on the noise level: It is small for zero noise and strong noise and takes its maximum at an intermediate noise intensity. Again, the curve "degree of regularity versus noise intensity" looks like an oscillator resonance curve "output amplitude versus driving frequency". In the case of the system (4.27), the phenomenon can be explained by the coincidence of the two characteristic times: an activation time (the mean time needed to excite the system from the stable point, i.e. to get strong enough noise shock) and an excursion time (the mean time needed to return from an excited state to the stable state) (Pikovsky and Kurths, 1997). Therefore, the phenomenon is called "coherence *resonance*".

Thus, some non-zero noise may provide the most coherent output signal. It was quite unexpected finding which appeared fruitful to explain many observations. Similarly to stochastic resonance, the concept of coherence resonance was further extended, e.g., as doubly stochastic coherence (Zaikin et al., 2003), spatial (Sun et al., 2008b) and spatiotemporal (Sun et al., 2008a) coherence resonances. It is widely exploited in neuroscience, in particular, coherence resonance was observed in neuron models (e.g. Lee et al., 1998; Volkov et al., 2003a) and central nervous system (e.g. Manjarrez et al., 2002).

To summarise, a possible constructive role of noise for functioning of natural non-linear systems and its exploitation in new technical devices is currently a widely debated topic in very different fields of research and applications.

References

Anishchenko, V.S., Neiman, A.B., Moss, F., Schimansky-Geier, L.: Stochastic resonance: Noise enhanced order. Physics – Uspekhi. **42**, 7–38 (1999)

Benzi, R., Parisi, G., Sutera, A., Vulpiani, A.: Stochastic resonance in climate change. Tellus. **34**, 10–16 (1982)

Benzi, R., Sutera, A., Vulpiani, A.: The mechanism of stochastic resonance. J. Phys. A. **14**, L453–L457 (1981)

Box, G.E.P., Jenkins, G.M.: Time series analysis. Forecasting and Control. Holden-Day, San Francisco (1970)

Bulsara, A., Hänggi, P., Marchesoni, F., Moss, F., Shlesinger, M. (eds.): Stochastic resonance in physics and biology. J. Stat. Phys. **70**, 1–512 (1993)

Collins, J.J., Chow, C.C., Capela, A.C., Imhoff, T.T.: Aperiodic stochastic resonance. Phys. Rev. E. **54**, 5575–5584 (1996)

Cordo, P., Inglis, J.T., Verschueren, S., Collins, J.J., Merfeld, D.M., et al.: Noise in human muscle spindles. Nature **383**, 769–770 (1996)

Douglass, J.K., Wilkens, L., Pantazelou, E., Moss, F.: Noise enhancement of information transfer in crayfish mechanoreceptors by stochastic resonance. Nature 365, 337–339 (1993)

Ermentrout, G.B., Galan, R.F., Urban, N.N.: Reliability, synchrony and noise. Trends Neurosci. 31, 428–434 (2008)

Gammaitoni, L., Hänggi, P., Jung, P., Marchesoni, F.: Stochastic Resonance. Rev. Mod. Phys. 70, 223–287 (1998)

Gihman, I.I., Skorohod, A.V.: The Theory of Stochastic Processes. Springer, Berlin (1974)

Golitsyn, G.S.: White noise as a base for explanation of many laws in nature. In: Gaponov-Grekhov, A.V., Nekorkin, V.I. (eds.) Nonlinear Waves – 2002. Institute of Applied Physics RAS, pp. 117–132. Nizhny Novgorod, (in Russian) (2003)

Greenwood, P.E., Ward, L.M., Russell, D.F., Neiman, A., Moss, F.: Stochastic resonance enhances the electrosensory information available to paddlefish for prey capture. Phys. Rev. Lett. **84**, 4773–4776 (2000)

Kendall, M.G., Stuart, A.: The Advanced Theory of Statistics,. vol. 2 and 3. Charles Griffin, London (1979)

Lee, S.-G., Neiman, A., Kim, S.: Coherence resonance in a Hodgkin-Huxley neuron. Phys. Rev. E. **57**, 3292–3297 (1998)

Levin, J.E., Miller, J.P.: Broadband neural encoding in the cricket cercal sensory system enhanced by stochastic resonance. Nature **380**, 165–168 (1996)

Longtin, A.: Stochastic resonance in neuron models. J. Stat. Phys. **70**, 309–327 (1993)

Malakhov, A.N.: Fluctuations in self-oscillatory systems. Nauka, Moscow, (in Russian) (1968)

Mannella, R. Numerical integration of stochastic differential equations. In: Vazquez, L., Tirando, F., Martin, I. (eds.) Supercomputation in Nonlinear and Disordered Systems: Algorithms, Applications and Architectures, pp. 100–129. World Scientific, Singapore (1997)

McDonnell, M.D., Abbott, D.: What is stochastic resonance? Definitions, misconceptions, debates, and its relevance to biology. PLoS Comput. Biol. **5**, e1000348 (2009)

Moss, F., Ward, L., Sannita, W.: Stochastic resonance and sensory information processing: a tutorial and review of application. Clin. Neurophysiol. **115**, 267–281 (2004)

Neiman, A., Silchenko, A., Anishchenko, V., Schimansky-Geier, L.: Stochastic resonance: Noise-enhanced phase coherence. Phys. Rev. E. **58**, 7118–7125 (1998)

Nicolis, C.: Solar variability and stochastic effects on climate. Sol. Phys. **74**, 473–478 (1981)

Nicolis, C.: Stochastic aspects of climatic transitions-response to a periodic forcing. Tellus. **34**, 1–9 (1982)

Nicolis, C.: Long-term climatic transitions and stochastic resonance. J. Stat. Phys. **70**, 3–14 (1993)

Nikitin, N.N., Razevig, V.D.: Methods of digital modelling of stochastic differential equations and estimation of their errors. Comput. Math. Math. Physics. **18**(1), 106–117, (in Russian) (1978)

Oksendal, B.K.: Stochastic Differential Equations: An Introduction with Applications. Springer, Berlin (1995)

Pikovsky, A.S., Kurths, J.: Coherence resonance in a noise-driven excitable system. Phys. Rev. Lett. **78**, 775–778 (1997)

Risken, H.: The Fokker – Planck Equation. Springer, Berlin (1989)

Russell, D.F., Wilkens, L.A., Moss, F.: Use of behavioural stochastic resonance by paddlefish for feeding. Nature **402**, 291–294 (1999)

Rytov, S.M., Kravtsov Yu.A., Tatarskiy, V.I.: Introduction to statistical radiophysics (part II). Nauka, Moscow (1978). Translated into English: Springer, Berlin (1987)

Siegert, S., Friedrich, R., Peinke, J.: Analysis of data sets of stochastic systems. Phys. Lett. A. **243**, 275–280 (1998)

Silchenko, A., Hu, C.-K.: Multifractal characterization of stochastic resonance. Phys. Rev. E. **63**, 041105 (2001)

Silchenko, A., Kapitaniak, T., Anishchenko, V.: Noise-enhanced phase locking in a stochastic bistable system driven by a chaotic signal. Phys. Rev. E. **59**, 1593–1599 (1999)

Stratonovich, R.L.: Topics in the theory of random noise. Gordon and Breach, New York (1967)

Sun, X., Lu Q, Kurths, J.: Correlated noise induced spatiotemporal coherence resonance in a square lattice network. Phys. A. **387**, 6679–6685 (2008a)

Sun, X., Perc, M., Lu, Q., Kurths, J.: Spatial coherence resonance on diffusive and small-world networks of Hodgkin–Huxley neurons. Chaos. **18**, 023102 (2008b)

Timmer, J.: Parameter estimation in nonlinear stochastic differential equations. Chaos, Solitons Fractals. **11**, 2571–2578 (2000)

Volkov, E.I., Ullner, E., Kurths, J.: Stochastic multiresonance in the coupled relaxation oscillators. Chaos. **15**, 023105 (2005)

Volkov, E.I., Stolyarov, M.N., Zaikin, A.A., Kurths, J.: Coherence resonance and polymodality in inhibitory coupled excitable oscillators. Phys. Rev. E. **67**, 066202 (2003a)

Volkov, E.I., Ullner, E., Zaikin, A.A., Kurths, J.: Oscillatory amplification of stochastic resonance in excitable systems. Phys. Rev. E. **68**, 026214 (2003b)

Volkov, E.I., Ullner, E., Zaikin, A.A., Kurths, J.: Frequency-dependent stochastic resonance in inhibitory coupled excitable systems. Phys. Rev. E. **68**, 061112 (2003c)

Volkov, I.K., Zuev, S.M., Ztvetkova, G.M.: Random Processes. N.E. Bauman MGTU, Ìoscow, (in Russian) (2000)

Wentzel' A.D.: Course of the Theory of Random Processes. Nauka, Moscow (1975). Translated into English: A course in the theory of stochastic processes. McGraw-Hill Intern., New York (1981)

Wiesenfeldt, K., Moss, F.: Stochastic resonance and the benefits of noise: from ice ages to crayfish and SQUIDs. Nature **373**, 33–36 (1995)

Zaikin, A., Garcia-Ojalvo, J., Bascones, R., Ullner, E., Kurths, J. Doubly stochastic coherence via noise-induced symmetry in bistable neural models. Phys. Rev. Lett. **90**, 030601 (2003)

Part II
Modelling from Time Series

Chapter 5
Problem Posing in Modelling from Data Series

Creation of models on the basis of observations and investigation of their properties is, in essence, the main contents of science.

"System Identification. Theory for the User" (Ljung, 1991)

5.1 Scheme of Model Construction Procedure

Despite infinitely many situations, objects and purposes of modelling from observed data, one can single out common stages in a modelling procedure and represent them with a scheme (Fig. 5.1). The procedure is started with consideration of available information about an object (including previously obtained experimental data from the object or similar ones, a theory developed for the class of objects under investigation and intuitive ideas) from the viewpoint of modelling purposes, with acquisition and preliminary analysis of data series. It ends with the use of an obtained model for solution to a concrete problem. The procedure is usually iterative, i.e. it is accompanied by multiple returns to a starting or an intermediate point of the scheme and represents a step-by-step approach to a "good" model.

Model structure[1] is formed at the key stage 2. This stage is often called structural identification. Firstly, one selects a type of equations. Below, we speak mainly of finite-dimensional deterministic models in the form of discrete maps

$$\mathbf{x}_{n+1} = \mathbf{f}(\mathbf{x}_n, \mathbf{c}) \tag{5.1}$$

or ordinary differential equations

$$d\mathbf{x}/dt = \mathbf{f}(\mathbf{x}, \mathbf{c}), \tag{5.2}$$

where $\mathbf{x}$ is the D-dimensional state vector, $\mathbf{f}$ is the vector-valued function, $\mathbf{c}$ is the P-dimensional parameter vector, n is the discrete time and t is the continuous time. Secondly, a kind of functions (scalar components of $\mathbf{f}$) is specified. Thirdly, one establishes a relationship between dynamical variables (components of $\mathbf{x}$) and

---

[1] A model structure is a parameterised set of models (Ljung, 1991).

B.P. Bezruchko, D.A. Smirnov, *Extracting Knowledge From Time Series*, Springer Series in Synergetics, DOI 10.1007/978-3-642-12601-7_5,
© Springer-Verlag Berlin Heidelberg 2010

Fig. 5.1 Typical scheme of
an empirical modelling
procedure

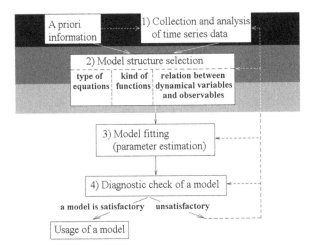

observed quantities η. Dynamical variables may coincide with the components of η. In a more general case, the relationship is specified in the form $\eta = \mathbf{h}(\mathbf{x})$, where **h** is called a *measurement function*. Moreover, one often introduces a random term $\zeta\,(\eta = \mathbf{h}(\mathbf{x}) + \zeta)$ to allow for a *measurement noise*. To make a model more realistic, one often incorporates random terms called *dynamical noise* into the Eq. (5.1) or (5.2).

Specification of a model structure is the most complicated and creative stage of the modelling procedure. After that, it remains just to determine concrete values of the parameters **c** (the stage 3). Here, one often speaks of *parameter estimation* or *model fitting*. To estimate parameters, one usually searches for an extremum of a *cost function*, e.g. minimum of the sum of squared deviations of a model realisation from the observed data. If necessary, one performs preliminary processing of the observed data series, such as filtering, numerical differentiation or integration. This is mainly a technical stage of numerical calculations. However, it requires the choice of an appropriate principle for parameter estimation and a relevant technique for its realisation.

One must also make a decision at the stage of model testing (the stage 4). Typically, model "quality" is checked with the use of a test part of an observed series specially reserved for this purpose. Depending on modelling purposes, one distinguishes between two types of problems: "cognitive identification" (the purpose is to get an adequate[2] model) and "practical identification" (there is a practical task to be solved with a model, e.g. a forecast). Accordingly, one either performs *validation (verification) of a model* in respect of the object properties of interest or checks *model efficiency* for achievement of a practical goal. If a model is found satisfactory (adequate or efficient), it is used further. Otherwise, it is returned to previous stages of the scheme (Fig. 5.1) for revision.

---

[2] *Adequacy* is understood as a correspondence between properties of a model and an object, i.e. a correct reproduction of original properties by a model (Philosophic dictionary, 1983, p. 13).

In particular, a researcher can return even to the stage of data collection (the stage 1) and ask for new data. It is appropriate to note that the data may not only be collected and analysed at this stage but also be pre-processed in different ways, e.g., to reduce measurement errors and fill possible gaps. This is especially important in geophysics, where the large field of *data assimilation* has appeared and evolved into a mature discipline (Anderson and Willebrand, 1989; Brassieur and Nihoul, 1994; Ghil and Malanotte-Rizzoli, 1991; Ide et al., 1997; Malanotte-Rizzoli, 1996). Data assimilation combines different estimation and modelling methods, where the Kalman filters occupy one of the central places (Bouttieur and Courtier, 1999; Evensen, 2007; Houtekamer and Mitchell, 1998; Robinson and Lermusiaux, 2000a, b).

5.2 Systematisation in Respect of A Priori Information

Background under the scheme in Fig. 5.1 changes from black (darkness of ignorance) to white reflecting a degree of prior uncertainty in modelling. The least favourable is a situation called "black box" when information about possibly adequate model structure is lacking so that one must start a modelling procedure from the very top of the above scheme. The more is known about how the model should look like (i.e. the lower is a "starting position" at the scheme), the more probable is a success. A "box" becomes "grey" and even "transparent". The latter means that a model structure is completely known a priori.

Methods for preliminary analysis of a time series, which can give useful information about a model structure and simplify a modelling problem, are discussed in Chap. 6. One can never avoid problems of the lower levels of the scheme, which are inevitably faced by a researcher overcoming the structural identification stage. Therefore, Chaps. 7 and 8 deal with the simplest situation, where everything is known about a model except for the concrete values of its parameters. It corresponds to the white background in Fig. 5.1.

Depending on the origin of a time series, two qualitatively different situations emerge in respect of the formulation of the modelling problem. The first one is when observations are a realisation of a certain mathematical model (a set of equations) obtained with a numerical technique. It is this situation where the term "reconstruction of equations" is completely appropriate. In such a case, model validation is much simpler since one can compare modelling results with the "true" original equations and their solutions. Besides, one may formulate theoretical conditions for the efficiency of modelling techniques for different classes of systems. The second situation is when a time series results from measurements of a certain real-world process so that a unique true model does not exist (Chap. 1) and one cannot assure success of modelling. One can only wonder at "inconceivable efficiency of mathematics" if a "good" model is achieved.[3]

---

[3] In addition, one can consider laboratory experiments as an intermediate situation. Strictly speaking, they represent real-world processes (not numerical simulations), but they can be manipulated

Further, we consider various techniques for model construction. To explain and illustrate them, we use mainly the former of the two situations, i.e. we reconstruct equations from their numerical solutions with various noises introduced to make a situation more realistic. We discuss different modelling problems in Chaps. 8, 9 and 10 according to the following "hierarchy":

(i) *Parameter estimation.* Structure of model equations is specified completely from physical or other conceptual considerations. Only parameter values are unknown. This problem setting is the simplest one and can be called "transparent box". However, essential difficulties can still arise due to a big number of unknown parameters and unobserved (hidden) dynamical variables.

(ii) *Reconstruction of equivalent characteristics.* A model structure is known to a significant degree from physical considerations. Therefore, one does not have to search for a multivariate function **f**. One needs to find only some of its components, which are univariate or bivariate functions (possibly, non-linear). We call them "equivalent characteristics". The term is borrowed from radiophysics but often appropriate in physics, biology and other fields.

(iii) *Black box reconstruction.* Since a priori information is lacking, one looks for a function **f** in a universal form. Solution to this problem is most often called "reconstruction of equations of motion". This is the most complicated situation.

Transition from the setting (i) to the setting (iii) is gradual. Many situations can be ordered according to the degree of prior uncertainty (darkness of the grey tone or non-transparency). However, it is not a linear ordering since not all situations can be readily compared to each other and recognised as "lighter" or "darker". For instance, it is not easy to compare information about the presence of a sinusoidal driving and a certain symmetry of orbits in a phase space.

5.3 Specific Features of Empirical Modelling Problems

5.3.1 Direct and Inverse Problems

It is often said that getting model equations from an observed time realisation belongs to the class of *inverse problems*. The term "inverse problem" is used in many mathematical disciplines. Inverse problems are those where input data and sought quantities switch their places as compared with some habitual basic problems which are traditionally called "direct".

---

in different ways and even constructed so as to correspond well to some mathematical equations. Thus, one may know a lot about appropriate mathematical description of such processes. It extends one's opportunities of successful modelling and acute model validation.

As a rule, direct problems are those whose formulations arise first (logically) after creation of a certain mathematical apparatus. Usually, one has regular techniques to solve direct problems, i.e. they are often relatively simple. An example of a direct problem is the Cauchy problem (a direct problem of dynamics): to find a particular solution to an ordinary differential equation, given initial conditions. Then, getting a set of ODEs whose particular solution is a given function is an inverse problem of dynamics.

It is appropriate to speak of inverse problems in experimental data analysis when one must determine parameters of a mathematical model of an investigated process from measured values of some variables. For instance, in spectroscopy and molecular modelling one considers the following problem: to determine a geometrical configuration of molecules of a certain substance and compute the corresponding parameters from an observed absorption spectrum. This is an inverse problem, while a direct one is to compute an absorption spectrum, given a model of a molecule.

5.3.2 Well-posed and Ill-posed Problems

Inverse problems are often ill-posed in the strict sense described below.

Let us express a problem formulation as follows: to find an unknown quantity Y (a solution) given some input data X. A problem is called *stable with respect to input data* if its solution depends on input data continuously $Y = \Phi(X)$, i.e. a solution changes weakly under a weak variation in X. A problem is called *well-posed according to Hadamard* if the three conditions are fulfilled: (i) a solution exists, (ii) a solution is unique, (iii) the problem is stable with respect to input data.

The first two conditions do not require comments. The third one is important from a practical point of view, since data are always measured with some errors. If a weak variation in the input data leads to a drastic change in a solution to a problem, then one cannot assure reliability of the solution so that usefulness of such a solution is doubtful. The third condition requires an answer which changes weakly under a weak variation in input data.

If at least one of the three conditions is violated, a problem is called *ill-posed according to Hadamard*. Of course, one should always tend to formulate well-posed problems. Therefore, ill-posed problems were out of mathematicians' interests for a long time (Tikhonov and Arsenin, 1974). It was thought that they did not make "physical sense", had to be reformulated, etc. However, as time went by, such problems more and more often emerged in different fields of practice. Therefore, special approaches to their solutions began to be developed such as regularisation techniques and construction of quasi-solutions.

Below, we often discuss well-posedness of modelling problems. Ill-posedness of a problem is not a "final verdict". It does not mean that "everything is bad". Thus, even the problem of differentiation is ill-posed[4] but differentiation is widely used.

---

[4] Significant difficulties in numerical estimation of derivatives from a time series (Sect. 7.4.2) are related to ill-posedness of the differentiation problem (Tikhonov and Arsenin, 1974). A concrete

Though it would be preferable to deal with well-posed problems, even a model obtained under ill-posed formulation can appear quite appropriate for practical purposes. It is possible, e.g., if ill-posedness are based on the non-uniqueness of a solution, but one manages to select a certain solution from a multitude of them, which gives satisfactory results. Thus, in molecular modelling the authors of Gribov et al. (1997) stress principal ill-posedness of the arising problems (namely, getting model parameters from an absorption spectrum or a diffraction pattern) and the necessity to consider models obtained under various experimental settings as mutually complementary and only partially reflecting object properties.

There is another widespread approach to solve an ill-posed problem. Let a solution Y^* exist for input data X^*. Let the data X^* be known with a certain error and denote such a "noise-corrupted" input as X. Strictly speaking, the problem may not have a solution for the input data X. Then, one seeks for a quantity $Z = \Phi(X)$, which is close to a solution in some sense, while the mapping Φ is continuous and $\Phi(X) \to Y^*$ for $\|X - X^*\| \to 0$. Such a quantity Z is called a *quasi-solution*.

In practice, ill-posedness of a problem of modelling from a time series is often eliminated due to ad hoc additional assumptions or a priori information about a model structure, which helps to choose appropriate basis functions (Chaps. 7, 8, 9 and 10). Also, there are more general procedures providing well-posedness of a problem. They consist of construction of the so-called regularising functional (Tikhonov and Arsenin, 1974; Vapnik, 1979, 1995).

A typical example of the latter situation is the problem of approximation of a dependence $Y(X)$ based on a finite "learning" sample (Sect. 7.2.1). It is quite easy to construct a curve passing via each experimental point on the plane (X, Y). However, there is an infinite multitude of such curves differing from a constructed one by arbitrary oscillations *between the points*. Each of the curves provides a minimal (zero) value of an empirical mean-squared approximation error and represents in this sense an equitable solution to the problem. The number of solutions is reduced if one imposes constraints on the acceptable value of inter-point oscillations. It is done with a regularising functional (Sect. 7.2.3). Thus, a check for ill-posedness of a problem is important, since it helps to select the most efficient way for its solution or even change the entire problem setting.

---

example is following. There is a continuously differentiable function $x_0(t)$, whose derivative is denoted as $dx_0(t)/dt = y_0(t)$. Let $x_0(t)$ be known with a certain error, i.e. the input data is a function $x(t) = x_0(t) + \xi(t)$, where $\xi(t)$ is a continuously differentiable function with $|\xi(t)| \leq \delta$. The input data $x(t)$ are very close to $x_0(t)$ in the sense of metrics L_∞. A derivative of the "noise-corrupted" function $x(t)$ is $dx(t)/dt = y_0(t) + d\xi(t)/dt$. Its deviation ε from $y_0(t)$ can be arbitrarily large in the same metrics. Thus, $dx(t)/dt = y_0(t) + \omega\delta\cos(\omega t)$ for $\xi(t) = \delta\sin(\omega t)$ so that the error $\varepsilon = \omega\delta$ can be arbitrarily large for arbitrarily small δ if ω is sufficiently large. In other words, the differentiation error can be arbitrarily large for arbitrarily low amplitude of "quickly oscillating noise $\xi(t)$". However, it is important how closeness in the space of input data and in the space of solutions is understood. If one regards close such input data that their difference $\xi(t)$ satisfies simultaneously two conditions – $|\xi(t)| \leq \delta$ and $|d\xi(t)/dt| \leq \delta$ (i.e. $\xi(t)$ is a slowly varying function) – then the differentiation problem gets well-posed.

5.3.3 Ill-conditioned Problems

For practical applications, it is important to mention a kind of problem which is well-posed from a theoretical viewpoint but whose solution is "quite sensitive" to weak variations in input data. "Quite sensitive" is not a rigorous concept. It is determined by a specific problem, but in general it means that for a small relative error δ in input data, a relative error ε in a solution is much greater: $\varepsilon = K \cdot \delta$, where $K >> 1$. Though a problem is theoretically stable with respect to input data (i.e. an error in a solution is infinitesimal for an infinitesimal error in input data), a solution error may appear very large for a *finite* small error in input data. Such problems are called *weakly* stable with respect to input data or *ill-conditioned* (Kalitkin, 1978; Press et al., 1988).

From a practical viewpoint, they do not differ from the ill-posed problems, given that all numerical calculations and representations of numbers are of finite precision. An example of an ill-conditioned problem is a set of linear algebraic equations, whose matrix is close to a degenerate one. Such a matrix is also often called ill-conditioned (Golub and Van Loan, 1989; Kalitkin, 1978; Press et al., 1988; Samarsky, 1982). Ill-conditioned problems often arise in construction of a "cumbersome" mathematical model from a time series. To solve them, one needs the same ideas and techniques as for the ill-posed problems.

References

Anderson, D., Willebrand, J. (eds.): Oceanic Circulation Models: Combining Data and Dynamics. Kluwer, Dordrecht (1989)

Bouttier, F., Courtier, P.: Data Assimilation Concepts and Methods. Lecture Notes, ECMWF. http://www.ecmwf.int/newsevents/training/lecture_notes (1999)

Brasseur P., Nihoul, J.C.J. (eds.): Data dissimilation: tools for modelling the ocean in a global change perspective. NATO ASI Series. Series I: Global Environ. Change **19** (1994)

Evensen, G.: Data Assimilation. The Ensemble Kalman Filter. Springer, Berlin (2007)

Ghil, M., Malanotte-Rizzoli, P.: Data assimilation in meteorology and oceanography. Adv. Geophys. **33**, 141–266 (1991)

Golub, G.H., Van Loan, C.F.: Matrix Computations, 2nd edn. Johns Hopkins University Press, Baltimore (1989)

Gribov, L.A., Baranov, V.I., Zelentsov, D.Yu.: Electronic-Vibrational Spectra of Polyatomic Molecules. Nauka, Moscow (in Russian) (1997)

Houtekamer, P.L., Mitchell, H.L.: Data assimilation using an ensemble Kalman filter technique. Mon. Wea. Rev. **126**, 796–811 (1998)

Ide, K., Courtier, P., Ghil, M., Lorenc, A.: Unified notation for data assimilation: Operational, sequential and variational. J. Meteor. Soc. Jpn. **75**, 181–189 (1997)

Kalitkin, N.N.: Numerical Methods. Nauka, Moscow, (in Russian) (1978)

Ljung, L.: System Identification. Theory for the User. Prentice-Hall, Engle Wood Cliffs, NJ (1991)

Malanotte-Rizzoli, P. (ed.): Modern Approaches to Data Assimilation in Ocean Modeling. Elsevier Science, Amsterdam (1996)

Press, W.H., Flannery, B.P., Teukolsky, S.A., Vetterling, W.T.: Numerical Recipes in C. Cambridge University Press, Cambridge (1988)

Robinson, A.R., Lermusiaux, P.F.J.: Overview of data assimilation. Harvard Rep. Phys. Interdisciplinary Ocean Sci. **62**, (2000a)

Robinson, A.R., Lermusiaux, P.F.J.: Interdisciplinary data assimilation. Harvard Reports in Physical. Interdisciplinary Ocean Sci. **63**, (2000b)

Samarsky, A.A.: Introduction to Numerical Methods. Nauka, Moscow, (in Russian) (1982)

Tikhonov, A.N., Arsenin, V.Ya. Methods for Solving Ill-Posed Problems. Nauka, Moscow (1974). Translated into English: Wiley, New York (1977)

Vapnik, V.N.: Estimation of Dependencies Based on Empirical Data. Nauka, Moscow (1979). Translated into English: Springer, New York, (1982)

Vapnik, V.N.: The Nature of Statistical Learning Theory. Springer, New York (1995)

Chapter 6
Data Series as a Source for Modelling

When a model is constructed from "first principles", its variables inherit the sense implied in those principles which can be general laws or derived equations, e.g., like Kirchhoff's laws in the theory of electric circuits. When an empirical model is constructed from a time realisation, it is a separate task to reveal relationships between model parameters and object characteristics. It is not always possible to measure all variables entering model equations either in principle or due to technical reasons. So, one has to deal with available data and, probably, perform additional data transformations before constructing a model. •

In this chapter, we consider acquisition and processing of informative signals from an original with the purpose to get ideas about opportunities and specificity of its modelling, i.e. the stage "collection and analysis of time series data" and "a priori information" at the top of the scheme in Fig. 5.1.

6.1 Observable and Model Quantities

6.1.1 Observations and Measurements

To get information necessary for modelling, one *observes* an object based on prior ideas or models at hand (Sect. 1.1). As a result, one gets qualitative or quantitative data. Qualitative statements may arise from pure *contemplation*, while quantitative information is gathered via *measurements* and *counting*. Counting is used when a set of discrete elements is dealt with, e.g., if one tries to register a number of emitted particles, to determine a population size and so forth. Measurement is a comparison of a measured quantity with a similar quantity accepted as a unit of measurement. The latter is represented by standards of various levels. When speaking of "observation", "observable quantities" and "observation results", one means a measurement or counting process, measured or counted quantities and resulting quantitative data, respectively (Mudrov and Kushko, 1976). It is widely accepted to omit the word "quantities" and speak of "observables". We denote observables by the letter η resembling a question mark to stress a non-trivial question about their possible relationships with model variables.

B.P. Bezruchko, D.A. Smirnov, *Extracting Knowledge From Time Series*, Springer Series in Synergetics, DOI 10.1007/978-3-642-12601-7_6,
© Springer-Verlag Berlin Heidelberg 2010

Any real-world system possesses an infinite set of properties. However, to achieve goals of modelling, it is often sufficient to consider a finite subset: *model variables* $x_1, x_2, \ldots, x_D$ and *model parameters* $c_1, c_2, \ldots, c_P$. The number of observables and model variables, as well as their physical meaning, may differ. Thus, the number k of observables $\eta_1, \eta_2, \ldots, \eta_k$ is usually less than a model dimension D. Quite often, observables have a meaning different from variables entering model equations. Anyway, observables are somehow related to the model variables. Such a relationship is determined by experimental conditions, accessibility of an original, its shape and size, lack of necessary equipment, imperfection of measurement tools and many other objective and subjective factors. For instance, opportunities of electric measurements inside a biological cell are restricted due to the small size of the latter (Fig. 6.1a). An oscillating guitar string is big and easily accessible so that one can measure velocity and coordinate of each small segment of this mechanical resonator (Fig. 6.1b). In contrast to that, an access to inner volume of a closed resonator can be realised only via holes in its walls, i.e. after partial destruction of an object (Fig. 6.1c).

Relationships between observables and model variables may be rather obvious. Both sets of quantities may even coincide. However, a simple relationship is more often lacking. In general, this question requires a special analysis as in the following examples: •

(i) Evolution of a biological population is often modelled with a map showing dependence of the population size at the next year x_{n+1} on its size at the current year x_n. If experimental data η are gathered by observers via direct counting, then *a model variable and an observable coincide*: $x = \eta$. However, if one observes only results of vital activity (traces on a land, dung, etc.) and tries to infer the population size from them indirectly, then one needs formulas for the recalculation or other techniques to reveal a *dependence between x and η*.

(ii) Physicists are well familiar with measurements of electromotive force (e.m.f.) E of a source with internal resistance r_i with the aid of a voltmeter (Fig. 6.2a, b). An object is a source of current, a model variable is e.m.f. ($x = E$) and an observable is a voltage U on the source clamps. To measure U, a voltmeter is connected to the source as shown in Fig. 6.2a. Resistance between the input clamps of a real-world voltmeter R_v (input resistance) cannot be infinite, therefore, after connection the loop gets closed and current I starts to flow. Then, the voltage on the source clamps reduces as compared with the case

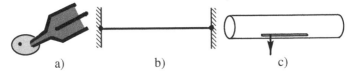

Fig. 6.1 Accessibility of different objects: (**a**) electric access to a cell via a glass capillary; (**b**) an open resonator; (**c**) a closed resonator

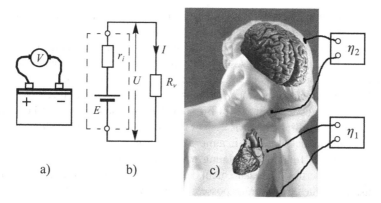

Fig. 6.2 Examples of characterising quantities and observables: (**a**) an experiment with a source of e.m.f.; (**b**) an equivalent scheme for the e.m.f. measurement; (**c**) observables in the analysis of brain and heart activity

without measurement device: $U = E - Ir_i = E\left(1 - r_i/(r_i + R_v)\right)$. Thus, the observable differs from the model variable, but *there is a unique functional dependence* between both quantities $\eta = f(x)$.

(iii) When electric potentials on a human skin are recorded, an observable η is a voltage between two points on a body surface. One of the points is taken as a reference (Fig. 6.2c). For an electroencephalogram (EEG) measurements, η is usually a voltage between points on a scalp and a ear; for electrocardiogram (ECG), it is a voltage between points on a chest and a leg. Even without special knowledge, one can see that the measured voltages are strongly transformed results of the processes occurring in ensembles of the brain cells or in the heart. Therefore, relationship between the observed potentials η and any model variables x characterising a human organism is a subject of special study. As mentioned in Sect. 1.6, a researcher is here in a position of passive observation of a complicated real-world process (Fig. 6.2c) rather than in a position of an active experimentation with a laboratory system (e.g. Fig. 6.2a). In particular, the greater difficulty of the passive case for modelling is manifested in a greater number of unclear questions. However, if a researcher models a potential recording itself, similar to kinematic description in mechanics, then a model variable and an observable coincide.

In any measurement, the results are affected by peculiarities of an experimental technique and device parameters, external regular influences and noises. A typical procedure for acquisition of experimental data is illustrated in Fig. 6.3: an outer curve bounds a modelling object, filled circles are sensors for measured quantities and triangles are devices (amplifiers, limiters, etc.) converting sensor data into observables η_i. Even if one precisely knows how a measured signal is distorted by the devices and makes corresponding corrections,[1] it is not possible in practice to get rid of interference (external regular or random influences).[2]

It is often feasible to suppress regular influences, while irregular ones called noises can only be reduced. Noises can be both of an external origin and inherent to an object. The former one is called *measurement* noise and the latter one is called *internal* or *dynamical* noise, since it affects the dynamics of an object. Noises are shown by the curves with arrows in Fig. 6.3. Measurement noise can be additive $\eta = f(x) + \xi$, multiplicative $\eta = f(x) \cdot \xi$ or enter the relationships between observables and model variables in a more complicated way.

The form of the function f relating observables and model variables is determined by the properties of measurement devices and transmission circuits. To avoid signal distortion, the devices must possess the following properties:

(i) A wide enough dynamic range $U_{max} - U_{min}$ allowing to transmit both high- and low-amplitude signals without violation of proportions. For instance, a dynamic range of a device shown in Fig. 6.4 is insufficient to transmit a high-amplitude signal.

(ii) A necessary bandwidth allowing to perceive and transmit all the spectral components (Sect. 6.4.2) of an input signal in the same way and ignore "alien" frequencies. Too large a bandwidth is undesirable since more interference and noise can mix in a signal. However, its decrease is fraught with a signal distortion due to the growth in the response time τ_u to a short pulse (a characteristic of the inertial properties of a device, Fig. 6.4c). Narrow frequency band of a converting device is often used for a purposeful filtering of a signal. Preserving only "low-frequency" spectral components (low-pass filtering) leads to

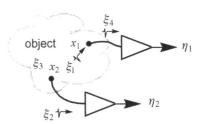

Fig. 6.3 A scheme illustrating data acquisition from an object under consideration: x_i are quantities, which characterise an object under study, and η_i are observables. *Filled circles* are sensors which send signals corrupted with noises ξ_i to measurement devices denoted by *triangles*

---

[1] For example, when e.m.f. is measured with a voltmeter, one can either allow for finiteness of R_v or use a classical no-current compensatory technique with a potentiometer (Kalashnikov, 1970, pp. 162–163; Herman, 2008) which is free of the above shortcoming.

[2] Thus, experiments show that in electric measurements (e.g., Fig. 6.2) a sensitive wideband device (e.g., a radio receiver) connected in parallel with a voltmeter or a cardiograph detects also noise, human speech and music, etc.

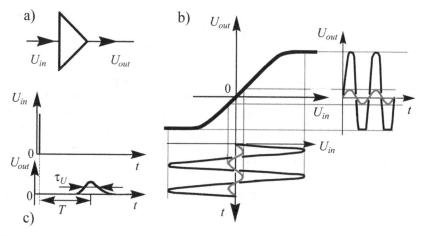

Fig. 6.4 Measurement device properties: (**a**) a device transmitting a signal; (**b**) its transfer characteristic $U_{out} = f(U_{in})$, a low-amplitude input signal (the *grey curve*) is transferred without distortions, while a high-amplitude one (the *black curve*) is limited, i.e. its peaks are cut off; (**c**) when a short pulse comes to an input of a device at a time instant $t = 0$, a response emerges at its output with a delay time T and is blurred due to finiteness of a response time τ_u

smoothing of a signal. High-pass filtering removes a constant non-zero component and slow *trends* (drifts).

(iii) A sufficiently high sensitivity. Internal noise level of a device should be small enough to give an opportunity to distinguish an informative signal at the device output confidently.

Questions of imperfection of measurement devices and non-trivial correspondence between model variables and observables relate also to statistical data used in modelling of social processes and humanitarian problems. Distortion of those data by people and historical time may be quite significant and uncontrolled. In modelling from such data, a researcher must be especially careful, having in mind that even electric measurements with sufficiently "objective" devices and noise reduction techniques often raise doubts and require special attention.

6.1.2 How to Increase or Reduce a Number of Characterising Quantities

If available observables are regarded as components of a state vector $\mathbf{x}(t)$, their number is often insufficient for a dynamical description of an original, i.e. for a unique forecast of future states based on a current one. There are several approaches to *increasing the number of model variables D*. Some of them are justified theoretically (Takens' theorems, Sect. 10.1.1), others rely on intuition and ad hoc ideas.

The simplest and most popular way is the *time-delay embedding*. According to it, components of the vector $\mathbf{x}(t)$ are the subsequent values of a scalar observable separated by a time delay τ: $x_1(t) = \eta(t)$, $x_2(t) = \eta(t + \tau)$, ..., $x_D(t) = \eta(t + (D-1)\tau)$.

According to the *successive differentiation technique*, temporal derivatives of an observable are used as dynamical variables: $x_1(t) = \eta(t)$, $x_2(t) = \mathrm{d}\eta(t)/\mathrm{d}t$, ..., $x_D(t) = \mathrm{d}^{D-1}\eta(t)/\mathrm{d}t^{D-1}$. However, it is hardly applicable, if the measurement noise is considerable (e.g. Fig. 7.8a, b): a negligible fringe on the plot of a slightly noise-corrupted cosine function $\eta(t) = \cos t + \xi(t)$ strongly amplifies under the differentiation so that an expected sinusoidal profile on the plot $\mathrm{d}\hat{x}/\mathrm{d}t$ versus t is not seen at all.

Noise can be reduced to some extent if one uses integrals of an observable as dynamical variables: $x_2(t) = \int\limits_0^t \eta(t')\mathrm{d}t'$, etc. Similarly, one can get a variable expressed via subsequent values of an observable via *weighted summation* as $x_2(t) = a_0\eta(t) + a_1\eta(t - \Delta t) + a_2\eta(t - 2\Delta t) + \ldots$, where a_k are weight coefficients.

One can also use a combination of all the above-mentioned approaches and other techniques to get time series of additional model variables (Sect. 10.1.2).

There are situations where the structure of a dynamical system is known, but computing the values of some dynamical variables directly from observables is impossible. Then, one speaks of "hidden" variables. In such a case, special techniques described in Sect. 8.2 may appear helpful.

In practice, it may also be necessary to reduce the number of observables if some of them do not carry information useful for modelling. It can be accomplished via the analysis of interrelations between observables and removal of the quantities, which represent linear combinations of the others. Furthermore, *dimensional and similitude methods* can be fruitful (Sena, 1977; Trubetskov, 1997): a spectacular historical example of a problem about fluid flow in a pipe, where one converts from directly measured dimensional quantities to a less number of their dimensionless combinations (similitude parameters), is considered in Barenblatt (1982).

6.2 Analogue-to-Digital Converters

To measure quantities of different nature, either constant or time-varying, one often tries to convert their values into electric voltages and currents. Electric signals can be easily transmitted from remote sensors and processed with the arsenal of standard devices and techniques. Previously, measured values were fixed by deviations of a galvanometer needle, paint traces on a plotter tape, and luminescence of an oscilloscope monitor. However, contemporary measurement systems usually represent data in a digital form with the aid of special transistor devices called *analogue-to-digital converters* (ADCs). The problem of analogue-to-digital conversion consists in the transformation of an input voltage at a measurement instant into a proportional number and, finally, in getting a discrete sequence of numbers. For a signal waveform to be undistorted, conditions of *Kotel'nikov's theorem* must be fulfilled: A continuous signal can be restored from a discrete sequence of its values only if a sampling frequency is at least twice as large as a maximal frequency which is present in its power spectrum (Sect. 6.4.2), i.e. corresponds to a non-zero component.

The principle of ADC functioning and the reasons limiting accuracy of the resulting data are illustrated in Fig. 6.5. The scheme realises the so-called parallel approach: An input voltage is compared simultaneously with n reference voltages. The number of reference voltages and the interval between the neighbouring ones are determined by the range of measured values and the required precision, i.e. the number of binary digits in output values. For the three-digit representation illustrated in the example and allowing to record eight different numbers including zero, one needs seven equidistant reference voltages. They are formed with the aid of a resistive divider of a reference voltage U_{ref}. The latter determines an upper limit of the measured voltages and is denoted $7U$ on the scheme.

A measured voltage U_{in} is compared to the reference levels with the aid of seven comparators k_i, whose output voltages take the values which are regarded in binary system equal to

- 1 if a voltage at the contact "+" exceeds a voltage at the contact "−",
- 0, otherwise.

Thus, if a measured voltage belongs to the interval $(5U/2, \ 7U/2)$, then the comparators with numbers from 1 to 3 are set to the state "1" and the comparators from 4 to 7 to the state "0". A special logical scheme (a priority encoder) converts

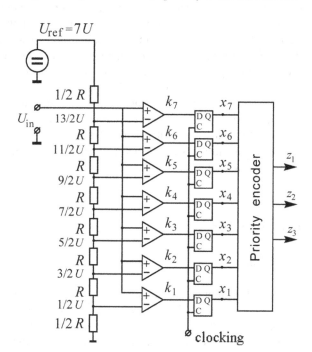

Fig. 6.5 A scheme of a three-digit ADC realising the parallel approach

those states into a binary number $z_1z_2z_3$ (011 in the example) or a corresponding decimal number (3 in the example). If a voltage varies in time, the priority encoder cannot be connected directly to outputs of the comparators, since it may lead to erroneous results. Therefore, one uses D triggers, shown by the squares "DQC" in Fig. 6.5, to save an instantaneous value of the voltage at outputs of the comparators and maintain it during a measurement interval. Measurement instants are dictated by a special clocking signal. If the latter is periodic, then one gets an equidistant sequence of binary numbers (a time series) at the encoder output.

The conversion of an analogue quantity into a several-digit number is characterised by a "quantisation error" equal to half an input voltage increment necessary to change the lowest order digit at the output. An eight-digit ADC has $2^8 = 256$ gradations ($\Delta x = x_{max}/256$), a 12-digit ADC has $2^{12} = 4096$ gradations ($\Delta x = x_{max}/4096$). If one performs an inverse transformation of the obtained number into a voltage with the aid of a digital-to-analogue converter, a quantisation error manifests itself as superimposed "noise". Besides, there are errors caused by the drift and non-linearity of the scheme parameters so that an overall error in the resulting observed values is determined by combinations of all the factors and indicated in a device certificate.

The parallel method for the realisation of the analogue-to-digital conversion is non-parsimonious, since one needs a separate comparator for each reference level. Thus, one needs 100 comparators to measure values ranging from 0 to 100 at a unit step. This number rises with the measurement resolution. Therefore, there have been developed and widely used approaches, which are better in this respect, e.g. *weighing* and *counting* techniques. Under the weighing technique, a result is obtained in several steps, since only a single digit of a binary number is produced at a single step. Firstly, one checks whether an input voltage exceeds a reference voltage of the highest order digit. If it does, the highest order digit is set equal to "1" and the reference voltage is subtracted from the input voltage. The remainder is compared with the next reference voltage and so on. Obviously, one needs as many comparison steps as many binary digits are contained in an output value. Under the counting technique, one counts a number of summations of the lowest order reference voltage with itself to reach an input voltage. If a maximal output value is equal to n, then one needs at most n steps to get a result. In practice, combinations of different approaches are often used.

6.3 Time Series

6.3.1 Terms

At an ADC output and in many other situations, data about a process under investigation are represented as a *finite sequence of values of an observed quantity corresponding to different time instants*, i.e. a *time series* $\eta(t_1), \eta(t_2) \ldots, \eta(t_N)$, where $t_1, t_2, \ldots, t_N$ are observation instants and their number N is called *time series length*. If a value of a single scalar quantity is measured at each time instant t_i, one speaks of a *scalar* time series. It is denoted as $\{\eta(t_i)\}_{i=1}^{N}$ or $\{\eta_i\}_{i=1}^{N}$, where $\eta_i = \eta(t_i)$. If k quantities $\eta_1, \ldots, \eta_k$ (Fig. 6.3) are measured simultaneously at each instant t_i, one speaks of a *vector* time series, since those quantities can be considered as components of a k-dimensional vector $\boldsymbol{\eta}$. A vector time series is denoted similarly: $\{\boldsymbol{\eta}(t_i)\}_{i=1}^{N}$ or $\{\boldsymbol{\eta}_i\}_{i=1}^{N}$. Thus, the notation η_i is used below in two different cases: (i) a scalar time series, where i is time index; (ii) a vector observable, where i means a coordinate number. Its meaning in each concrete case is determined by context unless otherwise stated.

Elements of a time series (scalars or vectors) are also called *data points*. A number of a point i is called *discrete time*. If time intervals between subsequent observation instants t_i are the same, $t_i - t_{i-1} = \Delta t$, $i = 2, \ldots, N$, then a time series is called *equidistant*, otherwise *non-equidistant*. One also says that the values are sampled uniformly or non-uniformly in time, respectively. An interval Δt between successive measurements is called *sampling interval* or discretisation interval. For a non-equidistant series, a sampling interval $\Delta t_i = t_i - t_{i-1}$ varies in time. In practice, one deals more often with equidistant series.

To check quality of a model constructed from a time series (Sects. 7.3, 8.2.1 and 10.4), one needs another time series from the same process, i.e. a time series which was not used for model fitting. Therefore, if the data amount allows, one distinguishes two parts in a time series $\{\eta_i\}_{i=1}^{N}$. One of them is used for model fitting and called a *training time series*. Another one is used for diagnostic check of a model and called a *test time series*.[3]

6.3.2 Examples

Time series from different fields of practice and several problems, which are solved with the use of such data, are exemplified below. The first problem is forecast of the future behaviour of a process.

6.3.2.1 Meteorology (a Science About Atmospheric Phenomena)

At weather stations, one measures hourly values of different quantities including air temperature, humidity, atmospheric pressure, wind speed at different heights, rainfall, etc. At some stations, measurements are performed for more than 100 years. Variations in the mentioned quantities characterised by the timescales of the order of several days are ascribed to *weather* processes (Monin and Piterbarg, 1997). Weather forecast is a famous problem of synoptic meteorology.

---

[3] In the field of artificial neural networks, somewhat different terminology is accepted. A test series is a series used to compare different empirical models. It allows to select the best one among them. For the "honest" comparison of the best model with an object, one uses one more time series called a *validation time series*. A training time series is often called a *learning sample*. However, we follow the terminology described in the text above.

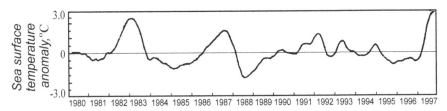

Fig. 6.6 Anomaly of sea surface temperature in eastern equatorial zone of Pacific Ocean (the values averaged over 5 months are shown). Sampling interval is $\Delta t = 1$ month (Keller, 1999)

Slower processes whose typical timescales exceed 3–4 months are called *climatic*. They are studied by *climatology*. Examples of their characteristics are sea surface temperature (a mean temperature of upper mixed layer, whose depth is about several dozens of meters, Fig. 6.6), sea level (at different coastal areas), thickness of ice, surface of ice cover, plant cover of the Earth, total monthly rainfall at a certain area, closed lakes level, etc. Besides, variations in many weather characteristics averaged over a significant time interval and/or a wide spatial area become climatic processes, e.g. mean monthly air temperature at a certain location, instantaneous value of temperature averaged over a 5° latitudinal zone, annual air temperature of the Northern Hemisphere. As for the spatial averaging, it is currently possible due to a global network of weather stations covering most of the surface of all continents. Distances between neighbouring stations are about 10–100 km.

6.3.2.2 Solar Physics

For a long time, a popular object of investigation is a time series of annual sunspot number (Fig. 6.7). This quantity is measured since telescope has been invented, more precisely, since 1610 (i.e. for almost 400 years). The process reflects magnetic activity of the Sun which affects, in particular, irradiation energy and solar wind intensity (Frik and Sokolov, 1998; Judd and Mees, 1995; Kugiumtzis et al., 1998; Yule, 1927) and, hence, leads to changes in the Earth's climate.

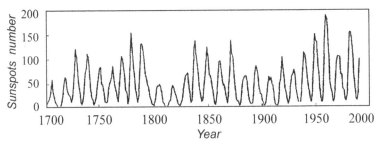

Fig. 6.7 Annual Wolf's numbers of sunspots. Sampling interval is $\Delta t = 1$ year. Eleven-year cycle of the Sun activity is noticeable

6.3.2.3 Seismology

Earthquakes occur in the block, cut by multiple ruptures, strongly non-uniform, solid shell of the Earth called lithosphere as a result of its deformation due to tectonic forces. General equations governing the process are still unknown (Sadovsky and Pisarenko, 1997). There are reasons to think that a seismic regime is affected by various physical and chemical processes including thermal, electric and others. In practice, one measures mechanical tension fields at different areas of the Earth's surface. Another form of data representation is time intervals between successive strong earthquakes $\Delta t_i = t_i - t_{i-1}$, where t_i is an instant of an ith shock. It can be interpreted as a non-equidistant time series. The values of Δt_i vary strongly between different observation areas and periods.

To study seismic activity, one uses equidistant time series as well. For instance, one measures a quantity proportional to the acceleration of the Earth's surface vibrations with the aid of seismographs (Koronovskii and Abramov, 1998).

6.3.2.4 Finance

For participants of events occurring at equity markets, it is important to be able to foresee changes in currency exchange rates, stock prices, etc. Those processes are affected by multiple factors and fluctuate quickly in a very complicated manner. Time series of currency exchange rates are often recorded at sampling interval of 1 h (and even down to 2 min). One often reports daily values of stock prices, see Figs. 6.8 and 6.9 and Box and Jenkins (1970); Cecen and Erkal (1996); Lequarre (1993); Makarenko (2003); Soofi and Cao (2002).

6.3.2.5 Physiology and Medicine

In this field, one often encounters the problem of diagnostics rather than forecast. Physiological signals reflecting activity of the heart, brain and other organs are one of the main sources of information for physicians in diagnosing. One uses electro-cardiograms (difference of potentials between different points at the chest surface, Fig. 6.10), electroencephalograms (potentials at the scalp), electrocorticograms (intracranial brain potentials), electromyograms (potentials inside muscles or on the

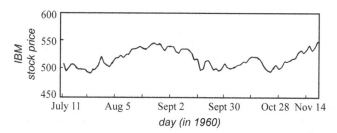

Fig. 6.8 Stock prices for the IBM company. Sampling interval $\Delta t = 1$ day; the values for the end of a day are shown (Box and Jenkins, 1970)

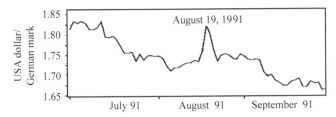

Fig. 6.9 Currency exchange rate between USA dollar and German mark in 1991 (Lequarre, 1993). Sampling interval $\Delta t = 1$ day

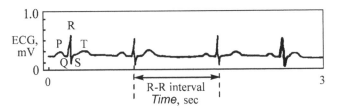

Fig. 6.10 Electrocardiogram (Rigney et al., 1993). A characteristic PQRST complex and the so-called $R - R$ *interval* are shown. Sampling interval $\Delta t = 1$ ms

skin), acceleration of finger oscillations for a stretched hand (physiological tremor), concentration of oxygen in the blood, heart rate, chest volume representing a respiration process, etc. A typical sampling interval is of the order of 1 ms.

It is important to be able to detect signs of a disease at early stages. For that, one might need quite sophisticated methods for the analysis of signals.

6.3.2.6 Transport

In this field as in many technical applications, a problem of automatic control often arises. For instance, one measures data representing simultaneously a course of a boat and a rudder turn angle (Fig. 6.11). Having those data, one can construct an automatic system allowing to make control of a boat more efficient, i.e. to reduce its wagging and fuel consumption (Ljung, 1991).

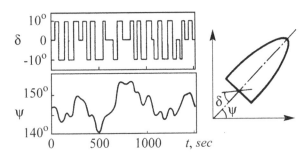

Fig. 6.11 Simultaneous time series of a rudder turn angle (random turns) and a course of a boat (Ljung, 1991). Sampling interval $\Delta t = 10$ s

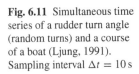

6.3.2.7 Hydrodynamics

Investigation of turbulent motion in a fluid is one of the oldest and most complex problems in non-linear dynamics. Chaotic regimes are realised in experiments with fluid between two cylinders rotating in opposite directions or in a mixer with a rotating turbine inside. The data are, for instance, time series of fluid velocity at a certain spatial location measured at typical sampling interval of 1 ms (Letellier et al., 1997).

6.3.2.8 Chemistry

Considerable attention of researchers is paid to chaotic behaviour in many chemical reactions. The data from those systems are presented in the form of time realisations of reagent concentrations, e.g. Ce^{IV} ions in the Belousov and Zhabotinsky reaction (Brown et al., 1994; Letellier et al., 1998b).

6.3.2.9 Laser Physics

Observed complex behaviour of a laser under periodic pulse pumping (Fig. 6.12) can be used to estimate some parameters of the laser and further to study a dependence of those parameters on external conditions including a temperature regime. The data are a time series of irradiation intensity measured with the aid of photodiodes.

6.3.2.10 Astrophysics

In fact, the only source of information about remote objects of the Universe (stars) is time series of their irradiation intensity. Those time series are collected with the aid of radio- and optical telescopes (Fig. 6.13).

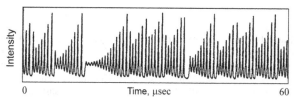

Fig. 6.12 Data from a ring laser in a chaotic regime (Hubner et al., 1993): irradiation intensity. Sampling interval $\Delta t = 40$ ns

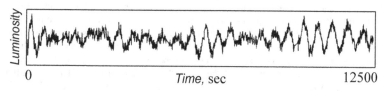

Fig. 6.13 Variations of luminance of a dwarf star PG-1159, $\Delta t = 10$ s (Clemens, 1993). The time series was recorded continuously during 231 h

The list of examples, problems and objects can be continued. Their number permanently rises. ADCs with $\Delta t = 10^{-8} - 10^{-9}$ s and data storage devices with memory sizes of hundreds of gigabytes are already easily accessible for a researcher. One still does not observe a saturation in the development of the devices for acquisition, storage and processing of time series. Even the above examples suffice to show that time realisations of motions in real-world systems are usually very complex and irregular. Yet, one now knows that complicated chaotic motions can be demonstrated even by simple non-linear dynamical systems so that the problem of modelling from a time series does not seem hopeless even though it requires a development of non-trivial techniques.

6.4 Elements of Time Series Analysis

6.4.1 Visual Express Analysis

Human capabilities to recognise visual images are so well developed that we can compete even with specialised computers in such activity. It is thought that a person gets about 70% of sensory information via the eyes. Visual analysis of data, if they are presented in a graphical form, can be very fruitful in modelling. It can give an idea about an appropriate form of model functions and kind and dimensionality of model equations. The most natural step is visual assessment of time realisations of a process $\eta(t)$, see Fig. 6.14 (left panels).

One should consider time realisations of sufficient length in order that peculiarities of motion allowing identification of a process could manifest themselves. Thus, for *periodic motions* (Fig. 6.14a), such a peculiarity is complete repeatability of a process with some period T. A motion is *quasi-periodic* (Fig. 6.14b) if there are

→

Fig. 6.14 Time realisations of regular (**a, b**) and irregular (**c–h**) processes and their periodograms (formulas are given in Sect. 6.4.2). (**a**) Periodic process: variations in voltage on a semiconductor diode in a harmonically driven *RL* diode circuit (provided by Prof. E.P. Seleznev). (**b**) Quasi-periodic process: variations in voltage on a non-linear element in coupled generators with quadratic non-linearity which individually exhibit periodic self-sustained oscillations (provided by Prof. V.I. Ponomarenko). (**c**) Narrowband stationary process. (**d**) Narrowband process with non-stationarity in respect of expectation. (**e**) Narrowband process with non-stationarity in respect of variance. The data on the panels **c, d, e** are signals from an accelerometer attached to a hand of a patient with Parkinson's disease during spontaneous tremor epochs (provided by the group of Prof. P. Tass, Research Centre Juelich, Germany). (**f**) Wideband stationary process: anomaly of the sea surface temperature in Pacific ocean ($5\,°N - 5\,°S$, $170\,°W - 120\,°W$), the data are available at http://www.ncep.noaa.gov. (**g**) Wideband process with non-stationarity in respect of expectation: variations in global surface temperature of the Earth. An anomaly of the GST (i.e. its difference from the mean temperature over the base period 1961–1990) is shown (Lean et al., 2005). (**h**) Wideband process with the signs of non-stationarity in respect of variance: variations in the global volcanic activity quantified by the optical depth of volcanic aerosol (Sato et al., 1993)

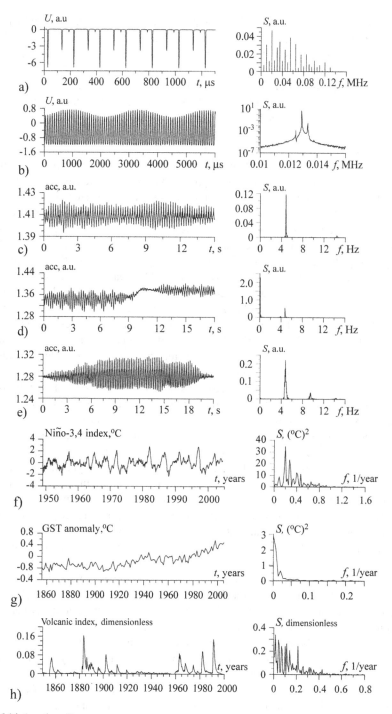

Fig. 6.14 (continued)

two or more characteristic timescales (i.e. periods of harmonic components) whose ratio T_i / T_j is an irrational number. Periodic and *quasi-periodic motions* are called *regular motions* in contrast to the cases illustrated in Fig. 6.14c–h which are depleted of obvious regularity, i.e. represent *irregular motions*.

Figure 6.14c, f shows *stationary processes* whose statistical characteristics do not change in time, while Fig. 6.14d, e, g, h shows temporal profiles looking more like *non-stationary processes* (see Sects. 4.1.3 and 6.4.4). In simple cases, non-stationary motions are recognised by eye if qualitatively or quantitatively different stages can be distinguished in their time realisations.

By considering the distance between successive maxima or minima in a time realisation and the shape of the temporal profile, one can estimate basic frequency of a signal and even a frequency range covered by its significant harmonic components (see Sect. 6.4.2). For instance, the distance between most pronounced minima is practically constant (about 200 μs) for the voltage variations shown in Fig. 6.14a; the distance between successive maxima for the accelerometer signal fluctuates stronger (Fig. 6.14c) and for the climatic process shown in Fig. 6.14f, it fluctuates much stronger. Therefore, a relative width of the peak in the power spectrum corresponding to a basic frequency is smallest for the electric signal (practically, a discrete line at 5 kHz, Fig. 6.14a), somewhat greater for the accelerometer signal (the peak at 5 Hz, Fig. 6.14c), and much greater for the climatic process (the smeared peaks at 0.2 and 0.33 1/year, Fig. 6.14f), i.e. the process in Fig. 6.14f is most wideband of these three examples. At the same time, the periodic electric signal exhibits a complicated time profile involving both flat intervals and voltage jumps (Fig. 6.14a). Therefore, its power spectrum exhibits many additional almost discrete lines at higher harmonics of the basic frequency: 10, 15, 20 kHz (an especially high peak) and so on up to 130 kHz. Non-stationarity often leads to an increase in the low-frequency components that is most clearly seen in Fig. 6.14d, g.

Another widespread approach to the visual assessment relies upon a procedure of *phase orbit reconstruction* when one shows the values of dynamical variables computed from an observable (Sects. 6.1.2 and 10.1.2) along the coordinate axes. Data points in such a space represent states of an object at successive time instants (Fig. 6.15). Possibilities of visual analysis of the phase portraits are quite limited. Without special tools, one can just consider two-dimensional projections of the phase portraits on a flat screen (Fig. 6.15b). Cycles are easily identified since they are represented by thin lines (the lower graph in Fig. 6.15b). A torus projection to a plane looks like a strip with sharp boundaries (the upper graph), which differs from a more "smeared" image of a chaotic attractor (the right graph). The pictures, which are more informative for the distinction between chaotic and quasi-periodic motions, are obtained with the aid of phase portrait sections, e.g. a stroboscopic section or a section based on the selection of extrema in the time realisations. Such a section for a torus is a closed curve (Fig. 6.15b, white line), for a cycle it is a point or several points and for a chaotic attractor it is a set of points with a complex structure. Analysis of phase portraits is more fruitful for the identification of complex non-periodic motions compared to the spectral analysis of the observed signals.

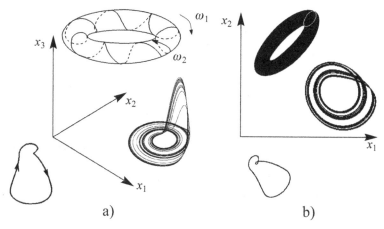

Fig. 6.15 Examples of different dynamical regimes. (**a**) Qualitative outlook of possible phase orbits in a three-dimensional phase space: a limit cycle (the *lower graph*), a torus (the *upper graph*) and a chaotic attractor (the *right graph*). (**b**) Projections to a plane representing typical pictures on an oscilloscope screen. The white line on the black background denote a projection of a two-dimensional section of the portrait

6.4.2 Spectral Analysis (Fourier and Wavelet Transform)

Most often, the term "spectral analysis" refers to Fourier transform, i.e. decomposition of a signal into harmonic components. However, in a generalised sense, spectral analysis is a name for any representation of a signal as a superposition of some *basis* functions. The term "spectrum" refers then to a set of those functions (components). Below, we briefly consider a traditional *Fourier analysis* and a more novel and "fashion" tool called *wavelet analysis* which is very fruitful, in particular, to study non-stationary signals.

6.4.2.1 Fourier Transform and Power Spectrum

This topic is a subject of multitude of books and research papers. It underlies such fields of applied mathematics as spectral analysis (Jenkins and Watts, 1968) and digital filters (Hamming, 1983; Rabiner and Gold, 1975). We only briefly touch on several points.

Firstly, let us recall that according to Weierstrass' theorem, any function $\eta = F(t)$ continuous on an interval $[a,b]$ with $F(a) = F(b)$ can be arbitrarily accurately represented by a trigonometric polynomial. The idea is readily realised in time series analysis. If a time series is equidistant and contains an even number of points N, sampling interval is Δt, $a = t_1$, $b = a + N \cdot \Delta t = t_N + \Delta t$, then one can show that an original signal $\{\eta(t_i)\}_{i=1}^{N}$ for all observation instants can be uniquely represented as the sum:

$$\eta(t_i) = a_0 + \sum_{k=1}^{N/2} a_k \cos(k\omega t_i) + \sum_{k=1}^{N/2-1} b_k \sin(k\omega t_i), \ i = 1, \ldots, N, \qquad (6.1)$$

where

$$\omega = \frac{2\pi}{b-a} = \frac{2\pi}{N\Delta t}.$$

Coefficients of the trigonometric polynomial (6.1) are expressed via the formulas

$$a_0 = \frac{1}{N}\sum_{i=1}^{N} \eta_i, \ a_{N/2} = \frac{1}{N}\sum_{i=1}^{N}(-1)^i \eta_i, \qquad (6.2)$$

$$a_k = \frac{2}{N}\sum_{i=1}^{N} \eta_i \cos(k\omega t_i), k = 1, \ldots, N/2 - 1, \qquad (6.3)$$

$$b_k = \frac{2}{N}\sum_{i=1}^{N} \eta_i \sin(k\omega t_i), k = 1, \ldots, N/2 - 1. \qquad (6.4)$$

The formulas (6.2), (6.3) and (6.4) converting the values of η_i into the coefficients a_k and b_k are called the direct *discrete Fourier transform* (DFT). The formula (6.1) providing calculation of η_i from a_k and b_k is called the inverse DFT.

Based on these transforms, one can construct an approximate description of an original signal, in particular, smooth it. For instance, higher frequencies corresponding to big values of k often reflect noise influence so that it is desirable to get rid of them. It can be accomplished in the simplest way if one zeros the corresponding coefficients a_k and b_k and performs the inverse DFT (6.1). Thereby, one gets a more gradually varying ("smoother") signal. This is a kind of a low-pass filter. A high-pass filter can be realised in a similar way by zeroing coefficients corresponding to small values of k. To get a band-pass filter, one zeros all the coefficients outside of a certain frequency band. These simple versions of digital filters are not the best ones (Hamming, 1983; Rabiner and Gold, 1975).

One can often get a sufficiently good approximation to a continuous function $F(t)$ over a finite interval with the aid of a trigonometric polynomial. At that, even trigonometric polynomials of quite a high order can be used, while the use of high-order algebraic polynomials leads to significant troubles (Sect. 7.2.3). If a function $F(t)$ is periodic, it can be approximated well by a trigonometric polynomial over the entire number axis. We stress that the trigonometric system of functions is especially useful and has no "competitors" for approximation of periodic functions.

Physical Interpretation as Power Spectrum

The mean-squared value of an observable η is proportional to physical power if η is an electric voltage or a current in a circuit. One can show that this mean-squared value is equal to the sum of the mean-squared values of the terms in the right-hand side of Eq. (6.1). In other words, the power is distributed *among frequencies*:

$$\frac{1}{N} \sum_{i=1}^{N} \eta_i^2 = \sum_{k=0}^{N/2} S_k, \tag{6.5}$$

where S_k is a power contained in a harmonic component with a frequency $k\omega$: $S_k = (a_k^2 + b_k^2)/2$ for $1 \leq k < N/2$ and $S_k = a_k^2$ for $k = 0, N/2$. Physically, an observed signal may represent a superposition of signals from several sources. If each of those sources demonstrates harmonic oscillations with its own frequency, then its intensity in the observed signal is reflected by the values of S_k at the corresponding frequency. The quantities S_k allow to detect different sources of oscillations and estimate their relative intensity. If each frequency corresponding to a considerable value of S_k is related to oscillations of a certain harmonic oscillator, then the number of considerable components is equal to the *number of degrees of freedom* involved in the process. Since the total power in a signal is represented as a set (spectrum) of components according to Eq. (6.5), the set of S_k is called "power spectrum" of the process $\eta(t)$. Strictly speaking, this is only an estimate of the power spectrum (see below).

The concept of the power spectrum is so easily defined only for a deterministic periodic function $\eta(t)$ with a period $2\pi/\omega$, since such a function is uniquely represented by a trigonometric Fourier series:

$$\eta(t) = a_0 + \sum_{k=1}^{\infty} [a_k \cos(k\omega t) + b_k \sin(k\omega t)], \tag{6.6}$$

whose coefficients are, in general, non-zero and expressed via the integrals of the original function $\eta(t)$.

However, even in this case, one must take into account that a finite set of coefficients in Eq. (6.1) obtained via the direct DFT from a time series is only an approximation to the theoretical spectrum. If the most part of the power is contained in relatively low frequencies (but higher than $\omega = 2\pi/(N\Delta t)$), then such an approximation is sufficiently accurate. Widely known is a phenomenon of frequency mimicry (masking) which is following. A model (6.1) includes maximal frequency of $\omega N/2 = \pi/\Delta t$, which is called Nyquist frequency. The period of the corresponding harmonic component is equal to the doubled sampling interval. Any components with the frequencies exceeding the Nyquist frequency would be linear combinations of the basis functions in Eq. (6.1) over the set of the observation instants t_i. If such components are introduced into a model, then the lower frequencies must be excluded to provide linear independence of the basis functions. In other words, higher frequency components cannot be distinguished from the combinations of the lower frequency components, e.g. $\cos\big((N/2+k)\,j\omega\Delta t\big) = \cos(\pi j + k\omega j \Delta t) = (-1)^j \cos(jk\omega\Delta t)$, where $k > 0$ and $t_j = j\Delta t$. It looks as if the former were masked by the latter.

The situation gets even more complex in the case of a non-periodic function $\eta(t)$. Such a function cannot be accurately represented by the series (6.6) over the entire number axis. However, under certain conditions (integrability over the entire number axis and smoothness), one can write down a similar representation in the form of the Fourier integral, i.e. replace discrete frequencies $k\omega$ in Eq. (6.6) by the continuous range of values:

$$\eta(t) = \int_0^\infty A(\omega)\cos(\omega t)d\omega + \int_0^\infty B(\omega)\sin(\omega t)d\omega, \tag{6.7}$$

$$A(\omega) = \frac{1}{\pi}\int_{-\infty}^\infty \eta(t)\cos(\omega t)dt, \quad B(\omega) = \frac{1}{\pi}\int_{-\infty}^\infty \eta(t)\sin(\omega t)dt. \tag{6.8}$$

The transforms (6.7) and (6.8) are called *continuous Fourier transforms* (the inverse and direct transforms, respectively). The above discrete transforms are their analogues. The energy[4] in a signal $\eta(t)$ is expanded into a continuous *energy spectrum* as $\int_{-\infty}^\infty \eta^2(t)dt = \int_0^\infty E(\omega)d\omega$, where $E(\omega) = A^2(\omega) + B^2(\omega)$.

Finally, let us consider the case where $\eta(t)$ is a realisation of a stationary random process. Typically, it is almost always non-periodic. Moreover, integrals (6.8) almost always do not exist, i.e. one cannot define A and B even as random quantities. Spectral contents of a process are then described via the *finitary Fourier transform*, i.e. for $\eta(t)$ over an interval $[-T/2, T/2]$ one gets

$$A_T(\omega) = \frac{1}{\pi}\int_{-T/2}^{T/2} \eta(t)\cos(\omega t)dt, \quad B_T(\omega) = \frac{1}{\pi}\int_{-T/2}^{T/2} \eta(t)\sin(\omega t)dt. \tag{6.9}$$

Further, one computes expectations of A_T, B_T and defines *power spectrum* as

$$S(\omega) = \lim_{T \to \infty} \frac{\langle A_T^2(\omega) + B_T^2(\omega)\rangle}{T}, \tag{6.10}$$

where angular brackets denote the expectation. The quantity on the left-hand side of (6.10) is power, since it represents energy divided by the time interval T. In this case, the values of S_k obtained with DFT are random quantities. The set of such values is a rough estimate of the power spectrum. In particular, it is not a consistent estimator since the probability density function for each S_k is proportional to that for the χ^2 distribution with two degrees of freedom so that the standard deviation of

---

[4] Not a power. Mean power equals zero in this case, since a signal must decay to zero at infinity to be integrable over the entire number axis.

S_k equals its mean and does not decrease with increasing time series length (Brockwell and Davis, 1987; Priestley, 1989). This set is called *periodogram* if all the components are multiplied by N, i.e. if one converts from power to energy. Several examples are presented in Fig. 6.14 (the right panels). To get an estimator with better statistical properties, it is desirable to average S_k over several realisations of the process or to "smooth" a single periodogram (Brockwell and Davis, 1987; Priestley, 1989).

Importance of the power spectrum concept is related to the fact that behaviour of many real-world systems in the low-amplitude oscillatory regimes is adequately described with harmonic functions. Such dynamical regimes are well known and understood in detail. They are observed everywhere in practice and widely used in technology, e.g. in communication systems. Linear systems (filters, amplifiers, etc.) are described in terms of the transfer functions, i.e. their basic characteristic is the way how the power spectrum of an input signal is transformed into the power spectrum of the output signal. The phase spectrum which is a set of initial phases of the harmonic components in Eq. (6.1) is also often important. Multiple peculiarities of the practical power spectrum estimation and filtering methods are discussed, e.g., in Hamming (1983), Jenkins and Watts (1968), Press et al. (1988), Rabiner and Gold (1975) and references therein.

An Example: Slowly Changing Frequency Contents and Windowed DFT

It is a widespread situation when a signal under investigation has a time-varying power spectrum. One of the simplest examples is a sequence of two sinusoidal segments with different frequencies:

$$\eta(t) = \begin{cases} \sin 2t, & -\pi \le t < 0, \\ \sin 4t, & 0 \le t < \pi. \end{cases} \tag{6.11}$$

We performed the analysis over the interval $[-\pi, \pi]$ from a time series of the length of 20 data points with the sampling interval $\Delta t = \pi/10$ and $t_1 = -\pi$ (Fig. 6.16a).

The signal can be described well with a trigonometric polynomial (6.1) containing many considerable components (Fig. 6.16a). However, more useful information can be obtained if the signal is divided into two segments (windows) and a separate trigonometric polynomial is fitted to each of them (Fig. 6.16b). This is a so-called windowed Fourier transform. In each window, one gets a spectrum consisting of a single significant component that makes physical interpretation of the results much easier. The windowed DFT reveals that frequency contents of the signal changes in time that cannot be detected with a single polynomial (6.1). Non-stationary signals are often encountered in practice and can be analysed with the windowed DFT. However, there exists a much more convenient contemporary tool for their analysis called "wavelet transform".

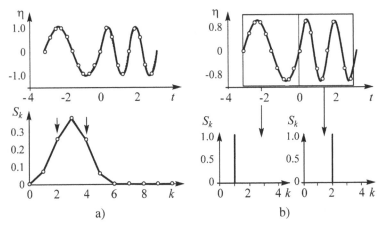

Fig. 6.16 An example of a non-stationary signal: (**a**) two segments of a sinusoid and a super-imposed plot of an approximating trigonometric polynomial (6.1). Power spectrum contains five significant components (the *bottom panel*). *Arrows* indicate frequencies of the two original sinu-soidal segments; (**b**) analogous plots for the windowed DFT. There is a single non-zero component in each spectrum. Their frequencies correspond to the values of k twice as small as in the *left panel* since the number of points in each window is twice as small as in the original time series

6.4.2.2 Wavelet Transform and Wavelet Spectrum

A very efficient approach to the analysis of functions $\eta = F(t)$ exhibiting pulses, discontinuities, breaks and other singularities is to use basis functions $\phi_k(t)$ called wavelets, which are well localised both in time and frequency domains. They have become an extremely popular tool during the last 20 years (see, e.g., the reviews Astaf'eva, 1996; Koronovskii and Hramov, 2003; Torrence and Compo, 1998 and references therein).

The term "wavelet" has been introduced in 1984 and become widely used. Many researchers call wavelet analysis "mathematical microscope" (Astaf'eva, 1996). Here, we do not go into strict mathematical definitions and explain only some basic points. Wavelet is a function $\phi(t)$, which

  (i) is well localised both in time domain (it quickly decays when $|t|$ rises) and frequency domain (its Fourier image is also well localised);

 (ii) has zero mean $\int\limits_{-\infty}^{\infty} \phi(t)\mathrm{d}t = 0$;

(iii) satisfies a scaling condition (a number of its oscillations does not change under variations in the timescale).

An example is the so-called DOG-wavelet[5] shown in Fig. 6.17:

$$\phi(t) = e^{-t^2/2} - 0.5e^{-t^2/8}. \tag{6.12}$$

---

[5] Difference of Gaussians, i.e. Gauss functions.

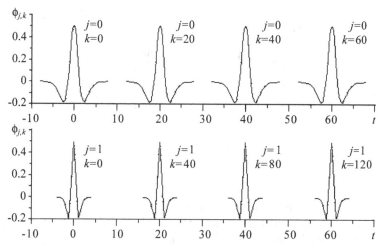

Fig. 6.17 An example of wavelet and construction of the set of basis functions via shifts (from *left* to *right*) and compressions (from *top* to *bottom*). Normalisation is not taken into account

Based on a certain wavelet function, one can construct a set of functions via its shifts and scaling transformations (compression and stretching along the t-axis). The functions obtained are conveniently denoted by two subscripts:

$$\phi_{j,k}(t) = 2^{j/2}\phi\left(2^j t - k\right), \quad -\infty < j, k < \infty, \tag{6.13}$$

where j and k are integers. Increasing j by 1 changes the scale along the time axis twice (compression of the function plot), while increasing k by 1 shifts the plot of the function $\phi_{j,k}$ by $k/2^j$ along the t-axis (Fig. 6.17). The normalising multiplier $2^{j/2}$ is introduced for convenience to preserve constant norm of $\phi_{j,k}$, i.e. the integrals of squared functions are all equal: $\left\|\phi_{j,k}\right\|^2 = \|\phi\|^2$.

The constructed set of the localised functions $\phi_{j,k}$ covers the entire t-axis due to shifts, compression and stretching. Under a proper choice of $\phi(t)$, the set is a basis in the space of functions, which are square summable over the entire axis. Strictly speaking, $\phi(t)$ is called wavelet only in this case (Astaf'eva, 1996). The condition is fulfilled for a wide class of functions including that presented in Fig. 6.17. Basis functions $\phi_{j,k}$ are often called wavelet functions. Since all of them are obtained via the transformations of $\phi(t)$, the latter is often called "mother wavelet".

Examples of mother wavelet are very diverse (Fig. 6.18). Numerical libraries include hundreds of them (http://www.wavelet.org). A widely used one is the complex *Morlet wavelet*

$$\phi(t) = \pi^{-1/4}e^{-t^2/2}(e^{-i\omega_0 t} - e^{-\omega_0^2/2}), \tag{6.14}$$

where ω_0 determines the number of its oscillations over the decay interval and the second term in parentheses is introduced to provide zero mean. Real part of the Morlet wavelet is shown in Fig. 6.18 for $\omega_0 = 6$.

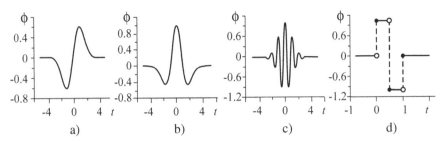

Fig. 6.18 Examples of wavelets: (**a**) WAVE wavelet; (**b**) "Mexican hat"; (**c**) Morlet wavelet (a real part); (**d**) HAAR wavelet

To construct an approximating function for a time series, one should select a finite number of terms from the infinite set of functions $\phi_{j,k}$. Wavelet-based approximation can be vividly illustrated with an example of *HAAR wavelet* (Fig. 6.18d) (Misiti et al., 2000). Let us consider an equidistant time series of length $N = 2^m$, where m is a positive integer, $\Delta t = 1/N$, $t_1 = 0$. As a set of basis functions, it is convenient to use the following functions from the entire set $\phi_{j,k}$ supplemented with a constant:

$$1,$$
$$\phi_{0,0}(t) \equiv \phi(t),$$
$$\phi_{1,0}(t) \equiv \phi(2t), \qquad\qquad \phi_{1,1}(t) \equiv \phi(2t - 1)$$
$$\phi_{2,0}(t) \equiv \phi(4t), \phi_{2,1}(t) \equiv \phi(4t - 1), \phi_{2,2}(t) \equiv \phi(4t - 2), \phi_{2,3}(t) \equiv \phi(4t - 3),$$
$$\cdots,$$
$$\phi_{m-1,0}(t) \equiv \phi(2^{m-1}t), \quad \cdots, \quad \phi_{m-1,2^{m-1}-1}(t) \equiv \phi(2^{m-1}t - 2^{m-1} + 1).$$
$$(6.15)$$

An original time series is precisely represented as the sum

$$\eta_i = c_0 + \sum_{j=0}^{m-1} \sum_{k=0}^{2^j - 1} c_{j,k} \phi_{j,k}, \qquad (6.16)$$

Coefficients corresponding to the terms with $j = m - 1$ depend only on the difference of the observed values at neighbouring time instants, i.e. on the oscillations with the period $2\Delta t$ corresponding to the Nyquist frequency. Since those functions describe only the smallest scale variations, the sum of them

$$D_1(t_i) = \sum_{k=0}^{2^{m-1}-1} c_{m-1,k} \phi_{m-1,k}(t_i), \qquad (6.17)$$

is said to describe the *first-level details*. The remaining component

$$A_1(t_i) = \eta_i - D_1(t_i) = c_0 + \sum_{j=0}^{m-2} \sum_{k=0}^{2^j-1} c_{j,k} \phi_{j,k}, \qquad (6.18)$$

is called the *first-level approximation*. The first-level approximation no longer contains variations with a period $2\Delta t$ (Fig. 6.19). Similarly, one can determine details in the first-level approximation and so on, which is realised via the basis functions with smaller j (Fig. 6.19). A general definition of the nth-level details and nth-level approximation is introduced analogously to Eqs. (6.17) and (6.18): details $D_n(t_i)$ are small-scale components of a signal and approximations $A_n(t_i)$ are larger scale components. Approximation of the last mth level is just a mean value of an original signal. Finally, an original signal is equal to the sum of its mean value and all details: $\eta_i = A_m(t_i) + D_m(t_i) + D_{m-1}(t_i) + \cdots + D_1(t_i)$.

The value of j determines the scale of the consideration: the greater the j, the smaller the scale. The value of k specifies a temporal point of consideration. To continue an analogy between a wavelet and a microscope, one can say that k is a focusing point, j determines its magnification and the kind of the mother wavelet is responsible for its "optical properties".

If only a small number of the wavelet coefficients $c_{j,k}$ appear significant, then the rest can be neglected (zeroed). Then, the preserved terms give a parsimonious and sufficiently accurate approximation to an original signal. In such a case, one says that the wavelet provides a compression of information since several wavelet coefficients can be stored instead of many values in the time series. If necessary, one can restore an original signal from those coefficients only with a small error. Wavelets are efficient to "compress" signals of different character, especially pulse-like ones. To compress signals, one can also use algebraic or trigonometric polynomials, but the field of wavelet applications appears much wider in practice (see, e.g., Frik and Sokolov, 1998; Misiti et al., 2000; http://www.wavelet.org).

Wavelet Analysis

Non-trivial conclusions about a process can be extracted from the study of its wavelet coefficients. This is a subject of the wavelet analysis in contrast to approximation and restoration of signal, which are the problems of synthesis. Above, we spoke of the *discrete wavelet analysis* since the subscripts j and k in the set of wavelet functions took discrete values. More and more careful interest is now paid to the *continuous wavelet analysis* when one uses continuous-valued "indices" of shift and scale (Astaf'eva, 1996; Koronovskii and Hramov, 2003; Torrence and Compo, 1998). The integral (continuous) *wavelet transform* of a signal $\eta(t)$ is defined by the expression

$$W(s,k) = \frac{1}{\sqrt{s}} \int_{-\infty}^{\infty} \eta(t) \cdot \phi\left(\frac{t-k}{s}\right) dt \equiv \int_{-\infty}^{\infty} \eta(t) \cdot \phi_{s,k}(t) \, dt, \qquad (6.19)$$

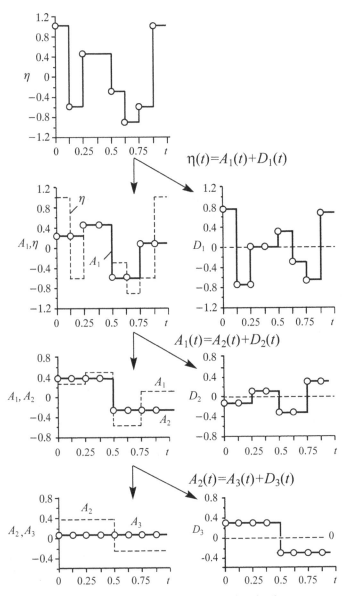

Fig. 6.19 Wavelet transform: approximations and details at various levels

where s and k are real numbers (continuous-valued parameters of scale and shift), wavelet functions are denoted as $\phi_{s,k}(t) = \left(1/\sqrt{s}\right) \phi\left((t-k)/s\right)$ and the parameter s is analogous to 2^{-j} in the discrete transform (6.13). The bivariate function $W(s,k)$ is called the *wavelet spectrum* of the signal $\eta(t)$. It makes vivid physical sense. A big value of $|W(s_1, k_1)|$ indicates that signal variations with the timescale s_1 around the time instant k_1 are intensive. Roughly speaking, the value of $|W(s, k_1)|$

at fixed k_1 shows frequency contents of the signal around the time instant k_1. If the values corresponding to small s are large, then small-scale (high-frequency) components are present. The values of $|W(s_1, k)|$ at fixed s_1 show how the intensity of the signal component corresponding to the timescale s_1 changes in time. Thus, the wavelet spectrum carries information both about frequency contents of the signal and its temporal localisation in contrast to the Fourier-based power spectrum, which provides information only about the frequency contents without any temporal localisation. Therefore, the wavelet spectrum is also called a time – frequency spectrum.

The *wavelet spectrum* satisfies "energy condition", which allows one to relate it to a decomposition of the signal energy in time and frequency domains:

$$\int\limits_{-\infty}^{\infty} \eta^2(t)\mathrm{d}t = \frac{1}{C_\phi} \int\limits_{-\infty}^{\infty} \int\limits_{-\infty}^{\infty} W^2(s, k)\frac{\mathrm{d}s\,\mathrm{d}k}{s^2}, \qquad (6.20)$$

where C_ϕ is a normalising coefficient depending on the kind of the mother wavelet. If the value of W^2 is integrated over time k, one gets a function of the timescale, which is called *global energy spectrum* or *scalogram*:

$$E_W(s) = \int\limits_{-\infty}^{\infty} W^2(s, k)\mathrm{d}k. \qquad (6.21)$$

It can be used for the global characterisation of the signal frequency contents along with the periodogram. Scalogram is typically a more accurate estimator of the power spectrum. It resembles a smoothed periodogram (Astaf'eva, 1996; Torrence and Compo, 1998).

A wavelet spectrum can be visualised as a surface in a three-dimensional space. More often, one uses a contour map of $|W(s, k)|$ or a two-dimensional map of its values on the plane (k, s) in greyscale, e.g., where the black colour denotes large values and the white colour indicates zero values. Of course, only an approximate computation of the integral (6.19) is possible in practice. To do it, one must specify a signal behaviour outside an observation interval $[a,b]$ (often, a signal is simply zeroed) that introduces artificial peculiarities called edge effects. How long intervals at the edges should be ignored depends on the mother wavelet used and on the timescale under consideration (Torrence and Compo, 1998). Let us illustrate performance of the wavelet analysis for the temporal profile shown in Fig. 6.20a: two sinusoidal segments with different frequencies (similar examples are considered, e.g., in Koronovskii and Hramov, 2003). DOG wavelet is used for a time series of the length of 1000 points over an interval $[-\pi, \pi]$ and zero padding outside the interval. The global Fourier-based spectrum does not reveal the "structure of non-stationarity". The wavelet spectrum clearly shows the characteristic timescales (Fig. 6.20b). In particular, it allows to distinguish low-frequency oscillations at the beginning of the time series; black spots correspond to the locations of the sinusoid

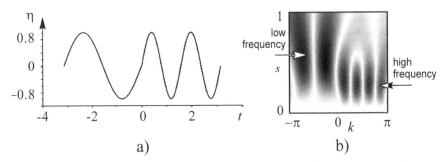

Fig. 6.20 Wavelet analysis: (**a**) two sinusoidal segments with different frequencies; (**b**) wavelet spectrum obtained with the aid of DOG wavelet

extrema and "instantaneous" period. Edge effects (which are, indeed, strong for a signal containing so small number of oscillations with a characteristic period) are not taken into account here for the sake of illustration simplicity.

Additional examples are given in Fig. 6.21 which presents wavelet spectra of the time series given in Fig. 6.14. For the computations, we have used the Morlet wavelet (6.14) with $\omega_0 = 6$. Edge effects for this wavelet function at the timescale s cover intervals of the width $s\sqrt{2}$ (Torrence and Compo, 1998). One can clearly see the basic periods as dark horizontal lines in Fig. 6.21a–e. Moreover, Fig. 6.21a exhibits additional structure related to the complicated temporal profile, which is seen in Fig. 6.14a. Decrease in the oscillation amplitude during the change in the mean value in Fig. 6.14d (the interval from the 9th to the 12th s) is reflected as a white "gap" in the horizontal black line in Fig. 6.21d. Figure 6.21e clearly shows the increase in the oscillation amplitude in the beginning and its decrease in the end. Figure 6.21h shows higher volcanic activity during the periods 1890–1910 and 1970–1990. One can see complicated irregular structures in Fig. 6.21f, g. Still, a characteristic timescale of about 60 months (5 years) can be recognised in Fig. 6.21f.

The wavelet analysis is extremely useful for the investigation of non-stationary signals containing segments with qualitatively different behaviour. It is efficient for essentially non-uniform signal (pulse-like, etc.) and signals with singularities (discontinuities, breaks, discontinuities in higher order derivatives), since it allows to localise singularities and find out their character. The wavelet spectrum exhibits a characteristic regular shape for fractal signals (roughly speaking, strongly jagged and self-similar signals): such signs appear inherent to many real-world processes. Moreover, one can analyse spatial profiles in the same way, e.g., the Moon relief (Misiti et al., 2000) exhibits a very complex shape with different scales related to the bombardment of the Moon with meteorites of various sizes. Wavelets are applied to data analysis in geophysics, biology, medicine, astrophysics and information processing systems, to speech recognition and synthesis, image compression, etc. Huge bibliography concerning wavelets can be found, e.g., in Astaf'eva (1996), Koronovskii and Hramov (2003), Torrence and Compo (1998) and at the website (http://www.wavelet.org).

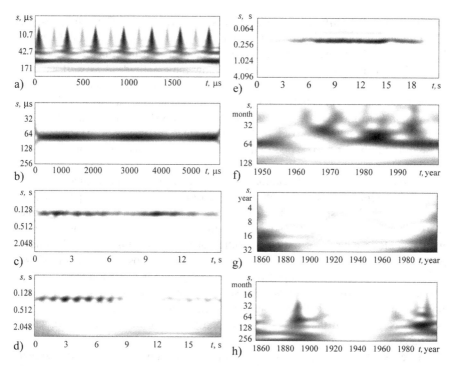

Fig. 6.21 Wavelet spectra for the signals shown in Fig. 6.14: (**a**) periodic electric signal (voltage on a diode); (**b**) quasi-periodic electric signal (voltage on a non-linear element); (**c**) stationary accelerometer signal; (**d**) non-stationary accelerometer signal in respect of the expectation; (**e**) non-stationary accelerometer signal in respect of the variance; (**f**) stationary climatic process (Niño-3.4 index); (**g**) non-stationary climatic process in respect of the expectation (variations in the global surface temperature); (**h**) non-stationary climatic process in respect of the variance (volcanic activity)

6.4.3 Phase of Signal and Empirical Mode Decomposition

It is very fruitful in multiple situations to consider a *phase of a signal*. Here, we discuss contemporary concepts of the phase. Roughly speaking, this is a variable characterising repeatability in a signal. It rises by 2π between any two successive maxima. Especial role of this variable is determined by its high sensitivity to weak perturbations of a system. Changes in the amplitude may require significant energy, while a phase can be easily changed by a weak "push".[6]

The term "phase" is often used as a synonym of the words "state" or "stage". In Sect. 2.1.3, we have discussed a state vector and a state space of a dynamical system and spoken of a phase orbit drawn by a state vector. The meaning of

---

[6] Thus, a phase of a pendulum oscillations (Fig. 3.5a) can be changed by holding it back at a point of maximal deflection from an equilibrium state without energy consumption.

the term "phase" is different in the field of signal processing. Thus, the phase of a harmonic signal $x(t) = A\cos(\omega t + \phi_0)$ is an argument of the cosine function $\phi = \omega t + \phi_0$. The phase ϕ determines the value of the cosine function. However, to specify a state completely, one needs to know the amplitude A as well, i.e. the phase is not a complete characteristic of a state. Apart from $\phi = \omega t + \phi_0$ called "unwrapped phase" (Fig. 6.22b, the upper panel), one uses a "wrapped" phase $\phi(t) = (\omega t + \phi_0)\bmod 2\pi$ defined only over the interval $[0, 2\pi)$ (Fig. 6.22b, the lower panel). The latter approach makes sense, since the values of the unwrapped phase differing by 2π correspond to the same states, the same values of the cosine function.

A vivid geometrical interpretation of the introduced concept of phase is possible if one represents a signal $x(t) = A\cos(\omega t + \phi_0)$ as a real part $\mathrm{Re}\,z(t)$ of a complex-valued signal $z(t) = A\,e^{i(\omega t + \phi_0)}$. Then, a vector $z(t)$ on the plane (x,y), where $x(t) = \mathrm{Re}\,z(t)$ and $y(t) = \mathrm{Im}\,z(t)$, rotates uniformly with a frequency ω along a circle of radius A centred at the origin (Fig. 6.22a). Its phase $\phi = \omega t + \phi_0$ is a rotation angle of $z(t)$ relative to the positive direction of the x-axis. To compute an unwrapped phase, one takes into account a number of full revolutions. Thus, such a phase is increased by 2π after each revolution. For a harmonic signal, it rises linearly in time at a speed equal to the angular frequency of oscillations. A plot of the wrapped phase is a piecewise linear function (a saw), Fig. 6.22b.

The concept of the phase originally introduced only for a harmonic signal was later generalised to more complicated situations. The most well-known and widespread generalisation to the case of non-harmonic signals is achieved via construction of an *analytic signal* (Gabor, 1946). The latter is a complex-valued signal, whose Fourier image has non-zero components only at positive frequencies. From an original signal $x(t)$, one constructs an analytic signal $z(t) = x(t) + iy(t)$, where $y(t)$ is the *Hilbert transform* of $x(t)$:

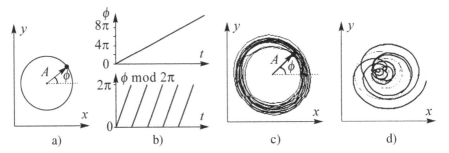

Fig. 6.22 Phase definition: (**a**) an orbit on the plane (x, y) for a harmonic signal $x(t)$, y is the Hilbert transform of x, A and ϕ are its amplitude and phase, respectively; (**b**) unwrapped and wrapped phases of a harmonic signal versus time; (**c**) an orbit on the complex plane for a non-harmonic narrowband signal, its amplitude and phase introduced through the Hilbert transform are shown; (**d**) the same illustration for a wideband signal, whose phase is ill-defined

$$y(t) = \text{P.V.} \int_{-\infty}^{\infty} \frac{x(t')dt'}{\pi(t-t')}, \tag{6.22}$$

P.V. denotes the Cauchy principal value of the improper integral. The phase is defined as an argument of the complex number z, i.e. as a rotation angle of the radius vector on the plane (x, y). This approach is widely used in radio-physics and electrical engineering (Blekhman, 1971, 1981; Pikovsky et al., 2001). For a harmonic signal $x(t) = A\cos(\omega t + \phi_0)$, the conjugated signal is $y(t) = \sin(\omega t + \phi_0)$ and the phase coincides with the above definition. For an "almost sinusoidal" signal, one observes rotation of the vector z not strictly along a circle but "almost" along a circle (Fig. 6.22c). The phase increases on average at the speed equal to the mean angular frequency of the oscillations.

The phase makes clear physical sense for oscillatory signals with a pronounced main rhythm (Anishchenko and Vadivasova, 2004; Pikovsky et al., 2001). For complicated irregular signals, an analytic signal constructed via the Hilbert transform may not reveal a rotation about a well-defined centre (Fig. 6.22d). Then, the above formal definition of the phase is, as a rule, useless. However, if the observed signal is a combination of a relatively simple signal from a system under investigation and a superimposed interference, then one can try to extract a simpler signal and determine its phase as described above. Let us consider two basic approaches to the extraction of a simple signal: band-pass filtering and empirical mode decomposition.

6.4.3.1 Band-pass Filtering

The simplest approach is to use a band-pass filter (Sect. 6.4.2), which let pass only components from a small neighbourhood of a certain selected frequency. If the frequency band is not wide, then one gets a signal with a pronounced main rhythm, whose phase is easily defined via the Hilbert transform (Fig. 6.22c). However, what frequency band should be used? Does a filtered signal relate to the process under investigation or is it just an artificial construction? One can answer such questions only taking into account additional information about the system under investigation. On the one hand, the frequency band should not be too narrow: In the limit case, one gets a single sinusoid whose phase is well defined but does not carry any interesting information. On the other hand, the frequency band should not be too wide, since then there would not be a rotation of the vector z about a single centre, i.e. repeatability to be described by the phase would not exist.

Another widespread opportunity to get an analytic signal is a complex wavelet transform (Lachaux et al., 2000):

$$z(t) = \frac{1}{\sqrt{s}} \int_{-\infty}^{\infty} x(t')\Phi^*\left(\frac{t'-t}{s}\right) dt' \tag{6.23}$$

at a fixed timescale s realised with the Morlet wavelet (6.14): $\Phi(t) = \pi^{-1/4}\left[\exp(-i\omega_0 t) - \exp\left(-\omega_0^2/2\right)\right]\exp(-t^2/2)$. This is equivalent to the band-pass filtering of an original signal with the subsequent application of the Hilbert transform. Namely, the frequency band is of the width $\Delta f/f = 1/4$ and centred at the frequency $f \approx 1/s$ at $\omega_0 = 6$. Edge effects are less prominent under the use of the wavelet transform (6.23) than for many other ways of filtering.

6.4.3.2 Empirical Mode Decomposition

Apart from linear filtering, one can use other options. A technique for the decomposition of a signal into the so-called "empirical modes" recently introduced in Huang et al. (1998) has become more and more popular during the last years. This is a kind of adaptive non-linear filtering. The phase of each empirical mode is readily defined, e.g., as a variable linearly rising by 2π between subsequent maxima or via the Hilbert transform. For that, each "mode" should be a zero-mean signal whose maxima are positive and minima are negative, i.e. its plot $x(t)$ inevitably intersects the abscissa axis ($x = 0$) between each maximum and minimum of a signal. The technique is easily implemented and does not require considerable computational efforts. A general algorithm is as follows:

(i)   To find all extrema in a signal $x(t)$.
(ii) To interpolate between the minima and get a lower envelope $e_{\min}(t)$. For instance, the neighbouring minima can be interconnected by straight line segments (linear interpolation). Analogously, an upper envelope $e_{\max}(t)$ is obtained from the maxima of a signal.
(iii) To compute the mean $m(t) = (e_{\max}(t) + e_{\min}(t))/2$.
(iv) To compute the so-called details $d(t) = x(t) - m(t)$. The meaning of the term is analogous to that used in the wavelet analysis (Sect. 6.4.2). The quantity $m(t)$ is called a *remainder*.
(v)   To perform the steps (i)–(iv) for the obtained details $d(t)$ and get new details $d(t)$ and a new remainder $m(t)$ (sifting procedure) until they satisfy two conditions: (1) the current remainder $m(t)$ is close to zero as compared with $d(t)$, (2) the number of extrema in $d(t)$ equals the number of its zeroes or differs from it by 1. One calls the resulting details $d(t)$ an "empirical mode" $f(t)$ or an "intrinsic mode function".
(vi) To compute a remainder, i.e. the difference between a signal and an empirical mode $m(t) = x(t) - f(t)$. To perform the steps (i)–(v) for the remainder $m(t)$ instead of the original signal $x(t)$. The entire procedure is stopped if $m(t)$ contains too few extrema.

The process is illustrated in Fig. 6.23 with a signal representing a sum of a sinusoid and two periodic signals with triangular profiles and different periods. The period of the first triangle wave is greater than that of the sinusoid and the period of the second triangle wave is less than that of the sinusoid. As a result of the above procedure, the original signal is decomposed into the sum of three components

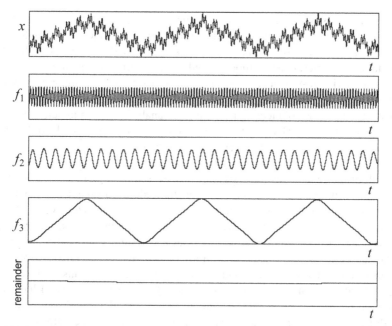

Fig. 6.23 Empirical mode decomposition (Huang et al., 1998). The *upper panel* shows an original signal. The next three panels show empirical modes. The first of them is a periodic triangular wave with a small period, the second one is a sinusoid and the third one is a triangular wave with larger period. The *lowest panel* shows a remainder, which exhibits a single extremum

(empirical modes) and a remainder, which is close to zero. An advantage of this technique over band-pass filtering is its adaptive character: it distinguishes modes based on the properties of a signal without the use of a *pre-selected* frequency band. In particular, it is more efficient in coping with non-stationary signals.

6.4.4 Stationarity Analysis

In general, *stationarity* of a process with respect to some property means constancy of that property in time. Definitions of wide-sense and narrow-sense stationarity are given in Sect. 4.1.3. Besides such a statistical stationarity related to the constancy of the distribution laws or their moments, one singles out dynamical stationarity, meaning constancy of an evolution operator (see also Sect. 11.1).

Majority of time series analysis techniques are based on the assumption of stationarity of an investigated process. However, multitude of real-world signals, including physiological, financial and others, look non-stationary. The latter results from processes, whose characteristic timescales exceed an observation time, or external events, which lead to changes in dynamics, e.g. adaptation in biological systems. Many characteristics calculated from a non-stationary time series appear meaningless or unreliable.

A lot of efforts were devoted to the problem of testing for stationarity. Previously, if *non-stationarity* was detected, a time series was rejected as useless for any further analysis or divided into segments sufficiently short to be considered as quasi-stationary. Later, many authors started to use information about the character of non-stationarity to study a process. There are situations when temporal variations in the properties of a process represent the most interesting contents of a time series. For instance, a purpose of electroencephalogram analysis is often to detect changes in the brain state. Such changes occur between different stages of sleep, between epileptic seizures and normal brain activity, and so on.

To check a process for stationarity based on a time series more or less reliably, the length of the time series must significantly exceed all timescales of interest. If components with characteristic timescales of the order of a time series length are present, then a process is typically recognised as non-stationary. However, a process may often be regarded stationary even if a time series length is considerably less than characteristic timescales of slow processes in a system. For instance, heart rate of a human being under relaxed conditions is as a rule homogeneous over time intervals of the order of several minutes. However, longer segments of data reveal new peculiarities arising due to slow biological rhythms. Since a usual 24-h electrocardiogram recording covers just a single cycle of a circadian (daily) rhythm, it is more difficult to consider it as stationary than longer or shorter recordings.

To detect non-stationarity, one uses the following basic approaches:

(i) Computation of a certain characteristic in moving window, i.e. in subsequent segments of a fixed length (it looks like a window for the consideration moves along the time axis). If the characteristic changes weakly and does not exhibit pronounced trends, then a time series is regarded stationary with respect to that characteristic, otherwise it is non-stationary. Examples of non-stationarity with respect to mean and variance are shown in Fig. 6.14d, e, g, h. In statistics there have been developed special techniques to test for stationarity with respect to the mean (Student's t-test, a non-parametric shift criterion, inversion criterion), the variance (Fisher's criterion, scattering criterion) and univariate distribution functions (Wilcoxon test) (Kendall and Stuart, 1979; Pugachev, 1979, 1984; von Mises, 1964). Based on the theory of statistical hypotheses testing, one may deny stationarity with respect to those characteristics at a given significance level (i.e. with a given probability of random error).

(ii) Comparison of characteristics in different time windows. One uses such characteristics as different statistical measures (criteria of χ^2, Cramer and Mises, Kolmogorov and Smirnov) (Kendall and Stuart, 1979; von Mises, 1964) and non-linear dynamics measures [cross-correlation integral (Cenis et al., 1991), cross-prediction error (Schreiber, 1997, 1999), distance between vectors of a dynamical model coefficients (Gribkov and Gribkova, 2000)]. Some approaches of such type are illustrated in Sect. 11.1.

One more recent approach, which often appears useful for the treatment of non-stationary signals (e.g. to detect drifts of parameters), is based on the analysis of *recurrences* in a reconstructed phase space (Facchini et al., 2005; Kennel, 1997; Rieke et al., 2002). *Recurrence plot* widely used in the *recurrence analysis* (Eckmann et al., 1987) is a diagram which shows time instants of close returns of an orbit to different states of a system. It is a very convenient tool to visualise a dynamical structure of a signal. Furthermore, recurrences in the space of model coefficients may be used to directly characterise non-stationarity and select quasi-stationary intervals of an observed signal (Sect. 11.1). Recurrence plots were introduced in Eckmann et al. (1987) and extended in many works, see the dissertation (Marwan, 2003) and the review (Marwan et al., 2007) for a detailed consideration. It is also worthwhile to note that the recurrence analysis extends possibilities of estimating dimensions, Lyapunov exponents and other dynamical invariants of a system (Marwan et al., 2007), which can be used in empirical modelling to select a model structure and validate a model.

6.4.5 Interdependence Analysis

Above, we have considered techniques for a scalar time series analysis. If one observes a vector time series, e.g., simultaneous measurements of two quantities $x(t)$ and $y(t)$, then opportunities to address new questions emerge. It is often important to reveal interdependence between $x(t)$ and $y(t)$ to get ideas about the presence and character of coupling between sources of the signals. Such information can be used in modelling as well. Thus, if there is a unique relationship between $x(t)$ and $y(t)$, then it is sufficient to measure only one of the quantities, since the other one does not carry new information. If there is certain interdependence, which is not unique, then it is reasonable to construct a model taking into account interaction between two sources of signals.

There are different approaches to the analysis of interdependencies. Historically, the first tools were *cross-correlation* and *cross-spectral analysis*. They are developed within the framework of mathematical statistics and attributed to the so-called linear time series analysis. *Cross-covariance function* is defined as covariance (Sect. 4.1.2) of x and y at time instants t_1 and t_2:

$$K_{x,y}(t_1, t_2) = E\left[(x(t_1) - E[x(t_1)])\,(y(t_2) - E[y(t_2)])\right]. \qquad (6.24)$$

For a stationary case, it depends only on the interval between the time instants:

$$K_{x,y}(\tau) = E\left[(x(t) - E[x])\,(y(t + \tau) - E[y])\right]. \qquad (6.25)$$

According to the terminology accepted mainly by mathematicians, normalised cross-covariance function is called *cross-correlation function* (CCF). The latter is a *correlation coefficient* between x and y. For a stationary case, the CCF reads

$$k_{x,y}(\tau) = \frac{E\left[(x(t) - E[x])\,(y(t + \tau) - E[y])\right]}{\sqrt{\mathrm{var}[x]\mathrm{var}[y]}}, \qquad (6.26)$$

where var$[x]$ and var$[y]$ are the variances of the processes $x(t)$ and $y(t)$. It always holds true that $-1 \le k_{x,y}(\tau) \le 1$. An absolute value of $k_{x,y}(\tau)$ reaches 1 in case of deterministically linear dependence $y(t+\tau) = \alpha\, x(t)+\beta$, while $k_{x,y}(\tau) = 0$ for statistically independent processes $x(t)$ and $y(t)$. However, if the processes are related uniquely, but non-linearly, then the CCF can be equal to zero (e.g., for $y(t) = x^2(t)$ and symmetric distribution of x about zero) and "overlooks" the presence of interdependence. Therefore, one says that the CCF *characterises a linear dependence* between signals. To estimate the CCF, one uses a usual formula for an empirical moment (Sect. 2.2.1).

There are multiple modifications and generalisations of the CCF. Thus, to characterise an interdependence between components of signals at a given frequency, rather than a total interdependence, one uses cross-spectral density and coherence (a normalised cross-spectral density). However, the estimation of the cross-spectrum and coherence is connected to greater difficulties compared to the estimation of an individual power spectrum (Bloomfield, 1976; Brockwell and Davis, 1987; Priestley, 1989). DFT-based estimators analogous to the periodogram estimator of the power spectrum have even worse estimation properties compared to the latter due to an estimation bias (Hannan and Thomson, 1971), observational noise effects (Brockwell and Davis, 1987), large estimation errors for the phase spectrum at small coherence (Prietley, 1989), etc. (Timmer et al., 1998). Cross-wavelet analysis further generalises characterisation of an interdependence by decomposing it in the time – frequency domain (Torrence and Compo, 1998). Analogous to the cross-spectrum and Fourier coherence, the cross-wavelet spectrum and wavelet coherence are estimated with greater difficulties, compared to the individual wavelet power spectrum (Maraun and Kurths, 2004).

To reveal non-linear dependencies, one uses generalisations of the correlation coefficient including Spearman's index of cograduation (Kendall and Stuart, 1979), correlation ratio (Aivazian, 1968; von Mises, 1964), mutual information function (Fraser and Swinney, 1986) and others.

The corresponding approaches are developed in non-linear dynamics along two directions. The first idea is to analyse mutual (possibly, non-linear) dependencies between state vectors $\mathbf{x}(t)$ and $\mathbf{y}(t)$ reconstructed from a time series via time-delay embedding or in other way (Sects. 6.1.2 and 10.1.2). The techniques rely upon the search for nearest neighbours in state spaces (Arnhold et al., 1999; Pecora et al., 1995) or construction of mutual predictive models (Schiff et al., 1996; Schreiber, 1999). If there is a unique dependence between simultaneous values of the state vectors $\mathbf{x}(t) = \mathbf{F}(\mathbf{y}(t))$, one speaks of *generalised synchronisation* (Rulkov et al., 1995; Pecora et al., 1995; Boccaletti et al., 2002). Fruitful approaches to quantification and visualisation of non-linear interrelations from observed data are based on the recurrence analysis (Sect. 6.4.4) and include cross-recurrence plots (Groth, 2005; Marwan and Kurths, 2002, 2004; Zbilut et al., 1998) and joint recurrence plots (Marwan et al., 2007).

The second idea is to analyse *only* the phases of observed signals. Since the phase is a very sensitive variable, an interaction between oscillatory systems often manifests itself as an interdependence between their phases, while the amplitudes may remain uncorrelated. If for two coupled self-sustained oscillators their unwrapped phase difference is constant $\phi_x(t) - \phi_y(t) = $ const, then one says that *phase synchronisation* takes place. This is a thresholdless phenomenon, i.e. it can be observed for arbitrarily weak coupling between systems if their own oscillation frequencies are arbitrarily close to each other (Afraimovich et al., 1989; Pikovsky et al., 2001) and noise is absent. If even weak noise is present, then the phase difference cannot be strictly constant and one considers a softened condition $\left|\phi_x(t) - \phi_y(t) - \text{const}\right| < 2\pi$. This is definition of 1:1 synchronisation. There also exists a higher order $m{:}n$ synchronisation defined by the condition $\left|m\phi_x(t) - n\phi_y(t) - \text{const}\right| < 2\pi$. For a considerable noise level, even the softened condition of the phase difference boundedness can be fulfilled only over a finite time interval. Then, one speaks of an effective synchronisation if that time interval significantly exceeds oscillation periods of both systems.

One introduces different numerical characteristics of phase interdependence often called *coefficients of phase synchronisation*. The most widespread among them is the so-called *mean phase coherence* (it has several names):

$$R_{m,n} = \sqrt{\left\langle \cos\left(m\phi_x(t) - n\phi_y(t)\right)\right\rangle^2 + \left\langle\sin\left(m\phi_x(t) - n\phi_y(t)\right)\right\rangle^2}, \qquad (6.27)$$

where angle brackets denote temporal averaging. It is equal to unity when the phase difference is constant (phase synchronisation) and to zero when each system exhibits oscillations with its own frequency independently of the other one. In the case of non-strict (e.g. due to noise) phase locking, the quantity $R_{m,n}$ can take an intermediate value and characterise a "degree of interdependence" between the phases. An example of efficient application of such a phase analysis to a medical diagnostic problem is given in Pikovsky et al. (2000).

In Chap. 12 and 13, we describe several techniques allowing to reveal and characterise "directional couplings" and their applications.

6.5 Experimental Example

This chapter is devoted to techniques and problems emerging at the starting stage of the modelling procedure and to acquisition and preliminary analysis of a time series (see Fig. 5.1). Instead of a summary, where we could say that probability of successful modelling rises with the amount and accuracy of prior knowledge about an object, let us discuss a real-world example of data acquisition and modelling. An object is a familiar circuit discussed above: a source of e.m.f. and resistors connected to it (Fig. 6.2a, b). However, a semiconductor sample (InSb, antimonide of indium) is included into it instead of the resistor R_v.[7] Under the room temperature, InSb is a conductor; therefore, experiments are carried out in liquid nitrogen at its boiling

---

[7] This narrow-band-gap semiconductor characterised by a large mobility of charge carriers is promising in respect of the increase in the operating speed of semiconductor devices.

Fig. 6.24 Experimental
set-up: (**a**) the scheme; (**b**) the
current I via a sample versus
the voltage U on a sample

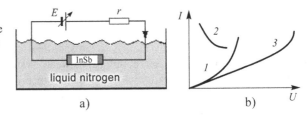

temperature of $-77\,°C$ (Fig. 6.24). Despite seeming simplicity of the object (at least, in comparison with living systems), we have selected it for an illustration due to diversity of its possible motions, mathematical tools needed for their description and difficulties in data acquisition depending on the modelling purposes.

What complexity can one expect from an oblong piece of a substance with two contacts connected to a source of constant e.m.f.? Depending on experimental conditions and exploited devices, one can observe diversity of processes ranging from a trivial direct current to oscillations at ultrahigh frequencies and even irradiation in the millimetre range of wavelengths. Registering of processes with characteristic frequencies ranging from 1 to 10^{12} Hz requires usage of different devices and analogue-to-digital converters. Moreover, starting from frequencies of about several gigahertz, digitising and, hence, modelling from time series are still technically impossible. Further, mathematical modelling of different phenomena mentioned above requires application of various tools ranging from algebraic equations to partial differential equations.

Under low voltages U at the contacts of the sample, it behaves like a usual resistor, i.e. one observes a direct current I and processes are appropriately modelled by an algebraic relationship, i.e. Ohm's law $I = U/R$, where R is a parameter meaning the resistance of the sample. The characterising quantities U and I are easily measured and can serve as observables. Linearity of their interdependence is violated with the rise in U due to the heating of the sample, whose conductance then rises as seen from the branch 1 of the characteristic (Fig. 6.24b).

With further increase in I and the heating intensity, boiling of the liquid transits from the bubble-boiling to the film[8]-boiling regime. This is reflected by the branch 2 on the dependency of the mean current on the mean voltage. Moreover, this is accompanied by the transition of the system "sample – source of e.m.f. – cooling liquid" to an oscillatory regime. Its characteristic frequencies range from less than 1 Hz to radio frequencies (dozens of kilohertz to several megahertz). Thus, one can use an ADC to record a time series. Oscillations at lower frequencies are determined by the arousal of multitude of bubbles at the surface, while higher frequencies are determined by the reactivity of the wires connected to the negative resistance of

---

[8] Liquid nitrogen in a thermos is under boiling temperature. At a low heat flow from the sample, small bubbles arise at its surface. They cover the entire surface at a more intensive heating. Thus, a vapour film is created, which isolates the sample from the cooling liquid.

the sample. Those phenomena can be modelled with stochastic differential equations. Yet, such a model contains quantities which are not directly related to the observables: heat flow from the sample, temperature dependence of its conductance, reactivity of the feed circuit.

A situation which is even more complex for observations and modelling arises if one tries to decrease heating influence and turns from a continuous power supply to a pulse regime when the voltage on the sample is supplied only during a short interval so that it has enough time to be cooled considerably until the next pulse. At that, without heat destruction of the sample and change in the boiling regime, one can achieve the voltages (the branch 3 in Fig. 6.24b) sufficient for a shock breakdown to start at local areas of the sample and created "pieces" of plasma to become a source of microwave radiation. To observe such processes, one needs a special equipment and microwave receivers, while the use of the current $I(t)$ and the voltage $U(t)$ as observables gets inefficient. The reason is that it is not clear how characterising quantities entering a DE-based model of the above oscillatory mechanism are related to such observables. More appropriate model variables would be the field strength in the sample and the concentration of the charge carriers. It is important to take into account a dependence of the drift velocity on the field strength and so on. An example of differential equations modelling such a dynamical regime is given in Bezruhcko and Erastova (1989). A more detailed modelling of the considered seemingly simple system requires to use non-linear partial differential equations.

This example cannot be regarded as an exclusive one in practice.

References

Afraimovich, V.S., Nekorkin, V.I., Osipov, G.V., Shalfeev, V.D.: Stability, structures, and chaos in nonlinear synchronisation networks. Gor'ky Inst. Appl. Phys. RAS, Gor'ky, (in Russian) (1989)

Aivazian, S.A.: Statistical Investigation of Dependencies. Metallurgiya, Moscow, (in Russian) (1968)

Anishchenko, V.S., Vadivasova, T.Ye.: Relationship between frequency and phase characteristics of chaos: two criteria of synchronization. J. Commun. Technol. Electron. **49**(1), 69–75 (2004)

Arnhold, J., Lehnertz, K., Grassberger, P., Elger, C.E.: A robust method for detecting interdependences: application to intracranially recorded EEG. Physica D. **134**, 419–430 (1999)

Astaf'eva N.M.: Wavelet analysis: basic theory and some applications. Phys. Uspekhi. **39**, 1085–1108 (1996)

Barenblatt, G.I.: Similarity, Self-similarity, Intermediate Asymptotics, Gidrometeoizdat, Leningrad (1982). Translated into English: Scaling, Self-Similarity, and Intermediate Asymptotics. Cambridge University Press, Cambridge (1996)

Bezruhcko, B.P., Erastova, E.N.: About possibility of chaotic solutions in model of narrow-band-gap semiconductor in ionisation by collision regime. Sov. Phys. *Semicon*duct. **23**(9), 1707–1709 (in Russian), (1989)

Blekhman, I.I.: Synchronisation in Nature and Technology. Nauka, Moscow (1981). Translated into English: ASME Press, New York (1988)

Blekhman, I.I.: Synchronisation of Dynamic Systems. Nauka, Moscow, (in Russian) (1971)

Bloomfield, P.: Fourier Analysis of Time Series: An Introduction. Wiley, New York (1976)

Boccaletti, S., Kurths, S., Osipov, G., Valladares, D., Zhou, C.: The synchronization of chaotic systems. Phys. Rep. **366**, 1–52 (2002)

Box, G.E.P., Jenkins, G.M.: Time Series Analysis. Forecasting and Control. Holden-Day, San Francisco (1970)

Brockwell, P.J., Davis, R.A.: Time Series: Theory and Methods. Springer, Berlin (1987)

Brown, R., Rulkov, N.F., Tracy, E.R.: Modeling and synchronizing chaotic systems from experimental data. Phys. Lett. A. **194**, 71–76 (1994)

Cecen, A.A., Erkal, C.: Distinguishing between stochastic and deterministic behavior in high frequency foreign exchange rate returns: Can non-linear dynamics help forecasting? Int. J. Forecasting. **12**, 465–473 (1996)

Čenis, A., Lasiene, G., Pyragas, K.: Estimation of interrelation between chaotic observable. Phys. D. **52**, 332–337 (1991)

Clemens, J.C.: Whole Earth telescope observations of the white dwarf star PG 1159-035 (data set E). In: Gerschenfeld, N.A., Weigend, A.S. (eds.) Time Series Prediction: Forecasting the Future and Understanding the Past. SFI Studies in the Science of Complexity, Proc. V. XV, pp. 139–150. Addison-Wesley, New York (1993)

Eckmann, J.-P., Kamphorst, S.O., Ruelle, D.: Recurrence plots of dynamical systems. Europhys. Lett. **5**, 973–977 (1987)

Facchini, A., Kantz, H., Tiezzi, E.: Recurrence plot analysis of nonstationary data: the understanding of curved patterns. Phys. Rev. E. **72**, 021915 (2005)

Fraser, A.M., Swinney, H.L.: Independent coordinates for strange attractors from mutual information. Phys. Rev. A. **33**, 1131–1140 (1986)

Frik, P., Sokolov, D.: Wavelets in astrophysics and geophysics. Computerra, vol. 8. Available at http://offline.computerra.ru/1998/236/1125/ (in Russian) (1998)

Gabor, D.: Theory of communication. J. Inst. Elect. Eng. (London). **93**, 429–459 (1946)

Gribkov, D., Gribkova, V.: Learning dynamics from nonstationary time series: analysis of electroencephalograms. Phys. Rev. E. **61**, 6538–6545 (2000)

Groth, A.: Visualization of coupling in time series by order recurrence plots. Phys. Rev. E. **72**, 046220 (2005)

Hamming, R.W.: Digital Filters, 2nd edn. Prentice-Hall, Englewood Cliffs, NJ (1983)

Herman, S.L.: Delmar's Standard Textbook of Electricity. Delmar Publishers, San Francisco (2008)

Huang, N.E., Shen, Z., Long, S.R.: The empirical mode decomposition and the Hilbert spectrum for nonlinear and non-stationary time series analysis. Proc. R. Soc. Lond. A. **454**, 903–995 (1998)

Hubner, U., Weiss, C.-O., Abraham, N.B., Tang, D.: Lorenz-like chaos in $NH_3 - FIR$ lasers (data set A). In: Gerschenfeld, N.A., Weigend, A.S. (eds.) Time Series Prediction: Forecasting the Future and Understanding the Past. SFI Studies in the Science of Complexity, Proc. V. XV, pp. 73–104. Addison-Wesley, New York (1993)

Jenkins, G., Watts, D.: Spectral Analysis and Its Applications. Holden-Day, New York (1968)

Judd, K., Mees, A.I. On selecting models for nonlinear time series. Phys. D. **82**, 426–444 (1995)

Kalashnikov, S.G.: Electricity. Nauka, Moscow, (in Russian) (1970)

Keller, C.F.: Climate, modeling, and predictability. Phys. D. **133**, 296–308 (1999)

Kendall, M.G., Stuart, A.: The Advanced Theory of Statistics, vols. 2 and 3. Charles Griffin, London (1979)

Kennel, M.B.: Statistical test for dynamical nonstationarity in observed time-series data. Phys. Rev. E. **56**, 316–321 (1997)

Koronovskii, A.A., Hramov, A.E.: Continuous Wavelet Analysis in Application to Nonlinear Dynamics Problems. College, Saratov (2003)

Koronovskii, N.V., Abramov, V.A.: Earthquakes: causes, consequences, forecast. Soros Educ. J. **12**, 71–78, (in Russian) (1998)

Kugiumtzis, D., Lingjaerde, O.C., Christophersen, N.: Regularized local linear prediction of chaotic time series. Phys. D. **112**, 344–360 (1998)

Lachaux, J.P., Rodriguez, E., Le Van Quyen, M., et al.: Studying single-trials of phase synchronous activity in the brain. Int. J. Bif. Chaos. **10**, 2429–2455 (2000)

Lean, J., Rottman, G., Harder, J., Kopp, G.: Source contributions to new understanding of global change and solar variability. Solar Phys. **230**, 27–53 (2005)

Lequarre, J.Y.: Foreign currency dealing: a brief introduction (data set C). In: Gerschenfeld, N.A., Weigend, A.S. (eds.) Time Series Prediction: Forecasting the Future and Understanding the Past. SFI Studies in the Science of Complexity, Proc. V. XV, pp. 131–137. Addison-Wesley, New York (1993)

Letellier, C., Le Sceller, L., Gouesbet, G., et al.: Recovering deterministic behavior from experimental time series in mixing reactor. AIChE J. **43**(9), 2194–2202 (1997)

Letellier, C., Macquet, J., Le Sceller, L., et al. On the non-equivalence of observables in phase space reconstructions from recorded time series. J. Phys. A: Math. Gen. **31**, 7913–7927 (1998)

Ljung, L.: System Identification. Theory for the User. Prentice-Hall, New York (1991)

Makarenko, N.G.: Embedology and neuro-prediction. Procs. V All-Russian Conf. "Neuroinformatics-2003". Part 1, pp. 86–148, Moscow, (in Russian) (2003)

Maraun, D., Kurths, J.: Cross wavelet analysis: significance testing and pitfalls. Nonlin. Proc. Geophys. **11**, 505–514 (2004)

Marwan, N., Kurths, J.: Nonlinear analysis of bivariate data with cross recurrence plots. Phys. Lett. A. **302**, 299–307 (2002)

Marwan, N., Kurths, J.: Cross recurrence plots and their applications. In: Benton, C.V. (ed.) Mathematical Physics Research at the Cutting Edge, pp. 101–139. Nova Science Publishers, Hauppauge (2004)

Marwan, N., Romano, M.C., Thiel, M., Kurths, J.: Recurrence plots for the analysis of complex systems. Phys. Rep. **438**, 237–329 (2007)

Marwan, N.: Encounters with Neighbours – Current Developments of Concepts Based on Recurrence Plots and Their Applications. Ph.D. Thesis. University of Potsdam, Potsdam (2003)

Misiti, R. et al. Wavelet Toolbox. User Guide for MatLab. The second edition (2000)

Monin, A.S., Piterbarg, L.I.: About predictability of weather and climate. In: Kravtsov, Yu.A. (ed.) Limits of Predictability, pp. 12–39. TsentrCom, Moscow, (in Russian) (1997)

Mudrov, V.L., Kushko, V.L.: Methods of Measurement Processing. Sov. Radio, Moscow, (in Russian) (1976)

Pecora, L.M., Carroll, T.L., Heagy, J.F.: Statistics for mathematical properties of maps between time series embeddings. Phys. Rev. E. **52**, 3420–3439 (1995)

Pikovsky, A.S., Rosenblum, M.G., Kurths, J.: Synchronisation. A Universal Concept in Nonlinear Sciences. Cambridge University Press, Cambridge (2001)

Press, W.H., Flannery, B.P., Teukolsky, S.A., Vetterling, W.T.: Numerical Recipes in C. Cambridge University Press, Cambridge (1988)

Priestley, M.B.: Spectral Analysis and Time Series. Academic, London (1989)

Pugachev, V.S.: Theory of Probabilities and Mathematical Statistics. Nauka, Moscow, (in Russian) (1979)

Pugachev, V.S.: Probability Theory and Mathematical Statistics for Engineers. Pergamon Press, Oxford (1984)

Rabiner, L.R., Gold, B.: Theory and Applications of Digital Signal Processing. Prentice Hall, New York (1975)

Rieke, C., Sternickel, K., Andrzejak, R.G., Elger, C.E., David, P., Lehnertz, K.: Measuring nonstationarity by analyzing the loss of recurrence in dynamical systems. Phys. Rev. Lett. **88**, 244102 (2002)

Rigney, D.R., Goldberger, A.L., Ocasio, W.C., et al.: Multi-channel physiological data: Description and analysis (data set B). In: Gerschenfeld, N.A., Weigend, A.S. (eds.) Time Series Prediction: Forecasting the Future and Understanding the Past. SFI Studies in the Science of Complexity, Proc. V. XV, pp. 105–129. Addison-Wesley, New York (1993)

Rulkov, N.F., Sushchik, M.M., Tsimring, L.S., Abarbanel, H.D.I.: Generalized synchronization of chaos in directionally coupled chaotic systems. Phys. Rev. E. **51**, 980–994 (1995)

Sadovsky, M.A., Pisarenko, V.F.: About time series forecast. In: Kravtsov, Yu.A. (ed.) Limits of Predictability, pp. 158–169. TsentrCom, Moscow (in Russian) (1997)

Sato, M., Hansen, J.E., McCormick, M.P., Pollack, J.B.: Stratospheric aerosol optical depths, 1850–1990. J. Geophys. Res. **98**, 22987–22994 (1993)

Schiff, S.J., So, P., Chang, T., Burke, R.E., Sauer, T.: Detecting dynamical interdependence and generalized synchrony through mutual prediction in a neural ensemble. Phys. Rev. E. **54**, 6708–6724 (1996)

Schreiber, T.: Detecting and analyzing nonstationarity in a time series using nonlinear cross predictions. Phys. Rev. Lett. **78**, 843–846 (1997)

Schreiber, T.: Interdisciplinary application of nonlinear time series methods. Phys. Rep. **308**, 3082–3145 (1999)

Sena, L.A.: Units of Physical Quantities and Their Dimensions. Nauka, Moscow (1977). Translated into English: Mir Publ., Moscow (1982)

Soofi, A.S., Cao, L. (eds.): Modeling and Forecasting Financial Data: Techniques of Nonlinear Dynamics. Kluwer, Dordrecht (2002)

Timmer, J., Lauk, M., Pfleger, W., Deuschl, G.: Cross-spectral analysis of physiological tremor and muscle activity. I. Theory and application to unsynchronized elecromyogram. Biol. Cybern. **78**, 349–357 (1998)

Torrence, C., Compo, G. A prectical guide to wavelet analysis. Bull. Amer. Meteor. Soc. **79**, 61–78 (1998)

Trubetskov, D.I.: Oscillations and Waves for Humanitarians. College, Saratov, (in Russian) (1997)

von Mises, R.: Mathematical Theory of Probability and Statistics. Academic Press, New York (1964)

Yule, G.U.: On a method of investigating periodicities in disturbed series, with special reference to Wolfer's sunspot numbers. Phil. Trans. R. Soc. London A. **226**, 267–298 (1927)

Zbilut, J.P., Giuliani, A., Webber, C.L.: Detecting deterministic signals in exceptionally noisy environments using cross-recurrence quantification. Phys. Lett. A. **246**, 122–128 (1998)

Chapter 7
Restoration of Explicit Temporal Dependencies

In the simplest formulation, modelling from a time series is considered as restoration of an explicit temporal dependence $\eta = f(t, \mathbf{c})$, where f is a certain function and $\mathbf{c}$ is the P-dimensional vector of model parameters. Such a problem setting is considered in the theory of function approximation (Akhieser, 1965) and mathematical statistics (Aivazian, 1968; Hardle, 1992; Seber, 1977). It can be interpreted as drawing a curve *through* experimental data points on the plane (t, η) or *near* those points (Fig. 7.1). A capability of solving this problem determines to a significant extent the success of modelling in more complex situations discussed in Chaps. 8, 9 and 10.

Below, two different formulations of the problem are discussed. Under the first formulation (Sects. 7.1 and 7.4.2), connection between the quantities t and η, i.e. a function $f(t, \mathbf{c})$, is a priori known up to the value of the parameter vector $\mathbf{c}$. The values of the parameters are of interest if they make physical sense and cannot be measured directly. The problem is *to estimate the parameters* as accurately as possible. Under the second formulation (Sects. 7.2 and 7.4.1), the purpose of modelling is to predict the value of η at a given time instant t, i.e. one needs to find a function f, which provides the forecast with as small error as possible. The form of f is a priori unknown, i.e. this problem setting can be considered as a kind of "black box" modelling.

In this chapter, we introduce a number of ideas and terms important for the entire further consideration. The most popular techniques for parameter estimation are described and compared in Sect. 7.1. The concepts of approximation, regression, interpolation and extrapolation are introduced in Sect. 7.2. Selection of a model class and "model overfitting" problem are considered in Sect. 7.2 as well. Diagnostic check of a model is discussed in Sect. 7.3. Applications of the models $\eta = f(t, \mathbf{c})$ to prediction and numerical differentiation are described in Sect. 7.4.

7.1 Parameter Estimation

We start with the deterministic setting of the "transparent box" problem, where a model structure is a priori known and only concrete values of parameters are unknown. An original process reads as

B.P. Bezruchko, D.A. Smirnov, *Extracting Knowledge From Time Series*, Springer Series in Synergetics, DOI 10.1007/978-3-642-12601-7_7,
© Springer-Verlag Berlin Heidelberg 2010

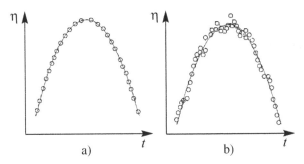

Fig. 7.1 Illustrations to the problem of drawing a curve of a given kind via experimental data points on a plane (*circles*): (**a**) no noise, it is simple to select a functional form for a curve; (**b**) data points do not lie precisely on a simple curve due to either random interference or complex character of the dependence so that the problem is to describe the observed dependence approximately with a simple function

$$\eta = f(t, \mathbf{c}_0), \qquad (7.1)$$

where the form of the function f is known and the value $\mathbf{c}_0$ of the parameter vector is unknown. To construct a model in the form $\eta = f(t, \mathbf{c})$, one should find such values of P components of the parameter vector $\mathbf{c}$ that the plot of f would go exactly through the experimental data points (t_i, η_i), $i = 1, \ldots, N$. The solution is found by solving the set of equations

$$\eta_i = f(t_i, c_1, \ldots, c_P), \ i = 1, \ldots, n, \qquad (7.2)$$

where n observation instants are chosen from the total set of N available instants. If f is linear with respect to the parameters, then the system (7.2) at $n = P$ has generically a unique solution, which can be found with any well-developed techniques (Golub and Van Loan, 1989; Press et al., 1988; Samarsky, 1982). Difficulties arise if matrix of the system is degenerate or ill-conditioned, i.e. the problem is ill-posed or ill-conditioned (Sect. 5.3). Then, one says that the parameters are *non-identifiable* from the given data set. Sometimes, such a difficulty can be eliminated if some observation instants t_i in Eq. (7.2) are replaced by other instants. If the number of data points N in a time series is greater than P, the remaining $N - n$ data points can be used for model validation. If N is less than P, the problem has no unique solution, i.e. it is ill-posed.

In the case of non-linear dependence of the function f on the parameters $\mathbf{c}$, the set of equations (7.2) is solved with iterative numerical techniques (Dennis and Schnabel, 1983; Kalitkin, 1978; Press et al., 1988; Samarsky, 1982). For instance, according to the widely used Newton technique, one first makes a *starting guess* $\mathbf{c}^{(0)}$ for a sought parameter vector. Then, one gets a correction $\Delta\mathbf{c}^{(0)}$ by solving a linearised problem and, thereby, generates a new approximation $\mathbf{c}^{(1)} = \mathbf{c}^{(0)} + \Delta\mathbf{c}^{(0)}$. Such operations are iterated until the process converges to a solution $\hat{\mathbf{c}}$ with a given precision. One finds a single solution to Eq. (7.2) for a single starting guess. If the set of equations (7.2) has many solutions, one needs to find all of them to select the "true" value of $\mathbf{c}$. Finding all solutions is a principal difficulty, since there is no general technique assuring it, even though many practically efficient algorithms are developed (Dennis and Schnabel, 1983). The problem is not very complicated if the number of unknown quantities is small. Similar to the linear case, it is often sufficient to take $n = P$ or n slightly greater than P (depending on the character of non-linearity).

In a stochastic case, an original process reads as

$$\eta = f(t, \mathbf{c}_0) + \xi(t), \tag{7.3}$$

where ξ is a zero-mean random process (noise). Here, one speaks of getting a statistical estimate $\hat{\mathbf{c}}$ as close to the true value $\mathbf{c}_0$ as possible, rather than of the accurate determination of the parameter values. Usually, $\xi(t_i)$ are assumed to be independent identically distributed random quantities. A probability distribution law for ξ may be unknown. Then, one either assumes that the probability density function $p(\xi)$ coincides with a certain well-known law (e.g. Gaussian, Laplacian, uniform on a finite interval) or uses a universal estimation technique applicable to a wide range of distribution laws.

7.1.1 Estimation Techniques

A set of possible estimation techniques is infinite. This is a set of all functions (estimators), which get "input" data $(t_i, \eta_i), i = 1, \ldots, N$ and give a value of $\hat{\mathbf{c}}$ as "output" (Ibragimov and Has'minskii, 1979; Pugachev, 1979, 1984). Within this multitude, there are several widely used techniques, which are characterised by high effectiveness for a wide range of processes, with simplicity of implementation, etc. Below, we consider some of them. Methodologically, it is useful to distinguish between the two situations:

(i) The quantity t is non-random and observation instants are fixed. This situation is encountered most often. It has got even a special name "Gauss and Markov scheme" (Vapnik, 1979).
(ii) The quantity t is random and observation instants are selected independently according to some distribution law $p_0(t)$.

7.1.1.1 "Simple Averaging" Technique

Let the time series length N be such that the time series could be divided into M parts, each of which contains P data points. Let us proceed as if the noise were absent, i.e. solve Eq. (7.2) requiring exact equalities $\eta_i = f(t_i, \mathbf{c})$ for each kth part of the time series. Let us assume that the set (7.2) has a unique solution $\hat{\mathbf{c}}_k$ for each part of the time series. The final estimate is found via simple averaging:

$$\hat{\mathbf{c}} = \frac{1}{M} \sum_{k=1}^{M} \hat{\mathbf{c}}_k.$$

7.1.1.2 Statistical Moments Technique

Strictly speaking, it concerns the situation where t is a random quantity and $f(t, \mathbf{c})$ is an algebraic polynomial. Let us describe the technique with an example of an original process:

$$\eta = c_1 + c_2 t + c_3 t^2 + \xi(t). \tag{7.4}$$

First, one finds the expectations of both sides of Eq. (7.4). Recalling that $E[\xi] = 0$, one gets

$$E[\eta] = c_1 + c_2 E[t] + c_3 E[t^2]. \tag{7.5}$$

Such manipulations are justified only if the quantity t has a necessary number of finite moments (in particular, up to the second-order moments in the example considered). If the distribution $p_0(t)$ exhibits "heavy tails", i.e. decreases according to a power law for $t \to \infty$, then already the second moment of the quantity t may not exist.

Then, one multiplies both sides of Eq. (7.4) by η and again finds the expectations:

$$E[\eta^2] = c_1 E[\eta] + c_2 E[\eta t] + c_3 E[\eta t^2]. \tag{7.6}$$

Similarly, one multiplies both sides of Eq. (7.4) by t and finds the expectations:

$$E[\eta t] = c_1 E[t] + c_2 E[t^2] + c_3 E[t^3]. \tag{7.7}$$

If the values of the statistical moments $E[\eta]$, $E[\eta t]$ and others entering Eqs. (7.5), (7.6) and (7.7) were known, then one could precisely find exact values of the parameters c_1, c_2, c_3 by solving Eqs. (7.5), (7.6) and (7.7) with respect to c_1, c_2, c_3. In data analysis, the theoretical moments are unknown, but by replacing them with their estimators (the sample moments, see Sect. 2.2.1), i.e. substituting $\langle \eta \rangle = \frac{1}{N} \sum_{i=1}^{N} \eta_i$ instead of $E[\eta]$ and so on, one gets a set of equations

$$\begin{aligned} c_1 + c_2 \langle t \rangle + c_3 \langle t^2 \rangle &= \langle \eta \rangle, \\ c_1 \langle \eta \rangle + c_2 \langle \eta t \rangle + c_3 \langle \eta t^2 \rangle &= \langle \eta^2 \rangle, \\ c_1 \langle t \rangle + c_2 \langle t^2 \rangle + c_3 \langle t^3 \rangle &= \langle \eta t \rangle, \end{aligned} \tag{7.8}$$

and finds an estimate $\hat{\mathbf{c}}$ as its solution.

The technique can be generalised to the case when f is not an algebraic polynomial. The values of parameters should then be found by solving non-linear equations containing quantities of the type $\langle f(t, \mathbf{c}) \rangle$ rather than the sample moments. It is more problematic in practice. If t is non-random, interpretation of the values $\langle t \rangle$, $\langle t^2 \rangle$ and others as estimators of statistical moments is impossible. Still, the technique can be applied, but one should remember that the name "statistical moments" is no longer completely appropriate.

7.1.1.3 Maximum Likelihood Technique

In contrast to the two previous techniques, to apply the ML technique one must know the distribution law $p(\xi)$, probably up to the values of some parameters. The ML-principle is described in Sect. 2.2.1 in application to the estimation of parameters of a univariate distribution. Everything is analogous to the problem considered here. An observed time series $\{\eta_1, \ldots, \eta_N\}$ is a random vector. For non-random instants t_i and random quantities ξ_i independent of each other, the *likelihood function* takes the form

$$L(\mathbf{c}) = \prod_{i=1}^{N} p(\eta_i - f(t_i, \mathbf{c})). \tag{7.9}$$

It depends on the parameters $\mathbf{c}$. The ML principle is realised via maximisation

$$\ln L(\mathbf{c}) = \sum_{i=1}^{N} \ln p(\eta_i - f(t_i, \mathbf{c})) \rightarrow \max. \tag{7.10}$$

In other words, the value of the N-dimensional probability density function at the observed "point" $\{\eta_1, \ldots, \eta_N\}$ is maximised.

If t is random and characterised by a distribution law $p_0(t)$, then almost nothing changes: each multiplier in Eq. (7.9) is multiplied by $p_0(t_i)$ and under the condition that $p_0(t)$ is independent of $\mathbf{c}$, the ML estimators are found again from condition (7.10).

A concrete form of the likelihood function depends on the distribution law $p(\xi)$. In general, finding ML estimates is a problem of multidimensional non-linear optimisation, which is solved with iterative techniques (Dennis and Schnabel, 1983). For three widespread distribution laws $p(\xi)$, the ML technique reduces to other well-known techniques: the least squares technique, the least absolute values technique and the least maximal deviation technique.

7.1.1.4 Least Squares Technique

If ξ is distributed according to the Gaussian law $p(\xi) = \left(1 \big/ \sqrt{2\pi\sigma_\xi^2}\right) \exp\left(-\xi^2 \big/ 2\sigma_\xi^2\right)$, which is appropriate to describe measurement errors under stable conditions

(Vapnik, 1979), then the likelihood function reads as

$$\ln L(\mathbf{c}) = -\frac{N \ln\left(2\pi\sigma_\xi^2\right)}{2} - \frac{1}{2\sigma_\xi^2} \sum_{i=1}^{N} (\eta_i - f(t_i, \mathbf{c}))^2. \qquad (7.11)$$

Its maximisation over $\mathbf{c}$ is tantamount to the minimisation

$$S(\mathbf{c}) = \sum_{i=1}^{N} (\eta_i - f(t_i, \mathbf{c}))^2 \to \min. \qquad (7.12)$$

This is the least squares (LS) technique. Geometrically, it means that the parameter estimates are chosen so to minimise the sum of the *squared vertical distances* from the experimental data points on the plane (t, η) to the plot of $f(t, \mathbf{c})$, see Fig. 7.2.

7.1.1.5 Least Absolute Values Technique

If ξ is distributed according to the Laplacian law $p(\xi) = (1/2\Delta)\exp(-|\xi|/\Delta)$, which is appropriate to describe measurement errors under unstable conditions (Vapnik, 1979), then the likelihood function reads as

$$\ln L(\mathbf{c}) = -N \ln(2\Delta) - \frac{1}{\Delta} \sum_{i=1}^{N} |\eta_i - f(t_i, \mathbf{c})|. \qquad (7.13)$$

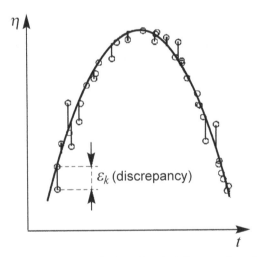

Fig. 7.2 Parameter estimation via minimisation of vertical distances from the experimental data points to the plot of a model function (some of these discrepancies are shown by *vertical lines*). One minimises either the sum of their squared values, the sum of their absolute values or their maximal value

Maximisation over $\mathbf{c}$ is equivalent to the minimisation of the sum of the absolute values of the vertical distances

$$S'(\mathbf{c}) = \sum_{i=1}^{N} |\eta_i - f(t_i, \mathbf{c})| \rightarrow \min. \tag{7.14}$$

7.1.1.6 Least Maximal Deviation Technique

If ξ is distributed uniformly on an interval of width Δ:

$$p(\xi) = \frac{1}{2\Delta}, \; -\Delta \leqslant \xi \leqslant \Delta,$$

which is appropriate to describe round-off errors of numerical computations (Vapnik, 1979), then the likelihood function reads as

$$L(\mathbf{c}) = \frac{1}{(2\Delta)^N} \prod_{i=1}^{N} \Theta(\Delta - |\eta_i - f(t_i, \mathbf{c})|), \tag{7.15}$$

where $\Theta(x) = 1$, $x > 0$ and $\Theta(x) = 0$, $x < 0$. Maximisation of Eq. (7.15) over $\mathbf{c}$ for unknown Δ is equivalent to the minimisation of the maximal vertical distance

$$S''(\mathbf{c}) = \max_{1 \leqslant i \leqslant N} |\eta_i - f(t_i, \mathbf{c})| \rightarrow \min. \tag{7.16}$$

Each of the techniques (7.12), (7.14) and (7.16) coincides with the ML technique only for the corresponding distribution law $p(\xi)$. Each of them may appear superior to the others even if the properties of $p(\xi)$ are varied in a certain way. Therefore, each of the three estimators has its own practical value, rather than being only a particular case of the ML approach.

7.1.2 Comparison of Techniques

Performance of the techniques can be assessed from different properties of the estimators. Let us compare the above techniques in the most popular way when the best estimator is that with the least value of the mean-squared error $E[\hat{c} - c_0]^2$. The latter is the sum of the estimator variance and squared bias as given by Eqs. (2.16) and (2.17) in Sect. 2.2.1:

$$E[\hat{c} - c_0]^2 = \sigma_{\hat{c}}^2 + (E[\hat{c}] - c_0)^2. \tag{7.17}$$

7.1.2.1 The Case of Linear Estimators

Let us start with a simple example where several compared estimators appear linear with respect to η_i (such estimators are called *linear*) and unbiased. To get such estimators, it is necessary for a function f to be linear with respect to c and for the properties of the noise ξ to be independent of time t. For instance, the conditions are satisfied for a process

$$\eta = c_0 t + \xi, \tag{7.18}$$

where ξ is distributed according to the Gaussian law with some variance σ_ξ^2. We apply the simple averaging, the statistical moments and the least squares techniques and compare the results.

Geometrical sense of the simple averaging technique is as follows. One draws a straight line through a point t_i, η_i and the origin and determines its angular coefficient $\hat{c}_i = \eta_i/t_i$. Then, one finds an averaged value over all $\hat{c}_i$ (Table 7.1). If some observation instants are close to zero, then the deviations of $\hat{c}_i$ from c_0 (equal to ξ_i/t_i) can be very large, which can lead to huge scattering of the estimator values. The estimator may not have even finite expectation and variance (Table 7.1). It is easy to show that the estimator is unbiased if it has a finite expectation.

Geometrical sense of the statistical moments technique is as follows. One finds the sample mean values of the coordinates $\langle t \rangle_N$ and $\langle \eta \rangle_N$. Then, one draws a straight line through the origin and the obtained "mean point" ("centre of mass") of the cloud of experimental points and takes its angular coefficient as an estimator $\hat{c}$. If its expectation is finite, it is unbiased. The variance of this estimator is less than that for the simple averaging estimator (Table 7.1, Fig. 7.3). The difference is especially large if some values of t are very close to zero.

As for the LS estimator, it has surely finite expectation and variance. It is unbiased and its variance is less than that for the two estimators above (Fig. 7.3).

For all the three techniques, the variance of the noise ξ can be estimated as the sample variance of the *model residual errors* $\hat{\sigma}_\xi^2 = S(\hat{c})/N$. *Residual errors* (also called just *residuals*) are the values $\varepsilon_i = \eta_i - f(t_i, \hat{c})$, where $\hat{c}$ is the parameter estimate obtained. In the case of Gaussian ξ, this estimator of σ_ξ^2 obtained from

Table 7.1 Results of the parameter estimation with different techniques for example (7.18). Variances of the estimators are shown for the case of random t. For non-random t, it is sufficient to replace its "expectation" with a sample mean

Techniques	Expression for $\hat{c}$	Variance of $\hat{c}$
Simple averaging	$(1/N) \sum_{i=1}^{N} \eta_i/t_i$	$\left(\sigma_\xi^2/N\right) E[1/t^2]$
Statistical moments	$\sum_{i=1}^{N} \eta_i \Big/ \sum_{i=1}^{N} t_i$	$\left(\sigma_\xi^2/N\right) E\left[1/\langle t \rangle_N^2\right]$
Least squares	$\sum_{i=1}^{N} \eta_i t_i \Big/ \sum_{i=1}^{N} t_i^2$	$\left(\sigma_\xi^2/N\right) E\left[1/\langle t^2 \rangle_N\right]$

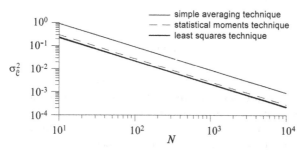

Fig. 7.3 Variances of different estimators versus the time series length in a double logarithmic scale for example (7.18) with $c_0 = 0.5$, random t distributed uniformly in the interval $[0.1, 1.1]$ and ξ distributed according to the Gaussian law with unit variance. The LS estimator has the least variance. The statistical moments-based estimator is the second best. The largest variance is observed for the simple averaging technique. The scaling law $\sigma_{\hat{c}}^2 \propto 1/N$ is the same for all the three estimators

the LS technique coincides with the ML estimator of σ_ξ^2. In all the three cases, the variance of $\hat{c}$ decreases with N. Recalling unbiasedness of the estimators, it means that the estimators become more accurate for greater N. The scaling law for the variance in the case of observation instants t_i randomly drawn from some distribution is $\sigma_{\hat{c}}^2 \propto 1/N$ (Fig. 7.3). However, it may differ for other selection of the observation instants t_i.

7.1.2.2 Asymptotic Properties

Let us consider properties of the estimators under the increase in sample size, $N \to \infty$, and start with the case where t is a random quantity with finite expectation and variance. Then, the ML estimator is an *asymptotically unbiased, asymptotically efficient* and *consistent estimator*. In other words, this is practically the best estimator for a sufficiently large N. The LS technique and statistical moments technique also provide consistent estimators under certain conditions (Ibragimov and Has'minskii, 1979; Vapnik, 1979).

If observation instants t_i are not random, the situation changes. In quite a typical case, the consecutive t_i are separated with a fixed time step $t_i = t_0 + i\Delta t$ so that an increase in the time series length N corresponds to an increase in the duration of the observation interval. Then, each new observation can make a contribution to the likelihood function (7.9) different from the previous observations. The likelihood function may well depend stronger and stronger on η_i with increasing i, i.e. the value of L in Eq. (7.9) becomes *non-stationary* with respect to i. This is reflected by the so-called *partial Fisher information*: In the case of a single scalar parameter, this is a value $I_i = \left(\partial f(t_i, c) \big/ \partial c \big|_{c=c_0} \right)^2$. The greater the I_i, the stronger the influence of an observation η_i on the value of the ML estimator. Such a strong influence takes place for the examples

$$\eta = \sin(c_0 t) + \xi \tag{7.19}$$

and

$$\eta = \sin(\exp(c_0 t)) + \xi, \tag{7.20}$$

where the single parameter is estimated from an equidistant time series with fixed Δt. In the case of Eq. (7.19), the ML estimator remains asymptotically unbiased, while its variance decreases *faster* than in the case of random t_i : $\sigma_{\hat{c}}^2 \propto 1/N^3$, i.e. the estimator is even more accurate for sufficiently large N. In the case of Eq. (7.20), the likelihood function loses smoothness and becomes too "jagged" at $N \to \infty$ (see, e.g., Fig. 8.3a in Sect. 8.1.2). Therefore, one can prove nothing about its asymptotic properties. At *finite* N, the variance of the estimator decreases very quickly with N, approximately via an exponential law (Pisarenko and Sornette, 2004). However, finding the global maximum of Eq. (7.9) is practically impossible for large N, since the cost function has too many local maxima. The same problems are encountered when one uses the techniques of the least squares, the least absolute values and the least maximal deviation.

7.1.2.3 Finite Sample Size

Taking into account the above consideration, one must often use only moderate time series lengths. However, in such a case the ML approach does not assure the best estimator properties. Therefore, a special attention is paid to the performance of the techniques for finite-size samples. For non-random t, Gaussian distribution of ξ and linear dependence of f on $\mathbf{c}$, the LS estimator (coinciding then with the ML estimator) is unbiased and efficient, i.e. the best one in the class of unbiased estimators. Under arbitrary distribution law for ξ, the LS technique (no longer equivalent to the ML technique) gives the best estimators in the narrower class of linear and unbiased estimators. However, there are non-linear biased estimators, which exhibit smaller error (7.17) than the LS estimators (Vapnik, 1979).

7.1.2.4 Optimal Estimation

In the simple example considered above, properties of the estimators do not depend on the true values of the parameters c_0. In general, such a dependence may take place: an estimator can exhibit small error for some value of c_0 and large errors for others. It is desirable to have a technique which gives the best estimates for any value of c_0 from a set of possible values of $\mathbf{C}$. However, such *uniformly the best* technique in general does not exist. There are estimators optimal in some sense and, hence, useful in a corresponding practical situation. Two popular ideas are minimax and Bayesian approaches.

Minimax Principle

It is often unknown which values of c_0 from a set $\mathbf{C}$ one can encounter in practice more often. If a large estimation error is extremely undesirable even for a single rarely encountered value of c_0, then one needs a technique providing the least worst-case error:

$$\sup_{c_0 \in C} \int (\hat{c} - c_0)^2 p(\hat{c} | c_0) d\hat{c} \to \min, \qquad (7.21)$$

where $p(\hat{c} | c_0)$ is the probability distribution density for an estimator $\hat{c}$ at a given c_0. One minimises an estimator error corresponding to the least favourable case. However, if such a value of c_0 is never encountered in practice, then the technique based on Eq. (7.21) appears "too careful" and gives an estimator, which is not the most accurate among others possible.

Bayesian Principle

As discussed in Sect. 2.2.1, true value of parameters c_0 can be considered as random quantities. Since some true values c_0 can be encountered more often than others, one may need a technique which gives minimal *average* estimator error, i.e.

$$\int (\hat{c} - c_0)^2 p(\hat{c} | c_0) p(c_0) \, d\hat{c} \, dc_0 \to \min. \qquad (7.22)$$

Such an approach rarely gives large errors (only for almost improbable c_0) and often gives small errors (for the most probable and, hence, most frequently encountered values of c_0).

7.1.2.5 Robust Estimation

The LS technique and the least absolute value technique are superior to the ML approach in the following sense. To apply the former two techniques, one does not have to know a priori the distribution law of the noise ξ. As mentioned above, these two techniques give optimal results if the noise is distributed according to Gaussian or Laplacian law, respectively. One may pose a question: Whether "good" properties of an estimator (e.g. small estimator error) are maintained under some variations in the noise distribution. If yes, which variations are allowable? If not, how can one construct an estimator stable (robust) with respect to some variations? This is a subject of the robust estimation theory, see, e.g. Vapnik (1979, 1995) and references therein.

Without going into rigorous formulations, we just indicate that both techniques are robust for sufficiently wide classes of distributions. Namely, the LS estimator is robust in the class of distributions, whose variance does not exceed a given finite value. The least absolute value estimator is robust in the class of distribution laws with $p(0) \geq \Delta > 0$, where Δ is a given finite number. One can construct other estimators which are robust in wider classes (Vapnik, 1995).

7.1.2.6 Concluding Remarks

For a sufficiently long time series, noise-free values of t_i, a priori known distribution law of the noise ξ and a sufficiently "good" likelihood function (so that one could find its global maximum), the best parameter estimators are given by the ML approach.

If the distribution law of the noise is unknown and the function f is linear with respect to the estimated parameters, the LS estimator and least absolute value estimator are the best in the corresponding classes of distributions. However, implementation of the LS technique is the simplest one, since an optimisation problem appears linear in contrast to other approaches. The LS technique is also often applied in the case of non-linear dependency of f on the parameters, even though one should then solve a more complicated problem of non-linear optimisation.

In total, the statistical moments technique is inferior to the LS technique with respect to the estimator accuracy. As a rule, the simple averaging technique is even worse. However, there are situations where the latter techniques have their own advantages (Pisarenko and Sornette, 2004).

We have only briefly touched on the questions of robust and optimal estimation, since the corresponding problems are more complicated and are still rarely formulated in modelling from time series.

7.2 Approximation

7.2.1 Problem Formulation and Terms

Similar to Sect. 7.1, we consider two situations: deterministic and stochastic. In the "deterministic" case, an original process is

$$\eta = F(t). \tag{7.23}$$

The form of the "true" function F is unknown so that one must find a model function f approximating $F(t)$ as accurately as possible over a given range of t. Approximation of $F(t)$ with another function $f(t)$ on an interval $[a, b]$ is the central problem in the theory of approximation (Akhieser, 1965). In a wide sense, *approximation* is a *replacement of an object with another one, which is in some sense close to the original.*[1] If a model function f obtained with the use of the observed

---

[1] An example. In the Ptolemaic astronomy, motions of celestial bodies relative to the Earth are approximated with combinations of motions along circumferences (epicycles), whose centres move along other circumferences around the Earth. To provide high accuracy in predictions of the future locations of planets on the coelosphere, which was important for sea navigation, a very sophisticated system of epicycles was developed. That work stimulated the development of the spherical geometry in the Middle Ages. When the geocentric system got an alternative, the heliocentric system of Copernicus, the latter one was inferior to its predecessor with respect to the prediction

data is applied to compute the values of η (i.e. to replace F) at intermediate time instants $t_1 < t < t_N$, then one speaks of *interpolation* of the dependency $\eta(t)$. If an approximating function is used outside the observed domain ($t < t_1$, $t > t_N$), then one speaks of *extrapolation* of the dependency $\eta(t)$.[2] In particular, extrapolation to the future ($t > t_N$) is the *forecast* in a narrow sense.

In the "stochastic" case, an original process is given as

$$\eta = F(t) + \xi(t), \tag{7.24}$$

where the form of the function F is a priori unknown and ξ is the zero-mean random quantity statistically independent of t, i.e. its properties do not depend on t. To characterise such a situation, one uses some additional concepts as compared with the deterministic case. Complete information about the original process is contained in the conditional probability density function $p_0(\eta|t)$ at a given t, but a "true" function $p_0(\eta|t)$ is a priori unknown. To get a model, one can aim at finding some law $p(\eta|t)$, which approximates $p_0(\eta|t)$ as accurately as possible. However, restoration of the entire distribution law from a finite sample is typically too problematic. Moreover, such a complete information is not always necessary. It is often enough to know the conditional mean $E[\eta|t]$ of the quantity η at a given t and the scattering of the values of η around $E[\eta|t]$ characterised by the conditional variance. The dependence of the conditional mean $E[\eta|t]$ on t is a deterministic function which is called *regression*.[3] For the case of Eq. (7.24), one gets $E[\eta|t] = F(t)$, since $E[\xi] = 0$, i.e. $F(t)$ is the regression. The regression predicts η at a given t most accurately, i.e. with the least mean-squared error:

$$\varepsilon^2(F) = \int (\eta - F(t))^2 p(\eta|t)d\eta = \min_f \varepsilon^2(f).$$

In empirical modelling, one usually aims at finding a function $f(t)$ approximating the "true" regression $F(t)$ as accurately as possible. This is called "restoration of regression" or "estimation of regression".

---

accuracy. With the further development of ideas and models of the heliocentric system, motions of planets got better described with it and it replaced the geocentric theory.

[2] Mathematical encyclopaedic dictionary gives a wider definition of interpolation as *an approximate or accurate finding of some quantity from its known separate values* (Mathematical dictionary, 1988). Under such a definition, interpolation covers even the notion of approximation. In the case of extrapolation, classical mathematics uses, vice versa, somewhat narrower meaning as *a continuation of a function outside its domain such that the continued function (as a rule, an analytic one) belongs to a given class*. We employ the definitions given in the main text, which are widely used by specialists in numerical analysis.

[3] The term was first used by English statistician F. Galton (1866). He studied how height of children Y depends on the height of their parents X and found the following. If the height of parents exceeds an average height of people by b, then the height of their children exceeds the average height by less than b. This phenomenon was called regression, i.e. "backward motion". Therefore, dependence of the conditional mean of Y on X was also called regression.

Thus, according to both formulations, one faces the problem of approximation of a function $F(t)$ based on the observed data. As a rule, a model is sought in a certain class $f(t, \mathbf{c})$ so that one must find the value of a P-dimensional parameter vector $\mathbf{c} = \hat{\mathbf{c}}$, which provides maximal closeness of the model function f to a true function F (Sect. 7.2.2).[4] At that, it is very important to select an optimal model size (Sect. 7.2.3) and an appropriate class of approximating functions (Sect. 7.2.4). Sometimes, one manages to do so based on the visual inspection of an observed signal and the selection of elementary model functions with similar plots. However, in general one performs approximation in a certain universal functional basis.

7.2.2 Parameter Estimation

The values of the model parameters $\mathbf{c}$ should be selected so to satisfy the condition $f(t, \hat{\mathbf{c}}) \approx F(t)$ in the best way. Closeness of functions is desirable not only at the observation instants $t_1, \ldots, t_N$ but also at *intermediate* instants and, sometimes, even at the past or future instants (providing the latter is very difficult). To characterise the closeness quantitatively, one introduces a measure of distance, i.e. some metrics ρ in the space of functions. The most popular choice of ρ is the mean-squared deviation with some weight function $w(t)$:

$$\rho(f, F) = \int (f(t, \mathbf{c}) - F(t))^2 w(t) dt. \qquad (7.25)$$

If t is a random quantity with the probability distribution $w(t)$, physical sense of such a metrics is a squared deviation of a model function $f(t, \mathbf{c})$ from $F(t)$ averaged over t. For the case of Eq. (7.24), the distance ρ is equal to the mean-squared deviation of $f(t, \mathbf{c})$ from $\eta(t)$ up to the variance of noise:

$$\iint (f(t, \mathbf{c}) - \eta(t))^2 w(t) p(\eta | t) dt \, d\eta = \iint (f(t, \mathbf{c}) - F(t) - \xi)^2 w(t) p(\xi) dt \, d\xi =$$

$$= \rho(f(t, \mathbf{c}), F(t)) + 2 \left(\int (f(t, \mathbf{c}) - F(t)) w(t) dt \right) \left(\int \xi p(\xi) d\xi \right) + \int \xi^2 p(\xi) d\xi =$$

$$= \rho(f(t, \mathbf{c}), F(t)) + \sigma_\xi^2. \qquad (7.26)$$

Other definitions of ρ and non-randomness of t are possible. However, all the considerations would remain similar for those different situations. Therefore, we confine ourselves with the measure (7.25).

---

[4] Even if the true regression $F(t)$ belongs to the selected model class $f(t, \mathbf{c})$, i.e. $F(t) = f(t, \mathbf{c}_0)$, restoration of $F(t)$ is not equivalent to the most accurate estimation of the parameters $\mathbf{c}_0$. The point is that the best estimate of the parameters in the "true" class $f(t, \mathbf{c})$ does not necessarily give the best approximation to F from a finite data sample, since the best approximation may be achieved in another class (Vapnik, 1979).

Thus, one needs to find the values of parameters minimising the functional (7.25). Since a "true" probability density $w(t)$ in Eq. (7.25) may be unknown, one must replace minimisation of Eq. (7.25) by minimisation of some functional, which can be evaluated from observed data and whose point of minimum is close to the point of minimum of Eq. (7.25). Similar to Eq. (7.12), such an *empirical* functional is the sample mean-squared value of the prediction error:

$$\varepsilon^2(\mathbf{c}) = S(\mathbf{c})/N = \frac{1}{N} \sum_{i=1}^{N} (\eta_i - f(t_i, \mathbf{c}))^2. \tag{7.27}$$

Model parameters $\hat{\mathbf{c}}$ are found via minimisation of Eq. (7.27). This is a kind of the so-called *empirical risk minimisation* (Vapnik, 1979, 1995). In our case, $\varepsilon^2(\mathbf{c})$ estimates the sum of the noise variance and an approximation error according to Eq. (7.26). If the class of functions $f(t, \mathbf{c})$ is selected appropriately so that it contains functions very close to F in the sense of ρ, then the quantity $\hat{\varepsilon}^2 = \varepsilon^2(\hat{\mathbf{c}})$ is almost unbiased estimator of the variance of the noise ξ.

As it is seen from expression (7.27), the technique to compute the parameter values is the ordinary LS technique, i.e. there are no technical differences in this respect between the problems considered in Sects. 7.1 and 7.2.

7.2.3 Model Size Selection, Overfitting and Ockham's Razor

Before discussing the choice of the class of functions f (Sect. 7.2.4), let us describe the selection of a model size for a given class. For that, let us consider the classical example of the algebraic polynomial approximation:

$$f(t, \mathbf{c}) = c_1 + c_2 t \ldots + c_{K+1} t^K. \tag{7.28}$$

The model size is usually understood as the number of free parameters of a model. Free parameters are those estimated from a time series without any imposed restrictions in the form of equalities. If all $P = K + 1$ coefficients in the model function (7.28) are free parameters, one comes to a well-known statistical problem in the field of polynomial regression: to select an optimal polynomial order K.

A theoretical justification for the use of an algebraic polynomial is given by the famous Weierstrass theorem, which states that any function continuous on a segment can be arbitrarily accurately uniformly approximated with an algebraic polynomial. Hence, theoretically one can achieve an arbitrarily small value of $\rho(f, F)$ with a polynomial of a sufficiently high order K.

7.2.3.1 Concepts of Underfitted and Overfitted Models

Which particular value of K should be selected in practice? A very small polynomial order is often unsuitable, since it does not allow to approximate a complicated dependence with high accuracy. In such a case, one says that a model is *underfitted*.

However, too large an order is also bad as shown below. It is important to construct a sufficiently *parsimonious model*.

A use-proven approach is to increase the polynomial order starting from zero and stop at the value of K, above which no significant improvement of a model is observed. There are several quantitative criteria: minimum of a test error $\hat{\varepsilon}^2_{test}$; saturation of an empirical error $\hat{\varepsilon}^2$; cross-validation and minimum of a cost function representing the sum of an empirical error $\hat{\varepsilon}^2$ and a certain penalty term (see below). All the criteria fight against ill-posedness of the problem. Ill-posedness manifests itself in that there are infinitely many model functions capable of accurate description of a *finite* set of data points on the plane (t, η). It is for this reason that one may not simply minimise $\hat{\varepsilon}^2$ to select K. The point is that $\hat{\varepsilon}^2$ is always a *non-increasing* function of K. This quantity equals zero when the number of polynomial coefficients equals the time series length N, i.e. the plot of such a polynomial goes *exactly* through the data points (t_i, η_i). However, such a model is typically very bad. It "learns" just to reproduce a concrete observed signal together with a superimposed random realisation of the noise ξ (Fig. 7.4a). Hence, on average it predicts new observations very inaccurately, since concrete values of ξ differ for new observations. Thus, such a model function $f(t, \hat{\mathbf{c}})$ possesses huge variance as an estimator of $F(t)$. Even if such $f(t, \hat{\mathbf{c}})$ is an unbiased estimator, its random error can be arbitrarily large. One says that such a model is not capable of *generalisation* of information. The model is *overfitted*, which is the main practical manifestation of the problem ill-posedness. Avoiding *model overfitting* is a key point in empirical modelling.

7.2.3.2 Minimum of the Test Error

Let a researcher have an additional time series from the same process: $t'_i, \eta'_i, i = 1, \ldots, N'$ (a test series). Then, a good criterion to select a polynomial order is to minimise the test approximation error

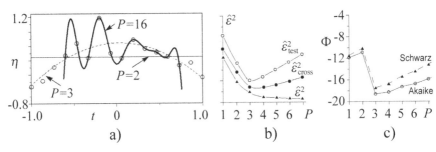

Fig. 7.4 Approximation of the quadratic function $F(t)$ from a time series of length $N = 16$ data points with polynomials of different orders: (**a**) plots of model functions for different orders $K = P - 1$. The thin line shows an underfitted model, the thick line shows an overfitted one and the dashed line shows an optimal model; (**b**) selection of an optimal order based on the prediction errors, a qualitative illustration; (**c**) Selection of an optimal order with the criteria of Schwarz and Akaike according to computations for the example shown in the panel (**a**)

$$\hat{\varepsilon}^2_{\text{test}} = \frac{1}{N'} \sum_{i=1}^{N'} \left(\eta'_i - f(t'_i, \hat{\mathbf{c}}) \right)^2, \tag{7.29}$$

where model coefficients $\hat{\mathbf{c}}$ are obtained from a training time series. Moreover, one can use the following additional consideration: for a non-overfitted model the values of $\hat{\varepsilon}^2$ and $\hat{\varepsilon}^2_{\text{test}}$ should be approximately the same. Minimisation of Eq. (7.29) seems to be the most reliable way to choose K (Fig. 7.4b, open circles), since it is based on the sample which is not used for the parameter estimation. Such criteria are called *out-of-sample criteria*.

7.2.3.3 Saturation of the "Training" Error

However, a test time series is not always at hand so that one must extract necessary information from a single training time series. The corresponding criteria are called *in-sample criteria*. As a rule, $\hat{\varepsilon}^2$ decreases with K. If for the values of K greater than some "threshold" value, the value of $\hat{\varepsilon}^2$ changes slowly with increasing K, i.e. a saturation takes place, such a threshold value can be selected as an optimal one (Fig. 7.4b, triangles). This approach often overestimates the value of K.

7.2.3.4 Cross-Validation Technique

It occupies an intermediate position between the two approaches described above. Its idea is as follows (Vapnik, 1979). One excludes from an available time series a single observation t_i, η_i and constructs a model from the resulting time series of the length $N - 1$. Let us denote it as $f(t, \hat{\mathbf{c}}_i)$. With this model, one computes prediction error $\hat{\varepsilon}_i$ for the excluded observation: $\hat{\varepsilon}_i = \eta_i - f(t_i, \hat{\mathbf{c}}_i)$. The same procedure is repeated for each observation in turn. Finally, one computes the mean-squared error $\hat{\varepsilon}^2_{\text{cross}} = \frac{1}{N} \sum_{i=1}^{N} \hat{\varepsilon}^2_i$ of such "cross-predictions". An optimal order K is obtained via the minimisation of $\hat{\varepsilon}^2_{\text{cross}}$ (Fig. 7.4b, filled circles). This approach may also overestimate a necessary value of K but it is more reliable than the criterion of $\hat{\varepsilon}^2$ saturation.

7.2.3.5 Minimum of the Cost Function with a Penalty Term

This approach involves a broad range of versions corresponding to different assumptions about the properties of investigated processes and different notions of an "optimal" model. We will briefly outline some of them, which are widely known. In a simple setting, such approaches to the automatic selection of the polynomial order rely on the minimisation of the cost function, which reads as

$$\Phi(P) = g_1(\hat{\varepsilon}^2) + g_2(P), \tag{7.30}$$

where g_1, g_2 are increasing functions of their arguments. The first term determines the contribution of the empirical error; the second one is the contribution of the model size. The corresponding techniques are developed within the theory of information and the theory of statistical estimation from different considerations often involving the ML principle. One may expect a minimum of Eq. (7.30) at an intermediate model size, since the empirical error is too large at small model sizes and the second term is too large at big model sizes.

The cost function $\Phi(P) = (N/2) \ln \hat{\varepsilon}^2 + P$ is called *Akaike criterion* (Akaike, 1974). It is obtained directly from the maximum likelihood principle. The first term is proportional to the negative likelihood of an optimal model with P parameters under the assumption that the model errors are independent Gaussian random quantities. The second term is heuristic: it is a simple penalty for a big model size but there is no rigorous proof of good statistical properties of the corresponding estimator. Thus, the Akaike criterion often overestimates the model size and cannot be asymptotically (i.e. in the case of a very large data set) optimal as shown in Schwarz (1978).

The cost function $\Phi(P) = (N/2) \ln \hat{\varepsilon}^2 + (P/2) \ln N$ is called *Schwarz criterion* (Schwarz, 1978). It is based on the additional assumption about general properties of prior distributions of the unknown parameters and on the Bayes' formalism (Sect. 2.2.1). It is justified better than the Akaike approach and exhibits much better statistical properties of the model size estimator. In particular, it does not systematically overestimate the model size (Schwarz, 1978). Due to this circumstance and the formal simplicity, the Schwarz criterion is quite often used in practice. Both approaches are illustrated in Fig. 7.4c, where any of them appears able to detect the correct model size $P = 3$. A real-world application of the Schwarz criterion is given in Sect. 13.3.

An information-theoretic-based approach was introduced by Jorma Rissanen (1989). The cost function, called a *description length*, reads as $\Phi(\mathbf{c}) = (N/2) \ln \hat{\varepsilon}^2(\mathbf{c}) + (P/2) \ln(2\pi e N/P) + P \ln \sqrt{\mathbf{c}^{\mathrm{T}} \mathbf{M} \mathbf{c}}$, where $\mathbf{c}$ is the P-dimensional vector of the model parameters and elements of the matrix $\mathbf{M}$ are given by (Rissanen, 1993)

$$M_{i,j}(\mathbf{c}) = \frac{1}{2} \frac{\partial^2 (\ln \hat{\varepsilon}^2(\mathbf{c}))}{\partial c_i \partial c_j}.$$

The idea behind this cost function is to minimise the "information" necessary to reproduce the observed data with a given model of the size P with parameters $\mathbf{c}$. Such "information" consists of two parts: model residual errors and model parameters. The length of the code needed to store the residual errors is given by the first term in the above expression for $\Phi(\mathbf{c})$, which is the negative likelihood $(N/2) \ln \hat{\varepsilon}^2(\mathbf{c})$. Specification of the model parameters to a given precision represents a "penalty" and is described by the other two terms. As distinct from the formulation (7.30), the minimisation of the description length does not require to minimise $\ln \hat{\varepsilon}^2(\mathbf{c})$ at each value of P, because the point of minimum of $\ln \hat{\varepsilon}^2(\mathbf{c})$ at a given P may correspond to a big value of the term $P \ln \sqrt{\mathbf{c}^{\mathrm{T}} \cdot \mathbf{M}(\mathbf{c}) \cdot \mathbf{c}}$ and, hence, yield a suboptimal value of the cost function Φ at a given P. In other respects, the use of the minimal description length principle is similar to the use of the cost functions (7.30). This approach has appeared very fruitful in practical problems, including construction of non-linear dynamical models (Judd and Mees, 1995, 1998), and currently has become more and more popular. Conceptually, it is directly related to the notion of *algorithmic complexity* of a process (Sect. 2.2.3).

A penalty term may depend not only on the model size P but also on some properties of a model. First, we should mention in this respect a very fruitful idea of *structural risk minimisation* (Vapnik, 1979, 1995), where the penalty term depends on the "capacity" (i.e. flexibility or generality) of the class of approximating functions corresponding to a given value of P. One of the advantages of this approach is that it is distribution free, i.e. no assumptions about the properties of the measurement and dynamical noises and prior distributions of the model parameters are needed. Secondly, a penalty term may be taken proportional to the mean-squared value of the second derivative $d^2 f/dt^2$ (see, e.g., Green and Silverman, 1994; Hastie and Tibshirani, 1990) to penalise fast fluctuations of the dependency $f(t)$. This idea is widely used in the construction of smoothing splines (Reinsch, 1967) and in other problems of non-parametric regression (Green and Silverman, 1994).

It is also worth noting that different cost functions of the form (7.30) and similar to them may give somewhat different optimal values of K in each concrete case. Therefore, one should take such results with caution (as rough guesses) in empirical modelling and perform a careful diagnostic check of models for several close values of K.

7.2.3.6 Ockham's Razor

All the approaches to the selection of the best model size (i.e. a minimal size sufficient for a satisfactory description of data) realise *the principle of parsimony* in modelling (Box and Jenkins, 1970; Judd and Mees, 1995). This is a concrete quantitative version of the general scientific and philosophical principle: "One should not multiply a number of entities needlessly". The principle is called *Ockham's razor*[5] even though similar statements were formulated already by Aristotle (see, e.g., Shalizi, 2003).

7.2.3.7 Model Structure Optimisation

Under any of the above approaches, a resulting model with a polynomial of an order K inevitably contains all monomials up to the order K inclusively. However, it is easy to imagine a situation when some of the intermediate monomials are superfluous. Thus, it would be desirable to exclude them from a model. This shortcoming arises due to the procedure used for the model selection: the terms are added in a *pre-defined* sequence.

---

[5] *Pluralitas non est ponenda sine necessitate.* William Ockham (1285–1349) is a famous English philosopher and logician.

In general, one needs a technique to search for an optimal model of a given size P, where all subsets of P functions – terms from a wider set of P_{max} functions are considered. This is very simple if the set of the basis functions is orthogonal over the observed sample. However, it can be very difficult in the opposite case, since a direct check of all combinations is practically impossible due to the *combinatorial explosion*, i.e. an astronomically big number of possible combinations even for moderately big values of P_{max} and P. There is a multitude of practically feasible simpler methods (Nakamura et al., 2004), some of which are just mentioned and others are described in some detail in Chap. 9:

(i) to add basis functions into a model one by one providing the fastest decrease of the empirical error $\hat{\varepsilon}^2$ (Judd and Mees, 1995);
(ii) to exclude the so-called "superfluous" terms from an originally large model one by one either providing the slowest increase of $\hat{\varepsilon}^2$ (Aguirre et al., 2001), or according to Student's criterion (Kendall and Stuart, 1979), or based on variability of coefficient values for different training series (Bezruchko et al., 2001).

The search for an optimal combination of the functions – terms is often called *model structure optimisation*.

7.2.4 Selecting the Class of Approximating Functions

7.2.4.1 Functional Basis

Any kind of functions can be used in modelling, but computations are simpler if the parameters enter the model function linearly:

$$f(t, \mathbf{c}) = c_1 \phi_1(t) + \ldots + c_P \phi_P(t), \qquad (7.31)$$

where ϕ_k are called *basis functions*. Typically, ϕ_k are elements of an infinite set of functions $\phi_k, k = 1, 2, \ldots$, allowing arbitrarily accurate approximation of any "sufficiently good" (e.g. continuous) function F. In other words, the set is a functional basis in the space of functions with certain properties. The function (7.31) is called a *generalised polynomial* with respect to the set of functions $\phi_1(t), \ldots, \phi_P(t)$. Its another name is a *pseudo-linear model*, where the prefix "pseudo" stresses that the dependence on the parameters is linear, rather than the dependence on the argument. In the case of Eq. (7.28), the standard polynomial basis is used for approximation, where basis functions are the monomials t^{k-1}.

The trigonometric set of basis functions $1, \cos \omega t, \sin \omega t, \cos 2\omega t, \sin 2\omega t, \ldots$ is also widely used. Then, in the case of a uniform sample, the least-squares problem is solved via the direct Fourier transform (Sect. 6.4.2). The values of the coefficients c_k even get physical meaning, since power spectrum components are expressed via c_k. The usage of a trigonometric polynomial for approximation allows to avoid big pips

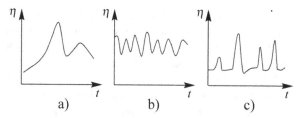

Fig. 7.5 Examples of signals which should be preferably approximated with (**a**) an algebraic polynomial; (**b**) a trigonometric polynomial; (**c**) wavelets

peculiar to a high-order algebraic polynomial. The former is the most efficient when repeatability is clearly seen in an observed signal (a signal is periodic or almost periodic), Fig. 7.5b. Algebraic polynomials are advantageous for gradually changing signals (even with an intricate profile) without periodicity, pulses and sharp changes, Fig. 7.5a.

One more popular kind of pseudo-linear models is a wavelet-based model, where basis functions are "pulses" of different width and with different locations on the time axis (Sect. 6.4.2). This class of functions is a flexible tool to describe the signals whose frequency contents change in time, signals of a pulse character (Fig. 7.5c), etc. Also, one can use combinations of different functional bases, e.g., Danilov and Safonov (1987).

7.2.4.2 Non-linear Parameter Dependence

Approximating functions may depend on their parameters in a non-linear way, e.g. radial, cylindrical and elliptic basis functions (Small and Judd, 1998) (Sect. 10.2.1), artificial neural networks (Makarenko, 2003) (Sects. 3.8 and 10.2.1) and ad hoc functions. In the case of non-linear parameter dependence, the minimisation problem (7.27) does not reduce to a linear set of equations. One should solve it with numerical iterative minimisation techniques. *Smooth optimisation* is widely applied (Dennis and Schnabel, 1983) including gradient descent, Newtonian techniques, Gauss and Newton technique, quasi-Newtonian techniques. However, a cost function may have many minima, while only one of them (the global one) gives the best parameter estimates. Depending on a starting guess, one can fall into the global minimum or one of the local ones. There are no universal ways to avoid this trouble. The local minima problem gets more difficult for greater number of estimated parameters. It is solved (without guarantees of success) via trials from different starting guesses or with sophisticated approaches of stochastic optimisation, e.g. genetic methods and simulated annealing (Crutchfield and McNamara, 1987).

7.2.4.3 Global and Local Models

If an approximating function is specified with a single formula in the entire range of argument values like a polynomial, then the approximation and the model are called *global* (Casdagli, 1989). All the above models are global. An alternative and often more effective way is to use a *local* (piecewise) approximating function, which is specified by different sets of the parameter values in different small domains of the argument values (Casdagli, 1989; Farmer and Sidorowich, 1987). The most popular examples are piecewise constant and piecewise linear approximations and cubic splines (Kalitkin, 1978; Press et al., 1988; Samarsky, 1982). Local models are advantageous for the description of less smooth dependencies with humps and discontinuities. However, they are more strongly affected by noise than the global models with small number of free parameters. Therefore, local models are superior in the case of large amount of data and low noise level for arbitrarily complicated non-linearity. Global models are better for smaller amount of data, higher noise level and not so "sharp" non-linearity.

7.2.4.4 Why a Certain Class of Model Functions Can Be Better

Theoretically, one can use any functional basis for the approximation of any sufficiently good function $F(t)$. In practice, different bases have their own advantages and shortcomings. The point is that to approximate a concrete dependence $F(t)$, one needs different model sizes in different bases. The best basis for the approximation of $F(t)$ gives the smallest size of the optimal model, i.e. a small number of the corresponding basis functions suffice. Thereby, one reduces random errors in the parameter estimates and the danger of model overfitting. Under an unlucky choice of the basis, an available amount of data may appear insufficient for the reliable estimation of the parameters of a model containing a big number of basis functions.

If a researcher, relying on the previous experience and prior knowledge about an object, chooses a special ad hoc model structure, there may appear an additional advantage that model parameters make physical sense. However, some physical meaning can be rarely extracted even from universal constructions. We note the "power interpretation" of the harmonic component amplitudes in a trigonometric polynomial and typically absent physical sense of an algebraic polynomial coefficients.

7.3 Model Validation

Basically, validation of models in the form of explicit time dependencies relies on studying *model residual errors*, i.e. the quantities $\varepsilon_i = \eta_i - f(t_i, \hat{\mathbf{c}})$. They are also called *discrepancies*. It is desirable to check properties of the residuals from a test time series, but one often has to deal only with a training one.

For the problem (7.1), where a model structure is known and there is no noise, all the residuals must be equal to zero up to the machine precision. In the presence of noise (7.3) or (7.24), one checks statistical properties of the residuals. Since one usually makes assumptions about the noise ξ when selecting a technique for the parameter estimation, model residuals must satisfy those assumptions to confirm relevance of the technique and reliability of the results. Most often, one assumes *statistical independence* of $\xi(t_i)$ and *normality* of ξ.

7.3.1 Independence of Residuals

There are different tests for independence such as criteria of ascending and descending series, criterion of series based on the sample median and criterion of squared sequential ratios (Aivazian, 1968; Hoel, 1971; von Mises, 1964). They are suitable for any distribution law of ξ. Here, we briefly consider only estimation of the auto-correlation function (ACF, Sect. 4.1.2) to check *uncorrelatedness of residuals*. This approach is directly applicable only for a uniform sampling. Uncorrelatedness is a weaker property than statistical independence, but both properties coincide for Gaussian distribution.

The ACF of the residuals can be estimated as

$$\hat{\rho}(n) = \sum_{i=1}^{N-n} \hat{\varepsilon}_i \hat{\varepsilon}_{i+n} \left/ \sum_{i=1}^{N} \hat{\varepsilon}_i^2 \right.,$$

where $n = 0, 1, \ldots, N_{\mathrm{ACF}} < N$. Typically, one uses $N_{\mathrm{ACF}} \le N/4$. For a sequence of independent quantities, the theoretical values of the ACF are $\rho_\xi(0) = 1$ and $\rho_\xi(n) = 0, n > 0$. Let us consider the plot $\hat{\rho}(n)$ (Fig. 7.6a, d) and estimate its deviation from the theoretical values. Under the assumption of independent residuals and sufficiently big N, the quantities $\hat{\rho}_\xi(n), n > 0$, are distributed approximately according to the Gaussian law and lie within an interval $\pm 2\hat{\sigma}_\xi/N$ with the probability of 0.95 (Box and Jenkins, 1970). The interval is shown by the dashed lines in Fig. 7.6a, d. Independence of the residuals for both examples is not denied, since not less than 95% of the estimated $\hat{\rho}$-values fall into the expected 95%-intervals.

7.3.2 Normality of Residuals

To test normality of the residual distribution, one can use such quantitative approaches as Kolmogorov and Smirnov test and criterion of χ^2 (Aivazian, 1968; Press et al., 1988; von Mises, 1964). We describe here only two graphical ways of visual inspection (Fig. 7.6).

The simplest way is to construct a histogram of residuals, i.e. an estimate of the probability density function. For that, one divides an entire range of $\hat{\varepsilon}_i$ into M subintervals (bins), which are typically of equal width. Then, a percentage of the observed residual values in each interval is counted. One draws a rectangle over each bin whose height is equal to the ratio of percentage of values within the bin to the bin width. Finally, one visually compares the histogram with a Gaussian probability density function, see two examples in Fig. 7.6b, e.

However, checking such "similarity" visually is not easy, especially taking into account that a histogram as an estimator of a probability density function exhibits large variance at small bin width: the reconstruction of a probability density from a sample is an ill-posed problem (Vapnik, 1979). More convenient is the visualisation of the plot on the so-called *normal probability paper* (Box and Jenkins, 1970). For

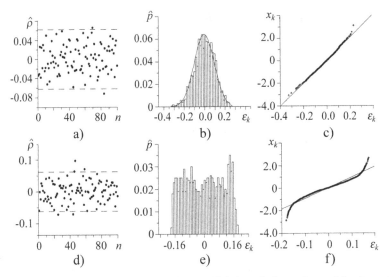

Fig. 7.6 Analysis of the residuals obtained after the LS fitting of a first-order model polynomial to a time series of length $N = 1000$ data points generated by the process $\eta(t) = 1 + t + \xi$ with normal (**a–c**) or uniform (**d–f**) distribution of the white noise ξ, $\sigma_\xi^2 = 0.01$. The *left* column shows the estimates of the ACF: uncorrelatedness is not denied, since the values of $\hat{\rho}$ fall outside of the 95% interval not more often than in 5% of cases. The *middle* column shows the histograms: only on the *upper panel* the histogram looks similar to a normal probability density function (in accordance with the original noise properties). The *right* column shows the plots on the normal probability paper: normality of the residuals is denied in the second case

that, one shows the observed values of $\hat{\varepsilon}_i$ along the abscissa axis and the corresponding[6] values of x_i calculated from the residual distribution function along the ordinate axis. If the residuals $\hat{\varepsilon}_i$ are distributed according to the Gaussian law with zero mean, then the points on such a plot lie on a straight line going through the origin. Its angular coefficient is determined by the variance of $\hat{\varepsilon}_i$. It is easier to assess visually whether the points lie on a straight line (Fig. 7.6c, f) rather than to assess the similarity between a histogram and a normal probability density function (Fig. 7.6b, e).

If a test time series is at hand, an additional sign of the model inadequacy is the difference between the sample residual variances for the training and test time series. Model inadequacy can be determined by its overfitting or underfitting, inappropriate

---

[6] Each value of $\hat{\varepsilon}_k$ corresponds to some value of the empirical distribution function for the residuals $\hat{\Phi}(\hat{\varepsilon}_k)$. The reconstruction of a distribution function is a well-posed problem (Vapnik, 1979). Thus, if all $\hat{\varepsilon}_i$ are pairwise different, then $\hat{\Phi}(\hat{\varepsilon}_k)$ can be estimated as the ratio of the number of values $\hat{\varepsilon}_i$, less than or equal to $\hat{\varepsilon}_k$, to their total number N. Let us denote the distribution function of the standard Gaussian law (zero mean and unit variance) as $\Phi_0(x)$. Let us denote x_k such a number that $\Phi_0(x_k) = \hat{\Phi}(\hat{\varepsilon}_k)$. x_k is unique since $\Phi_0(x)$ is a continuous and strictly monotonous function. x_k is related to $\hat{\varepsilon}_k$ as $x_k = \Phi_0^{-1}(\hat{\Phi}(\hat{\varepsilon}_k))$. The plot x_k versus $\hat{\varepsilon}_k$ is called a plot on the normal probability paper.

choice of the class of approximating functions, falling into local rather than into global extremum of the cost function, incorrect assumptions about the noise properties, and not the best technique for the parameter estimation. Finally, the model structure may appear irrelevant, e.g. sequential values of ξ for the original process (7.24) may depend on each other. Thus, one should find out which of the mentioned reasons take place and return to the corresponding stage of the modelling procedure (Fig. 5.1).

7.4 Examples of Model Applications

Despite simplicity of the models considered in this chapter, the skill in constructing them is important in modelling practice. More often, they play a role as elements of more complex problems but may have their own value as well. Let us consider their applications to the problems of the forecast (one of the central topics of the book) and numerical differentiation (which is important in construction of differential equations, Chaps. 8, 9 and 10).

7.4.1 Forecast

The problem is to predict the value of η at a given time instant t. It is necessary not only to give a *"point" prediction* $\hat{\eta}(t)$ but also to assess its possible error, i.e. indicate an interval within which a true value $\eta(t)$ lies with high probability. The latter is called an *interval prediction*. One often searches for such an interval in the form $\hat{\eta}(t) \pm \Delta\hat{\eta}(t)$. For simplicity, we consider the problem in the case of a single estimated parameter.

For the models considered in this chapter, one typically uses the quantity $\hat{\eta}(t) = f(t, \hat{c})$ as a point predictor. Its error is

$$e(t) \equiv \hat{\eta}(t) - \eta(t) = [f(t, \hat{c}) - f(t, c_0)] - \xi(t). \tag{7.32}$$

For a *pseudo-linear model* or a small error in a parameter estimate, one can rewrite the expression as

$$e(t) = k(t) \cdot (\hat{c} - c_0) - \xi(t), \tag{7.33}$$

where $k(t) = \partial f(t, c)/\partial c|_{c=c_0}$. The prediction error is characterised by the respective mean-squared value:

$$E[e^2(t)] = k^2(t) \cdot E[\hat{c} - c_0]^2 + \sigma_\xi^2. \tag{7.34}$$

For an unbiased parameter estimator, it follows from Eq. (7.33) that $\hat{\eta}(t)$ is an unbiased estimator of $\eta(t)$: $E[e(t)] = 0$. In other words, the predictor does not exhibit a systematic error. There is only a random error with the variance

$$\sigma_e^2(t) = E[e^2(t)] = k^2(t)\sigma_{\hat{c}}^2 + \sigma_\xi^2. \tag{7.35}$$

If the noise ξ in the original process is normal, then a 95% confidence interval for $\eta(t)$ is $\hat{\eta}(t) \pm 1.96\sigma_e^2$. This formula is often used as an approximation even if the distribution law of ξ is unknown.[7]

If the noise variance and the parameter estimator error are not large, then the prediction error (7.35) remains small while $k(t)$ is small. The latter holds true for any t if $f(t, c)$ is insensitive to variations in c. Otherwise, $k(t)$ and the prediction error may rise with time. For instance, in the case of $f(t, c) = \sin(ct)$, one gets $k(t) = t \cos(c_0 t)$. This is an unbounded function, therefore, an originally very small prediction error rises at larger t up to the scale of the observed oscillations.[8]

We note that estimating the prediction error is possible here due to the information about the properties of the noise ξ. In general, an "origin" of the model errors is unknown so that it is impossible to assess their probable values based on the same model which gives point predictions (Judd and Small, 2000, see also Sect. 10.3).

7.4.1.1 Interpolation and Extrapolation

Already in the previous example, one can see that forecast at distant time instants (*extrapolation*) can be much less accurate. It holds true even in the case of a priori known model structure. If a model structure is unknown, then to provide success-ful extrapolation one needs to estimate the parameter values accurately and to be assured that the selected form of the function f is suitable to describe a process for a wide range of t. The latter is typically not the case. Thus, a high-order algebraic polynomial usually extrapolates *very badly*.

Nevertheless, extrapolation is used for some practical applications. For instance, one uses models with a first- or second-order algebraic polynomial f to predict slow processes in econometrics. There, annual observations (e.g. profit of a factory) may be performed during several years and a 1-year-ahead (i.e. one-step-ahead)

---

[7] Instead of the variance σ_ξ^2 one may substitute its estimate $\hat{\sigma}_\xi^2$ into equation (7.35). The variance of the parameter estimator $\sigma_{\hat{c}}^2$ is usually proportional to σ_ξ^2 and inversely proportional to time series length N or higher degrees of N (Sect. 7.1.2). Formulas for a more general case of several estimated parameters $\mathbf{c}$ are as follows. For a pseudo-linear model (7.31), the covariance matrix for the parameter estimators is given by $Cov(\hat{\mathbf{c}}) = \hat{\sigma}_\xi^2 (\mathbf{A}^T\mathbf{A})^{-1}$, where $A_{j,k} = \sum_{i=1}^{N} \phi_j(t_i)\phi_k(t_i)$ (Gnedenko, 1950). Diagonal elements of $Cov(\hat{\mathbf{c}})$ are the variances of the parameter estimators $\hat{\sigma}_{\hat{c}_i}^2 = [Cov(\hat{\mathbf{c}})]_{ii}$. For a model non-linear with respect to the parameters, the covariance matrix estimator is obtained as an inverse Hessian of the likelihood function: $Cov(\hat{\mathbf{c}}) = \mathbf{H}^{-1}(\hat{\mathbf{c}})$, where $H_{ij}(\hat{\mathbf{c}}) = -\partial^2 \ln L(\hat{\mathbf{c}})/\partial c_i \partial c_j$.

[8] It is for the same reason that the parameter estimator variance for a process (7.19) decreases faster (as $1/N^3$) with N (Sect. 7.1.2). Since the function f is sensitive to variations in the parameter value, the cost function S in equation (7.12) is also sensitive to them, since the squared value of $\partial f(t, c)/\partial c|_{c=c_0}$ is the partial Fisher information determining the contribution of each new obser-vation to the likelihood function.

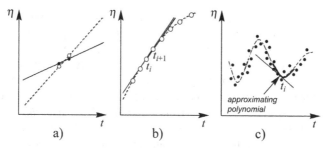

Fig. 7.7 Forecast and differentiation: (**a**) prediction of a linear dependence based on the two observed data points. Forecast between the points is sufficiently accurate (interpolation). Prediction error at distant time instants is large (extrapolation); (**b**) an illustration to the differentiation scheme (7.36). The *dashed line* is a plot of $F(t)$, the *thin solid line* is a true tangent, the *thick solid line* is a secant (7.36); (**c**) differentiation with the aid of an approximating polynomial (the *thick line*). The *thin line* is a tangent to a model polynomial which is close to the tangent to the original function

prediction is quite important. Due to slowness of the processes, the assumption of linear or quadratic character of temporal dependence over one more year looks plausible. In other words, extrapolation to the "nearest future" with a simple model appears reasonable.

As for the prediction at intermediate time instants (*interpolation*), it is often sufficiently reliable (Fig. 7.7a). The only important assumption is that an approximated dependence does not exhibit strong fluctuations between observation instants. It is often well justified. There is a huge literature about interpolation (Johnson and Riess, 1982; Kalitkin, 1978; Press et al., 1988; Samarsky, 1982; Stoer and Bulirsch, 1980), in particular, the following techniques are used:

(i) An interpolating algebraic polynomial (its plot goes exactly through all the experimental data points, i.e. a high polynomial order is used) is suitable for gradual dependencies and moderate time series length (roughly speaking, not more than about 30 data points),

(ii) Piecewise linear interpolation (data points at subsequent time instants are connected by straight lines, so that one obtains a polygonal line) is better for not very smooth dependencies and high sampling rates.

(iii) Cubic splines (through each pair of neighbouring data points one draws a cubic parabola so that the entire resulting curve is continuous together with its first and second derivatives) are rather efficient and quite a universal tool to approximate smooth dependencies at high sampling frequency (Samarsky, 1982; Stoer and Bulirsch, 1980; Press et al., 1988).

7.4.2 Numerical Differentiation

The values of model parameters can be of interest themselves, e.g. in the widely encountered problem of *numerical differentiation*. Let an original process be

$\eta(t_i) = x(t_i) + \xi(t_i)$, where $t_i = t_0 + i\Delta t$ and $x(t)$ is the smooth deterministic function. One needs to find the values of the derivative $dx(t_i)/dt$ from $\eta(t_i)$.

The most widespread approach consists of the approximation of the function $x(t)$ with a low-order algebraic polynomial in the vicinity of each time instant t_i of interest and estimation of the polynomial coefficients via the LS technique. The derivative of the model polynomial is taken as an estimator of the value of $dx(t_i)/dt$. This approach is also called *Savitzky – Golay filter* and *digital smoothing polynomial* (Press et al., 1988). If the sampling frequency is sufficiently high and the function $x(t)$ is sufficiently smooth, then approximation with an algebraic polynomial is justified by the Taylor theorem about expansion of a smooth function in a power series with respect to the deviations $t - t_i$ in the vicinity of t_i.

7.4.2.1 Differentiation in a Noise-Free Case

If $\xi = 0$, then one typically uses interpolating polynomials. Accuracy of the differentiation is determined by how well the dependence $x(t)$ can be approximated by a polynomial in the vicinity of the selected point t_i. The simplest scheme is to draw a straight line through the points t_i and t_{i+1} (the first-order polynomial, Fig. 7.7b). Then, a derivative estimator is

$$\frac{d\hat{x}(t_i)}{dt} = \frac{\eta_{i+1} - \eta_i}{t_{i+1} - t_i} = \frac{\eta_{i+1} - \eta_i}{\Delta t}. \tag{7.36}$$

Its error can be estimated with the use of the Taylor formula $\eta_{i+1} \approx \eta_i + (dx(t_i)/dt)\,\Delta t + (d^2x(t_i)/dt^2)\,\Delta t^2/2$. One gets a differentiation error $(d^2x(t_i)/dt^2)\,\Delta t/2$, which is proportional to Δt. In such a case, one says that a technique has the first order of accuracy. Decrease in Δt leads to more accurate derivative estimates in the absence of noise and computation errors. However, making Δt arbitrarily close to zero is not possible, since then ill-posedness of the numerical differentiation problem manifests itself (see below).

One can increase the order of accuracy by using higher order polynomials. For instance, via construction of the second-order polynomial from the observation instants t_{i-1}, t_i and t_{i+1} for a uniform sampling, one comes to the formula

$$d\hat{x}(t_i)/dt = \frac{\eta_{i+1} - \eta_{i-1}}{2\Delta t}. \tag{7.37}$$

It is tantamount to draw a straight line through the points at the instants t_{i-1} and t_{i+1}. Its order of accuracy is two since its error equals $(d^3x(t_i)/dt^3)\,\Delta t^2/6$. There are also schemes with higher orders of accuracy (Johnson and Riess, 1982; Press et al., 1988; Samarsky, 1982; Stoer and Bulirsch, 1980).

7.4.2.2 Ill-Posedness of the Problem

It is not allowable to take a very small value of Δt, e.g. in the scheme (7.36), for the following reason. Let the values η_i, η_{i+1} be known with small errors ξ_i, ξ_{i+1}

of the order of δ. Then, the derivative estimation error is dominated by the term $(\xi_{i+1} - \xi_i)/\Delta t \sim \delta/\Delta t$. For arbitrarily small error δ, one gets an arbitrarily large derivative estimation error in the case of sufficiently small Δt. It means ill-posedness of the problem: instability of a solution with respect to the input data (Sect. 5.3.2).

Ill-posedness is avoided if Δt is taken to be of the order of $\delta^{1/2}$ or greater. Under such a restriction, the derivative estimation error is bounded and tends to zero for $\delta \to 0$. This is an example of *regularisation* of an ill-posed problem.

7.4.2.3 Differentiation of a Noise-Corrupted Signal

A direct usage of the interpolating formulas (7.36) or (7.37) leads to the "noise amplification" and huge errors in the derivative estimator (Fig. 7.8a, b). To reduce the noise effect, a model algebraic polynomial is constructed in a sufficiently wide window $[t_{i-k_1\Delta t}, t_{i+k_2\Delta t}]$ with the aid of the LS technique (Fig. 7.7c). Efficiency of such an approach is shown in Fig. 7.8c. Since a derivative at a given time instant is equal to one of the polynomial coefficients, its root-mean-squared error can be estimated as the estimation error for the respective coefficient (Sect. 7.4.1).

The number of polynomial coefficients should be small as compared with the number of data points $k_1 + k_2 + 1$ in the time window $[t_{i-k_1\Delta t}, t_{i+k_2\Delta t}]$. The more data points are contained in the window, the less is the estimator variance, i.e. the higher is the accuracy of the differentiation. However, the interval $\max\{k_1, k_2\}\Delta t$ should not be too wide for an approximation with a low-order polynomial to be satisfactory. Thus, for a fixed polynomial order, there is an optimal window width, for which an error due to noise and an error due to model inconsistency are approximately similar. Increase in the polynomial order may increase the estimation accuracy, but only if one can use a wider time window. The latter circumstance depends on the character of the function $x(t)$ and the value of Δt. Under fixed Δt, there is some optimal value of the polynomial order providing the most accurate derivative estimator. In practice, it typically ranges from 1 to 3.

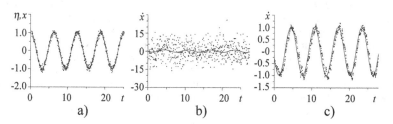

Fig. 7.8 Numerical differentiation of a signal $\eta(t) = \cos t + \xi(t)$ with Gaussian noise ξ, $\sigma_\xi^2 = 0.01$, $\Delta t = 0.01$: (**a**) an original time series, the *solid line* represents noise-free values $x(t)$; (**b**) differentiation according to scheme (7.37), the *solid line* represents true values of the derivative of $x(t)$; (**c**) differentiation with the second-order smoothing polynomial fitted to the windows covering 51 data points. The *solid line* shows true values of the estimated derivative. The estimation error is highly reduced

One can calculate the higher order derivatives as well by using the respective derivatives of the model polynomial. However, to estimate a high-order derivative, one must construct a high-order polynomial that often leads to large errors. Differentiation errors rise rapidly with the derivative order. Our experience shows that under typical practical settings, one manages to estimate reasonably well the derivatives up to the third order as the best case.

References

Aguirre, L.A., Freitas, U.S., Letellier, C., Maquet, J.: Structure-selection techniques applied to continuous-time nonlinear models. Physica D. **158**, 1–18 (2001)

Aivazian, S.A.: Statistical Investigation of Dependencies. Metallurgiya, Moscow, (in Russian) (1968)

Akaike, H.: A new look at the statistical identification model. IEEE Trans. Automatic Control. **19**, 716–723 (1974)

Akhieser, N.I.: Lectures on Theory of Approximation, 2nd edn. Nauka, Moscow (1965). 1st edn. is translated into English: Ungar Publishing Co., New York (1956)

Bezruchko, B.P., Dikanev, T.V., Smirnov, D.A.: Role of transient processes for reconstruction of model equations from time series. Phys. Rev. E. **64**, 036210 (2001)

Box, G.E.P., Jenkins, G.M.: Time Series Analysis. Forecasting and Control. Holden-Day, San Francisco (1970)

Casdagli, M.: Nonlinear prediction of chaotic time series. Physica D. **35**, 335–356 (1989)

Crutchfield, J.P., McNamara, B.S.: Equations of motion from a data series. Complex Syst. **1**, 417–452 (1987)

Danilov Yu.A., Safonov, V.L.: Usage of uniform function approximation in trigonometric basis to compute integrals and sums and to process experimental data. Preprint Institute Atomic Energy No. 4381/1. TsNIIAtomInform, Moscow, (in Russian) (1987)

Dennis, J., Schnabel, R.: Numerical Methods for Unconstrained Optimization and Nonlinear Equations. Prentice-Hall, Engle Wood Cliffs, NJ (1983)

Farmer, J.D., Sidorowich, J.J.: Predicting chaotic time series. Phys. Rev. Lett. **59**, 845–848 (1987)

Golub, G.H., Van Loan, C.F.: Matrix Computations. 2nd edn. Johns Hopkins University Press, Baltimore (1989)

Green, P.J., Silverman, B.W.: Nonparametric Regression and Generalized Linear Models. Chapman and Hall, London (1994)

Hardle, W.: Applied Nonparametric Regression. Cambridge University. Press, Cambridge (1992)

Hastie, T.J., Tibshirani, R.J.: Generalized Additive Models. Chapman and Hall, London (1990)

Hoel, P.G.: Introduction to Mathematical Statistics. 4th edn. Wiley, New York (1971)

Ibragimov, I.A., Has'minskii R.Z.: Asymptotic Theory of Estimation. Nauka, Moscow (1979). Translated into English Under the Title Statistical Estimation: Springer, New York (1981)

Johnson, L.W., Riess, R.D.: Numerical Analysis. 2nd edn. Addison-Wesley, Reading, MA (1982)

Judd, K., Mees, A.I. On selecting models for nonlinear time series. Physica D. **82**, 426–444 (1995)

Judd, K., Mees, A.I.: Embedding as a modeling problem. Phys. D. **120**, 273–286 (1998)

Judd, K., Small, M.: Towards long-term prediction. Phys. D. **136**, 31–44 (2000)

Kalitkin, N.N.: Numerical Methods. Nauka, Moscow, (in Russian) (1978)

Kendall, M.G., Stuart, A.: The Advanced Theory of Statistics, vol. 2 and 3. Charles Griffin, London (1979)

Makarenko, N.G.: Embedology and neuro-prediction. Procs. V All-Russian Conf. "Neuroinformatics-2003". Part 1, pp. 86–148. Moscow, (in Russian) (2003)

Nakamura, T., Kilminster, D., Judd, K., Mees, A. A comparative study of model selection methods for nonlinear time series. Int. J. Bif. Chaos. **14**, 1129–1146 (2004)

Pisarenko, V.F., Sornette, D.: Statistical methods of parameter estimation for deterministically chaotic time series. Phys. Rev. E. **69**, 036122 (2004)

Press, W.H., Flannery, B.P., Teukolsky, S.A., Vetterling, W.T.: Numerical Recipes in C. Cambridge University Press, Cambridge (1988)

Pugachev, V.S.: Probability Theory and Mathematical Statistics for Engineers. Pergamon Press, Oxford (1984)

Pugachev, V.S.: Theory of Probabilities and Mathematical Statistics. Nauka, Moscow, (in Russian) (1979)

Reinsch, C.H.: Smoothing by spline functions. Num. Math. **10**, 177–183 (1967)

Rissanen, J. A universal prior for integers and estimation by minimum description length. Ann. Stat. **11**, 416–431 (1993)

Rissanen, J.: Stochastic complexity in statistical inquiry. World Scientific, Singapore (1989)

Samarsky, A.A.: Introduction to Numerical Methods. Nauka, Moscow, (in Russian) (1982)

Schwarz, G.: Estimating the order of a model. Ann. Stat. **6**, 461–464 (1978)

Seber, G.A.F.: Linear Regression Analysis. Wiley, New York (1977)

Shalizi, C.R.: Methods and techniques of complex systems science: an overview, vol. 3, arXiv:nlin.AO/0307015 (2003). Available at http://www.arxiv.org/abs/nlin.AO/0307015

Small, M., Judd, K.: Comparisons of new nonlinear modeling techniques with applications to infant respiration. Phys. D. **117**, 283–298 (1998)

Stoer, J., Bulirsch, R.: Introduction to Numerical Analysis. Springer, New York (1980)

Vapnik, V.N.: Estimation of Dependencies Based on Empirical Data. Nauka, Moscow (1979). Translated into English: Springer, New York, (1982)

Vapnik, V.N.: The Nature of Statistical Learning Theory. Springer, New York (1995)

von Mises, R.: Mathematical Theory of Probability and Statistics. Academic Press, New York (1964)

Chapter 8
Model Equations: Parameter Estimation

Motions and processes observed in nature are extremely diverse and complex. Therefore, opportunities to model them with explicit functions of time are rather restricted. Much greater potential is expected from difference and differential equations (Sects. 3.3, 3.5 and 3.6). Even a simple one-dimensional map with a quadratic maximum is capable of demonstrating chaotic behaviour (Sect. 3.6.2). Such model equations *in contrast to explicit functions of time* describe how *a future state of an object depends on its current state* or how *velocity of the state change depends on the state itself*. However, a technology for the construction of these more sophisticated models, including parameter estimation and selection of approximating functions, is basically the same. A simple example: construction of a one-dimensional map $\eta_{n+1} = f(\eta_n, \mathbf{c})$ differs from obtaining an explicit temporal dependence $\eta = f(t, \mathbf{c})$ only in that one needs to draw a curve through experimental data points on the plane (η_n, η_{n+1}) (Fig. 8.1a–c) rather than on the plane (t, η) (Fig. 7.1). To construct model ODEs $d\mathbf{x}/dt = \mathbf{f}(\mathbf{x}, \mathbf{c})$, one may first get time series of the derivatives dx_k/dt ($k = 1, \ldots, D$, where D is a model dimension) via numerical differentiation and then approximate a dependence of dx_k/dt on $\mathbf{x}$ in a usual way. Model equations can be multidimensional, which is another difference from the construction of models as explicit functions of time.

For a long time, in empirical modelling of complex processes, one used *linear difference equations* containing noise to allow for irregularity (Sect. 4.4). The idea was first suggested in 1927 (Yule, 1927) and appeared very fruitful so that autoregression and moving average models became a main tool for the description of complex behaviour for the next 50 years.

Only in 1960–1970s, researchers widely realised that simple low-dimensional models in the form of non-linear maps or differential equations can exhibit complex oscillations even without noise influence. It gave a new impulse to the development of empirical modelling techniques, since arousal of powerful and widely accessible computers provided practical implementation of the ideas.

In this chapter, we consider a situation when an observed time series $\eta_i = h(\mathbf{x}(t_i))$, $i = 1, \ldots, N$, is produced by iterations of a map $\mathbf{x}_{n+1} = \mathbf{f}(\mathbf{x}_n, \mathbf{c})$ or integration of an ordinary differential equation $d\mathbf{x}/dt = \mathbf{f}(\mathbf{x}, \mathbf{c})$, whose structure is *completely known*. The problem is *to estimate parameters* $\mathbf{c}$ from the observed

B.P. Bezruchko, D.A. Smirnov, *Extracting Knowledge From Time Series*, Springer Series in Synergetics, DOI 10.1007/978-3-642-12601-7_8, © Springer-Verlag Berlin Heidelberg 2010

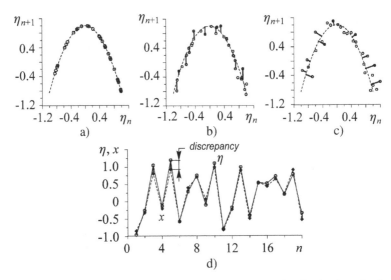

Fig. 8.1 Parameter estimation for the quadratic map (8.1) at $c_0 = 1.85$. *Circles* show the observed values: (**a**) no noise; the *dashed line* is an original parabola; (**b**) only dynamical noise is present; the *dashed line* is a model parabola obtained via minimisation of the mean-squared vertical distance (some of those distances are shown by *solid lines*); (**c**) only measurement noise is present; the *dashed line* is a model parabola obtained via minimisation of the mean-squared orthogonal distance; (**d**) rhombs show a model realisation which is the closest one to an original time series in the sense (8.4)

data. This is called "transparent box" problem (Sect. 5.2). To make the consideration more realistic, we add dynamical and/or measurement noise.

Such a problem setting is encountered in different applications and attracts serious attention. One singles out two main aspects of interest considered below:

(i) Parameter estimation *with a required accuracy* is important if the parameters cannot be measured directly due to experimental conditions. Then, the modelling procedure plays a role of "measurement device" (Butkovsky et al., 2002; Horbelt and Timmer, 2003; Jaeger and Kanrz, 1996; Judd, 2003; McSharry and Smith, 1999; Pisarenko and Sornette, 2004; Smirnov et al., 2005b) (see Sect. 8.1).

(ii) Parameter estimation *in the case of data deficit* is even more problematic. Such is a situation when one cannot get time series of all model dynamical variables x_k from the measured values of an observable η, i.e. some variables are "hidden" (Baake et al., 1992; Bezruchko et al., 2006; Breeden and Hubler, 1990; Parlitz, 1996; Voss et al., 2004; (see Sect. 8.2).

8.1 Parameter Estimators and Their Accuracy

Let us consider the estimation of a single parameter in a non-linear map from its noise-corrupted time realisation. The object is the quadratic map in a chaotic regime with unknown parameter $c = c_0$:

$$x_{n+1} = f(x_n, c_0) + \xi_n = 1 - c_0 x_n^2 + \xi_n, \, \eta_n = x_n + \zeta_n, \qquad (8.1)$$

where ξ_n, ζ_n are independent random processes. The first of them is a *dynamical noise* (since it influences the dynamics) and the second one is a *measurement noise* (since it affects only the observations).

If both noises are absent, one has $\eta_n = x_n$ and experimental data points on the plane x_n, x_{n+1} lie exactly on the sought parabola (Fig. 8.1a). Finding the value of c reduces to an algebraic equation whose solution is $\hat{c} = (1 - x_{n+1}) / x_n^2$. Hence, it is sufficient to use any two observations x_n, x_{n+1} with $x_n \neq 0$. As a result, a model coincides with the object up to the computation error.

If noise is present either in dynamics or measurements, one looks for a *parameter estimate* rather than for the precise value of the parameter. The most widely used estimation techniques are described in Sect. 7.1.1. Peculiarities of their application under the considered problem setting are as follows.

8.1.1 Dynamical Noise

Let the dynamical noise ξ_n in Eq. (8.1) be a sequence of statistically independent random quantities and identically distributed with a probability density $p(\xi)$. To estimate the parameter c, one can use the ML technique (Sects. 2.2.1, 7.1.1 and 7.1.2), which is the most efficient one under sufficiently general conditions (Ibragimov and Has'minskii, 1979; Pisarenko and Sornette, 2004). The likelihood function [see also Eqs. (2.26) and (7.10)] reads in this case as

$$\ln L(c) \equiv \ln p_N(\eta_1, \eta_2, \ldots, \eta_N \,|\, c) \approx \sum_{n=1}^{N-1} \ln p\,(\eta_{n+1} - f(\eta_n, c)). \qquad (8.2)$$

To apply the technique, one must know the distribution law $p(\xi)$, which is rarely the case. Most often, one assumes Gaussian noise so that the maximisation of Eq. (8.2) becomes equivalent to the so-called *ordinary least-squares technique*, i.e. to the minimisation

$$S(c) = \sum_{i=1}^{N-1} (\eta_{n+1} - f(\eta_n, c))^2 \to \min. \qquad (8.3)$$

It means that the plot of a model function on the plane (η_n, η_{n+1}) should go in such a way so as to minimise the sum of the squared *vertical distances* from it to the experimental data points (Fig. 8.1b).

As a rule, the error in the estimator $\hat{c}$ decreases with a time series length N. Under the considered problem setting, both the ML approach and the LS technique give asymptotically unbiased and consistent estimators. It can be shown that the estimator variance decreases as N^{-1} analogous to the examples in Sect. 7.1.2. The reason can be described in the same way: the terms in Eq. (8.3) are stationary with respect to i, i.e. a partial Fisher information is bounded.

The ordinary LS technique often gives an acceptable estimator accuracy even if the noise is not Gaussian (Sects. 7.1.1 and 7.1.2). Although one may apply other methods, e.g., the least absolute values technique, the LS technique is much easily implemented. An essential technical difficulty arises when the "relief" of the cost function (8.3) exhibits many local minima, which is often the case if f is non-linear with respect to $\mathbf{c}$. Then, the optimisation problem is solved in an iterative way with some starting guesses for the sought parameters (Dennis and Schnabel, 1983). Whether a global extremum is found depends on how "lucky" are the starting guesses i.e. how close they are to the true parameter values. In the example (8.1), f is linear with respect to c; therefore, the cost function S is quadratic with respect to c and exhibits the only minimum, which is easily found as a solution to a linear algebraic equation.

We note that if f is linear with respect to x, the model (8.1) is a linear first-order *autoregression model*. More general ARMA models involve a dependence of x_{n+1} on several previous values of x and ξ, see Eq. (4.13) in Sect. 4.4.

8.1.2 Measurement Noise

If only a measurement noise is present ($\eta_n = x_n + \zeta_n$), the estimation problem gets more complicated. This is because one aims at finding a dependence of x_{n+1} on x_n, where x_n is an "independent" variable whose observed values are noise corrupted [a confluent analysis problem, see Eq. (2.28) in Sect. 2.2.1.8).

8.1.2.1 Bias in the Estimator Obtained Via the Ordinary LS Technique

The bias is non-zero for an arbitrarily long time series, since the technique (8.3) is developed under the assumption of only a dynamical noise presence. It can be illustrated with the example (8.1), where one has

$$S(c) = \sum_{i=1}^{N-1} (\eta_{i+1} - f(\eta_i, c))^2 = \sum_{i=1}^{N-1} \left(x_{i+1} + \zeta_{i+1} - 1 + c(x_i + \zeta_i)^2 \right)^2 =$$
$$= \sum_{i=1}^{N-1} \left(cx_i^2 - c_0 x_i^2 + \zeta_{i+1} + 2cx_i\zeta_i + c\zeta_i^2 \right)^2.$$

Minimum of S over c can be found from the condition $\partial S/\partial c = 0$, which reduces to the form

$$\sum_{i=1}^{N-1}\left(cx_i^2 - c_0 x_i^2 + \zeta_{i+1} + 2cx_i\zeta_{i+1} + c\zeta_i^2\right) \cdot \left(x_i^2 + 2x_i\zeta_{i+1} + \zeta_i^2\right) = 0.$$

By solving this equation, one gets an estimator (McSharry and Smith, 1999)

$$\hat{c} = \frac{c_0\left(\sum_{i=1}^{N-1}x_i^4 + 2\sum_{i=1}^{N-1}x_i^3\zeta_i + \sum_{i=1}^{N-1}x_i^2\zeta_i^2\right) - \sum_{i=1}^{N-1}x_i^2\zeta_{i+1} - 2\sum_{i=1}^{N-1}x_i\zeta_i\zeta_{i+1} - \sum_{i=1}^{N-1}\zeta_i^2\zeta_{i+1}}{\sum_{i=1}^{N-1}x_i^4 + 6\sum_{i=1}^{N-1}x_i^2\zeta_i^2 + \sum_{i=1}^{N-1}x_i^4 + 4\sum_{i=1}^{N-1}x_i^3\zeta_i + 4\sum_{i=1}^{N-1}x_i\zeta_i^3}.$$

Under the condition $N \to \infty$, one can take into account statistical independence ζ_i of ζ_{i+1} and x_i and replace the sums like $\sum_{i=1}^{N-1} x_i^4$ (temporal averaging) by the integrals like $\int_{-\infty}^{\infty} \mu(x, c_0)x^4 dx \equiv \langle x^4\rangle$ (ensemble averaging). Here, $\mu(x, c_0)$ is an invariant measure for the map (8.1), i.e. a probability density function. At $c_0 = 2$ it can be found analytically: $\mu(x, 2) = 1/\pi\left(1 - x^2\right), -1 < x < 1$. Hence, one gets $\langle x^2\rangle = 1/2$, $\langle x^4\rangle = 3/8$ and $\langle x^n\rangle = 0$ for uneven n. Finally, at $c_0 = 2$, one comes to the asymptotic expression $\hat{c} = c_0\left(4\sigma_\zeta^2 + 3\right)\left/\left(8\langle\zeta^4\rangle + 24\sigma_\zeta^2 + 3\right)\right.$. This is a biased estimator. It underestimates the true value c_0, since the denominator is greater than the numerator in the above expression. Figure 8.2 shows the asymptotic value of $\hat{c}$ versus noise-to-signal ratio for Gaussian noise. It is close to the true value only under the low noise levels; its bias is less than 1% if the noise level is less

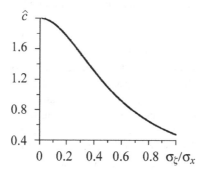

Fig. 8.2 Ordinary LS estimates of the parameter c in Eq. (8.1) versus the noise level at $N = \infty$ and $c_0 = 2$. Noise-to-signal ratio quantified as the ratio of standard deviations is shown versus the abscissa axis (McSharry and Smith, 1999)

than 0.05. The bias rises with the noise level. Analogous properties are observed for other noise distributions (Butkovsky et al., 2002; McSharry and Smith, 1999).

However, since the LS technique is simple for the implementation and can be easily used in the case of many estimated parameters in contrast to other methods, it is often applied in practice with an additional assumption of low measurement noise level.

8.1.2.2 Increasing Estimator Accuracy at High Measurement Noise Level

It is possible in part through the use of the *total least-squares technique* (Jaeger and Kanrz, 1996) when one minimises the sum of the squared orthogonal distances from the data points (η_n, η_{n+1}) to the plot of the function $f(x_n, c)$ (Fig. 8.1c). Thereby, one takes into account that deviations of the observed data points with coordinates (η_n, η_{n+1}) from the plot of the sought function $f(x_n, c_0)$ are induced by the noise influence on both coordinates. Therefore, the deviations may occur in any direction, not only vertically. The use of the orthogonal distances is justified in Jaeger and Kanrz (1996) as an approximate version of the ML approach.

However, a non-zero estimator bias is not fully eliminated under the use of the total LS technique (especially in the case of a very strong noise), since the latter is just an approximation to the ML technique. It may seem that a way out is to write down the likelihood function for the new situation "honestly", i.e. taking into account how the noise enters the observations. For a Gaussian noise, the problem reduces to the minimisation of the sum of squared deviations of a model *realisation* from *an observed time series* (Fig. 8.1d):

$$S(c, x_1) = \sum_{n=0}^{N-1} \left(\eta_{n+1} - f^{(n)}(x_1, c) \right)^2 \to \min, \qquad (8.4)$$

where $f^{(n)}$ is an nth iterate of the map $x_{n+1} = f(x_n, c)$, $f^{(0)}(x, c) = x$, and the initial model state x_1 is considered as an estimated quantity as well.

An orbit of a chaotic system is very sensitive to initial conditions and parameters. Therefore, the variance of the estimator obtained from Eq. (8.4) for a chaotic orbit rapidly decreases with N, sometimes even exponentially (Horbelt and Timmer, 2003; Pisarenko and Sornette, 2004). This is a desirable property, but it is achieved only if one manages to find the global minimum of Eq. (8.4). In practice, even for moderate values of N, the "relief" of S for a chaotic system becomes strongly "jagged" (Fig. 8.3a) so that it gets almost impossible to find the global minimum numerically (Dennis and Schnabel, 1983). To do it, one would need very "lucky" starting guesses for c and x_1. It is also difficult to speak of asymptotic properties of the estimators since the cost function gets non-smooth in the limit $N \to \infty$. Therefore, one develops modifications of the ML technique in application to the parameter estimation from a chaotic time series (Pisarenko and Sornette, 2004; Smirnov et al., 2005b).

Fig. 8.3 Cost functions for the quadratic map (8.1) at $c_0 = 1.85$ and $N = 20$. The *left panel* shows the cost function for the direct iterates (8.4), where $x_1 = 0.3$; the *right* one for the reverse iterates (8.5), where $x_N = f^{(N-1)}(0.3, c_0)$

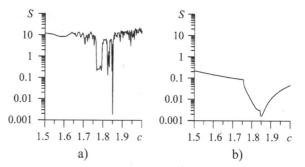

Thus, according to a piecewise technique, one divides an original time series into segments of a moderate length so that it is possible to find the global minimum of Eq. (8.4) for each of them and averages the obtained estimates. This is a reasonable approach, but a resulting estimator may remain asymptotically biased. Its variance decreases again as N^{-1}. Some ways to improve the estimator properties are described in Sect. 8.2.

Here, we note only an approach specific to one-dimensional maps (Smirnov et al., 2005b). It is based on the property that the only Lyapunov exponent of a one-dimensional chaotic map becomes negative under the time reversal so that an orbit gets much less sensitive to parameters and a "final condition". Therefore, one minimises a quantity

$$S(c, x_N) = \sum_{n=0}^{N-1} \left(\eta_{N-n} - f^{(-n)}(x_N, c) \right)^2 \to \min, \qquad (8.5)$$

where $f^{(-n)}$ is an nth iterate of the map $x_{n+1} = f(x_n, c)$ in reverse time, in particular $f^{(-1)}$ is an inverse function for f with respect to x. The plot of the cost function (8.5) looks sufficiently smooth for an arbitrarily long time series (Fig. 8.3b) so that it is not difficult to find its global minimum. At low and moderate noise levels (σ_ζ/σ_x up to 0.05–0.15), the error in the estimator (8.5) appears less than for the piecewise technique. Moreover, the expression (8.5) gives asymptotically unbiased estimates, whose variance typically scales as N^{-2} at weak noise. The high rate of error decrease is determined by the returns of a chaotic orbit to a small vicinity of the extremum of the function f (Smirnov et al., 2005b).

8.2 Hidden Variables

If the measurement noise level is considerable, the state variable x is often regarded "hidden" since its true values are, in fact, unknown. Variables are "even more hidden" if even their noise-corrupted values cannot be measured directly or computed from the observed data. This is often the case in practice. In such a situation, the parameter estimation is much more problematic than in the cases considered in Sect. 8.1. However, if one manages to solve the problem, then an additional opportunity to restore time series of hidden variables appears as a by-product. Then, a modelling procedure serves as a measurement device with respect to dynamical variables as well.

8.2.1 Measurement Noise

We illustrate the techniques with the example of the parameter estimation in ordinary differential equations without a dynamical noise. An object is the classical chaotic system – Lorenz system:

$$
\begin{aligned}
dx_1/dt &= c_1(x_2 - x_1), \\
dx_2/dt &= -x_2 + x_1(c_3 - x_3), \\
dx_3/dt &= c_2 x_3 + x_1 x_2,
\end{aligned}
\tag{8.6}
$$

for a "canonical" set of parameter values $c_1 = 10, c_2 = 8/3, c_3 = 46$. A noise-corrupted realisation of x_1 is considered as observed data, i.e. $\eta_n = x_1(t_n) + \zeta_n$, while the variables x_2 and x_3 are hidden. A model is constructed in the form (8.6) where all the three parameters c_k are regarded unknown.

8.2.1.1 Initial Value Approach

All the estimation techniques are based to a certain extent on the ideas like Eq. (8.4), i.e. one chooses such initial conditions and parameter values to provide maximal closeness of a model realisation to an observed time series in the least-squares sense. The direct solution to the problem like (8.4) is called an initial value approach (Horbelt, 2001; Voss et al., 2004). As indicated in Sect. 8.1.2, it is not applicable to a long chaotic time series. Improving the approach is not straightforward. Thus, a simple division of a time series into segments with subsequent averaging of the respective estimates gives a low accuracy of the resulting estimator. The reverse-time iterations are not suitable in the case of a multidimensional dissipative system.

8.2.1.2 Multiple Shooting Approach

The difficulties can be overcome in part with Bock's algorithm (Baake et al., 1992; Bock, 1981). It is also called a multiple shooting approach since one replaces the Cauchy initial value problem for an entire observation interval with a set of boundary value problems. Namely, one divides an original time series $\{\eta_1, \eta_2, \ldots, \eta_N\}$ into L non-overlapping segments of length M and considers model initial states $\mathbf{x}^{(i)}$ for each of them (i.e. at time instants $t_{(i-1)M+1}, i = 1, \ldots, L$) as estimated quantities, but *not as free parameters*. One solves the problem of conditional minimisation, which reads in the case of a scalar observable $\eta = h(\mathbf{x}) + \zeta$ ($\eta = x_1 + \zeta$ in our example) as

$$S\left(\mathbf{c}, \mathbf{x}^{(1)}, \mathbf{x}^{(2)}, \ldots, \mathbf{x}^{(L)}\right) =$$

$$= \sum_{i=1}^{L} \sum_{n=1}^{M} \left[\eta\left(t_{(i-1)M+n}\right) - h\left(\mathbf{x}\left(t_{(i-1)M+n} - t_{(i-1)M+1}, \mathbf{x}^{(i)}, \mathbf{c}\right)\right)\right]^2 \to \min,$$

$$\mathbf{x}\left(t_{iM+1} - t_{(i-1)M+1}, \mathbf{x}^{(i)}, \mathbf{c}\right) = \mathbf{x}^{(i+1)}, \quad i = 1, 2, \ldots, L-1. \tag{8.7}$$

The quantity $\mathbf{x}(t, \mathbf{x}^{(i)}, \mathbf{c})$ denotes a model realisation (a solution to model equations), i.e. a model state $\mathbf{x}$ at a time instant t for an initial state $\mathbf{x}^{(i)}$ and a parameter value $\mathbf{c}$. The first equation in Eq. (8.7) means minimisation of the deviations of a model realisation from the observed series over the entire observation interval. The second line provides "matching" of the segments to get finally a *continuous* model orbit over the entire observation interval. This matching imposes the conditions of the equality type on the L sought vectors $\mathbf{x}^{(i)}$, i.e. only one of the vectors can be regarded as a free parameter of the problem.

Next, one solves the problem with ordinary numerical iterative techniques using some *starting guesses* for the sought quantities $\mathbf{c}, \mathbf{x}^{(1)}, \mathbf{x}^{(2)}, \ldots, \mathbf{x}^{(L)}$. The starting guesses correspond, as a rule, to L non-matching pieces of a model orbit (Fig. 8.4b,

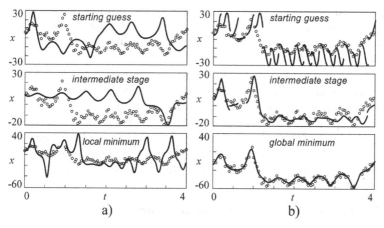

Fig. 8.4 Parameter estimation from a chaotic realisation of the coordinate $x = x_1$ of the Lorenz system (8.6), $N = 100$ data points, the sampling interval is 0.04, measurement noise is Gaussian and white with standard deviation of 0.2 of the standard deviation of the noise-free signal (Horbelt, 2001). (**a**) An initial value approach. Realisations of an observable (circles) and a corresponding model variable. The fitting process converges to a local minimum, where a model orbit and parameter estimates strongly differ from the true ones. (**b**) A multiple shooting approach. The fitting process converges to the global minimum, where a model orbit and parameter estimates are close to the true ones

upper panel). The situation is similar for intermediate values of the sought quantities during the iterative minimisation procedure, but the "discrepancies" should get smaller and smaller if the procedure converges at given starting guesses. Such a temporary admission of the model orbit discontinuity distinguishes Bock's algorithm (Fig. 8.4b) from the initial value approach (Fig. 8.4a) and provides greater flexibility of the former.

An example of the application of the two techniques to the parameter estimation for the system (8.6) is shown in Fig. 8.4. The multiple shooting technique "finds" the *global minimum*, while the initial value approach stops at a local one. This is quite a typical situation. However, the multiple shooting technique does not assure finding the global minimum. It only softens requirements to the "goodness" of starting guesses for the sought quantities (Bezruchko et al., 2006). For even longer chaotic time series, it also gets inefficient since the basic principle of closeness of a chaotic model orbit to the observed time series over a long time interval again leads to very strict requirements to starting guesses.

8.2.1.3 Modification of the Multiple Shooting Technique

As shown in Bezruchko et al. (2006), one can avoid some difficulties via allowing discontinuity of the resulting model orbit at several time instants within the observation interval, i.e. via ignoring several equalities in the last line of Eq. (8.7). In such a way it is easier to find the global minimum of the cost function S in Eq. (8.7).

Such a modification allows to use arbitrarily long chaotic time series, but the requital is that sometimes a model with an inadequate structure may be accepted as a "good" one due to its ability to reproduce short segments of the time series. Therefore, one should carefully select the number and the size of the continuity segments for the model orbit.

There is also an additional difficulty in the situation with hidden variables. Apart from lucky starting guesses for the parameters $\mathbf{c}$, it appears important to generate lucky starting guesses for the hidden variables (components of vectors $\mathbf{x}^{(i)}$) in contrast to very optimistic early statements (Baake et al., 1992). Quite often, one has to proceed intuitively or via blind trials and errors. However, useful information may be obtained sometimes through a preliminary study of the properties of model time realisations at several trial parameter values (Bezruchko et al., 2006).

8.2.1.4 Synchronisation-Based Parameter Estimation

A further improvement of the technique is based on the idea of synchronising a model by the observed time series. It was suggested in Parlitz (1996) and further elaborated in many works. Here, we briefly describe its main points, advantages and difficulties following the works Parlitz (1996) and Parlitz et al. (1996).

Let us consider a system of ODEs $d\mathbf{y}/dt = \mathbf{f}(\mathbf{y}, \mathbf{c}_0)$ with a state vector $\mathbf{y} \in R^D$ and a parameter vector $\mathbf{c}_0 \in R^P$ as an object of modelling. Parameter values $\mathbf{c}_0$ are unknown. An observable vector is $\mathbf{\eta} = \mathbf{h}(\mathbf{y})$. It may have a lower dimension than the state vector $\mathbf{y}$ (the case of hidden variables). A model is given by the equation

$d\mathbf{x}/dt = \mathbf{f}(\mathbf{x}, \mathbf{c})$. Let us assume that there is a unidirectional coupling scheme using the available signal $\boldsymbol{\eta}(t)$, which enables asymptotically stable synchronisation of the model ($\mathbf{x}$) by the object ($\mathbf{y}$), i.e. $\mathbf{x} \to \mathbf{y}$ as $t \to \infty$ if $\mathbf{c} = \mathbf{c}_0$. Thus, one can vary model parameters $\mathbf{c}$ and integrate model equations with somehow introduced input signal $\boldsymbol{\eta}(t)$ at each value of $\mathbf{c}$. If at some value $\mathbf{c} = \hat{\mathbf{c}}$ identical synchronisation between $\boldsymbol{\eta}(t)$ and the corresponding model realisation $\mathbf{h}(\mathbf{x}(t))$ (i.e. the regime $\boldsymbol{\eta}(t) = \mathbf{h}(\mathbf{x}(t))$) is achieved after some transient process, then the value $\hat{\mathbf{c}}$ should be equal to $\mathbf{c}_0$. If only an approximate relationship $\boldsymbol{\eta}(t) \approx \mathbf{h}(\mathbf{x}(t))$ holds true, then $\hat{\mathbf{c}}$ can be taken as an estimate of $\mathbf{c}_0$.

There are different implementations of the idea. One can introduce a measure of the discrepancy between $\boldsymbol{\eta}(t)$ and $\mathbf{h}(\mathbf{x}(t))$, e.g. their mean-squared difference after some transient, and minimise it as a function of $\mathbf{c}$ (Parlitz et al., 1996). This is a complicated problem of non-linear optimisation similar to those encountered under the multiple shooting approach. An advantage of the synchronisation-based estimation is that the minimised function often changes quite gradually with $\mathbf{c}$ and has a pronounced minimum at $\mathbf{c} = \mathbf{c}_0$ with a broad "basin of attraction", i.e. a starting guess for $\mathbf{c}$ does not have to be so "lucky" as under the multiple shooting approach. This is explained by the following property of many non-linear systems: If $\mathbf{c}$ is not equal to $\mathbf{c}_0$ but reasonably close to it, the identical synchronisation is impossible but there often occurs the generalised synchronisation (Sect. 6.4.5), where $\mathbf{x}$ is a function of $\mathbf{y}$ not very much different from $\mathbf{x} = \mathbf{y}$. Then, the discrepancy between $\boldsymbol{\eta}(t)$ and $\mathbf{h}(\mathbf{x}(t))$ changes smoothly in the vicinity of $\mathbf{c} = \mathbf{c}_0$.

It can be even more convenient to avoid minimisation of a complicated cost function by considering the parameters $\mathbf{c}$ as additional variables in model ODEs and update their values depending on the current mismatch between $\boldsymbol{\eta}(t)$ and $\mathbf{h}(\mathbf{x}(t))$ in the course of integration of the model ODEs (Parlitz, 1996). An example is again the chaotic Lorenz system

$$
\begin{aligned}
dy_1/dt &= \sigma(y_2 - y_1), \\
dy_2/dt &= c_{1,0}y_1 - c_{2,0}y_2 - y_1 y_3 + c_{3,0}, \\
dy_3/dt &= y_1 y_2 - b y_3,
\end{aligned}
\tag{8.8}
$$

with the parameters $c_{1,0} = 28, c_{2,0} = 1, c_{3,0} = 0, \sigma = 10, b = 8/3$ and an observable $\eta = h(\mathbf{y}) = y_2$. The following unidirectional coupling scheme and equations for the parameter updates were considered:

$$
\begin{aligned}
dx_1/dt &= \sigma(\eta - x_1), \\
dx_2/dt &= c_1 x_1 - c_2 x_2 - x_1 x_3 + c_3, \\
dx_3/dt &= x_1 x_2 - b x_3, \\
dc_1/dt &= (\eta - x_2)x_1, \\
dc_2/dt &= -(\eta - x_2)x_2, \\
dc_3/dt &= \eta - x_2.
\end{aligned}
\tag{8.9}
$$

Using a global Lyapunov function, the author has shown that at correct parameter values $c = c_0$, the model is synchronised by the signal $\eta(t)$ at all initial conditions. The system (8.9) tends to the regime $x(t) = y(t)$ and $c(t) = c_0$ at any values of $x_1(0), x_2(0), x_3(0), c_1(0), c_3(0)$ and any positive $c_2(0)$. Parameter estimates in this example appeared not very sensitive to the mismatch in the parameter σ. More general recommendations on the choice of the coupling scheme were also suggested from geometrical considerations in Parlitz (1996). Rather different equations for the parameter updates were suggested by several authors, e.g., Konnur (2003).

The synchronisation-based approach is theoretically justified for the noise-free case. Still, numerical experiments have shown its good performance when a moderate measurement noise is present. As it has been already mentioned, the technique can be readily used in the case of hidden variables. However, despite several important advantages of the approach, it may encounter its own significant difficulties. Firstly, an asymptotically stable synchronisation may not be achieved for any observable $\eta = h(y)$. This possibility depends on the system under study and the coupling scheme. Secondly, it is not always clear what coupling scheme should be used to assure synchronisation at $c = c_0$. Finally, it may be very important to select appropriate initial conditions in Eqs. (8.8) and (8.9), in other words, to select starting guesses for the model parameters and hidden variables. Further details can be found in Chen and Kurths (2007); Freitas et al. (2005); Hu et al. (2007); Huang (2004); Konnur (2003); Marino and Miguez (2005); Maybhate and Amritkar (1999); Parlitz (1996); Parlitz et al. (1996); Tao et al. (2004).

8.2.2 Dynamical and Measurement Noise

Estimating model parameters in the case of simultaneous presence of the dynamical and measurement noise is a more complicated task. However, there have been recently developed corresponding sophisticated techniques such as Kalman filtering-based methods (Sitz et al., 2002, 2004; Voss et al., 2004) and Bayesian approaches (Bremer and Kaplan, 2001; Davies, 1994; Meyer and Christensen, 2000). Below, we describe in some detail a recently suggested technique (Sitz et al., 2002) called *unscented Kalman filtering*.

Kalman filtering is a general idea which was originally developed for the state estimation in linear systems from observed noise-corrupted data (Kalman and Bucy, 1961). It is widely used, e.g., in data assimilation (Sect. 5.1) as well as in many other applications (Bar-Shalom and Fortmann, 1988). The idea has been generalised for the estimation of model parameters together with model states. It has been recently further generalised for the estimation of parameters and state vectors in non-linear systems (Sitz et al., 2002).

In the linear case, the model is assumed to be of the form

$$x_i = A \cdot x_{i-1} + \xi_i, \qquad (8.10)$$
$$\eta_i = B \cdot x_i + \zeta_i,$$

where $\mathbf{x}$ is a state vector of the dimension D, i is the discrete time, $\boldsymbol{\eta}$ is the vector of observables whose dimension may be different from that of $\mathbf{x}$, $\mathbf{A}$ and $\mathbf{B}$ are constant matrices, and $\boldsymbol{\xi}_n$ and $\boldsymbol{\zeta}_n$ are independent zero-mean Gaussian white noises with diagonal covariance matrices. The problem is to estimate the state $\mathbf{x}$ at a time instant n having observations $\boldsymbol{\eta}_i$ up to time n inclusively, i.e. a set $\mathbf{H}_n = \{\boldsymbol{\eta}_1, \boldsymbol{\eta}_2, \ldots, \boldsymbol{\eta}_n\}$. Let us denote such an estimator as $\hat{\mathbf{x}}(n|n)$. Kalman filter provides an optimal linear estimate, i.e. unbiased and with the smallest variance. Formally, the estimation procedure consists of the predictor and corrector steps. At the predictor step, one estimates the value $\hat{\mathbf{x}}(n|n-1)$, i.e. takes into account only the observations $\mathbf{H}_{n-1}$ up to time instant $n-1$. The optimal solution to such a problem is $\hat{\mathbf{x}}(n|n-1) = E[\mathbf{x}_n|\mathbf{H}_{n-1}]$. It can be written as

$$\hat{\mathbf{x}}(n|n-1) = E[\mathbf{A} \cdot \mathbf{x}_{n-1}|\mathbf{H}_{n-1}] = \mathbf{A} \cdot E[\mathbf{x}_{n-1}|\mathbf{H}_{n-1}] = \mathbf{A} \cdot \hat{\mathbf{x}}(n-1|n-1),$$
$$\hat{\boldsymbol{\eta}}(n|n-1) = E[\mathbf{B} \cdot \mathbf{x}_n|\mathbf{H}_{n-1}] = \mathbf{B} \cdot \mathbf{A} \cdot \hat{\mathbf{x}}(n-1|n-1). \tag{8.11}$$

The second equation in Eq. (8.11) will be used at the correction step below. The point estimators must be equipped with the confidence bands (see Sect. 2.2.1). It is known that due to the linearity of the system, the estimators are Gaussian distributed. Thus, their confidence bands are simply expressed via their covariance matrices whose optimal estimates read as

$$\mathbf{P}(n|n-1) = E[(\mathbf{x}_n - \hat{\mathbf{x}}(n|n-1)) \cdot (\mathbf{x}_n - \hat{\mathbf{x}}(n|n-1))^{\mathrm{T}}|\mathbf{H}_{n-1}],$$
$$\mathbf{P}_{\boldsymbol{\eta}\boldsymbol{\eta}}(n|n-1) = E[(\boldsymbol{\eta}_n - \hat{\boldsymbol{\eta}}(n|n-1)) \cdot (\boldsymbol{\eta}_n - \hat{\boldsymbol{\eta}}(n|n-1))^{\mathrm{T}}|\mathbf{H}_{n-1}], \tag{8.12}$$
$$\mathbf{P}_{\mathbf{x}\boldsymbol{\eta}}(n|n-1) = E[(\mathbf{x}_n - \mathbf{x}(n|n-1)) \cdot (\boldsymbol{\eta}_n - \hat{\boldsymbol{\eta}}(n|n-1))^{\mathrm{T}}|\mathbf{H}_{n-1}].$$

These matrices can be expressed via their previous estimates $\mathbf{P}(n-1|n-1)$, $\mathbf{P}_{\boldsymbol{\eta}\boldsymbol{\eta}}(n-1|n-1)$, $\mathbf{P}_{\mathbf{x}\boldsymbol{\eta}}(n-1|n-1)$ analytically for the linear system.

Now, the corrector step updates the predictor-step estimators taking into account the last observation $\boldsymbol{\eta}_n$ as follows:

$$\hat{\mathbf{x}}(n|n) = \hat{\mathbf{x}}(n|n-1) + \mathbf{K}_n \cdot (\boldsymbol{\eta}_n - \hat{\boldsymbol{\eta}}(n|n-1)),$$
$$\mathbf{P}(n|n) = \mathbf{P}(n|n-1) - \mathbf{K}_n \cdot \mathbf{P}_{\boldsymbol{\eta}\boldsymbol{\eta}}(n|n-1) \cdot \mathbf{K}_n^{\mathrm{T}},$$
$$\mathbf{K}_n = \mathbf{P}_{\mathbf{x}\boldsymbol{\eta}}(n|n-1) \cdot \mathbf{P}_{\boldsymbol{\eta}\boldsymbol{\eta}}^{-1}(n|n-1). \tag{8.13}$$

Thus, the corrections represent discrepancy between the predictor-step estimates and actual observations multiplied by the so-called Kalman gain matrix $\mathbf{K}_n$. Having Eqs. (8.11), (8.12) and (8.13), one can start from initial guesses $\hat{\mathbf{x}}(1|1)$ and $\mathbf{P}(1|1)$ and recursively get optimal state estimates for all subsequent time instants taking into account subsequent observations. Due to Gaussianity of the distributions, a 95% confidence band for the jth component of the vector $\mathbf{x}_n$ is given by $\hat{x}_j(n|n) \pm 1.96\sqrt{\mathbf{P}_{jj}(n|n)}$.

To apply the idea to a non-linear system $\mathbf{x}_{i+1} = \mathbf{f}(\mathbf{x}_i) + \boldsymbol{\xi}_i$ and $\boldsymbol{\eta}_i = \mathbf{h}(\mathbf{x}_i) + \boldsymbol{\zeta}_i$, one can either approximate non-linear functions with Taylor expansion or simulate the distribution of states and compute many model orbits to get an estimator $\hat{\mathbf{x}}(n|n)$ and its covariance matrix $\mathbf{P}(n|n)$. The latter idea appeared more fruitful in practice (Sitz et al., 2002). Its fast and convenient implementation includes the selection of the so-called sigma points $\mathbf{x}^{(1)}, \ldots, \mathbf{x}^{(2D)}$ specifying the distribution of states at a time instant $n-1$:

$$
\begin{aligned}
\mathbf{x}^{(j)}(n-1\,|\,n-1) &= \hat{\mathbf{x}}(n-1\,|\,n-1) + \left[\sqrt{D \cdot \mathbf{P}(n-1\,|\,n-1)}\right]_j, \\
\mathbf{x}^{(j+D)}(n-1\,|\,n-1) &= \hat{\mathbf{x}}(n-1\,|\,n-1) - \left[\sqrt{D \cdot \mathbf{P}(n-1\,|\,n-1)}\right]_j, \quad (8.14)
\end{aligned}
$$

where $j = 1, 2, \ldots, D$ and $\left[\sqrt{\cdot}\right]_j$ means jth column of the matrix square root. The sigma points are propagated through the non-linear systems giving

$$
\begin{aligned}
\mathbf{x}^{(j)}(n\,|\,n-1) &= \mathbf{f}(\mathbf{x}^{(j)}(n-1\,|\,n-1)), \\
\mathbf{y}^{(j)}(n\,|\,n-1) &= \mathbf{h}(\mathbf{x}^{(j)}(n\,|\,n-1)), \quad (8.15)
\end{aligned}
$$

where $j = 1, \ldots, 2D$. Now, their sample means and covariances define the predictor estimates as follows:

$$
\hat{\mathbf{x}}(n\,|\,n-1) = \frac{1}{2D} \sum_{j=1}^{2D} \mathbf{x}^{(j)}(n\,|\,n-1),
$$

$$
\hat{\boldsymbol{\eta}}(n\,|\,n-1) = \frac{1}{2D} \sum_{j=1}^{2D} \mathbf{y}^{(j)}(n\,|\,n-1),
$$

$$
\mathbf{P}_{\boldsymbol{\eta\eta}}(n\,|\,n-1) = \frac{1}{2D} \sum_{j=1}^{2D} (\mathbf{y}^{(j)}(n\,|\,n-1) - \hat{\boldsymbol{\eta}}(n\,|\,n-1)) \cdot (\mathbf{y}^{(j)}(n\,|\,n-1) - \hat{\boldsymbol{\eta}}(n\,|\,n-1))^{\mathrm{T}},
$$

$$
\mathbf{P}_{\mathbf{x}\boldsymbol{\eta}}(n\,|\,n-1) = \frac{1}{2D} \sum_{j=1}^{2D} (\mathbf{x}^{(j)}(n\,|\,n-1) - \hat{\mathbf{x}}(n\,|\,n-1)) \cdot (\mathbf{y}^{(j)}(n\,|\,n-1) - \hat{\boldsymbol{\eta}}(n\,|\,n-1))^{\mathrm{T}},
$$

$$
\mathbf{P}(n\,|\,n-1) = \frac{1}{2D} \sum_{j=1}^{2D} (\mathbf{x}^{(j)}(n\,|\,n-1) - \hat{\mathbf{x}}(n\,|\,n-1)) \cdot (\mathbf{x}^{(j)}(n\,|\,n-1) - \hat{\mathbf{x}}(n\,|\,n-1))^{\mathrm{T}}.
$$

$$(8.16)$$

The estimates (8.16) are updated via the usual Kalman formulas (8.13). This procedure is called unscented Kalman filtering (Sitz et al., 2002).

If parameter $\mathbf{a}$ of a system $\mathbf{x}_{i+1} = \mathbf{f}(\mathbf{x}_i, \mathbf{a}) + \boldsymbol{\xi}_i$ is unknown, then it can be formally considered as an additional state component. Moreover, it is convenient to consider noises $\boldsymbol{\xi}_i$ and $\boldsymbol{\zeta}_i$ as components of a joint state vector. Thus, joint equations of motion read as

$$\mathbf{x}_{i+1} = \mathbf{f}(\mathbf{x}_i, \mathbf{a}_i) + \boldsymbol{\xi}_i,$$
$$\mathbf{a}_{i+1} = \mathbf{a}_i, \qquad\qquad\qquad (8.17)$$
$$\boldsymbol{\xi}_{i+1} = \boldsymbol{\xi}_i,$$
$$\boldsymbol{\zeta}_{i+1} = \boldsymbol{\zeta}_i,$$

where the last three equations do not alter starting values of the additional state components. However, the estimates of these components change in time due to the correction step (8.13), which takes into account a new observation.

The technique can be easily generalised to the case of ordinary and stochastic differential equations, where numerical integration scheme would enter the first equation of Eq. (8.17) instead of a simple function $\mathbf{f}$.

We note that the procedure can be used in the case of hidden variables, which correspond to the situation where the dimension of the observable vector $\boldsymbol{\eta}$ is less than D. Examples with a scalar observable from two- and three-dimensional dynamical systems are considered in Sitz et al. (2002), where efficiency of the approach is illustrated. Thus, even in the case of the deterministically chaotic Lorenz system, the unscented Kalman filtering allowed accurate estimation of the three parameters from a scalar time realisation. Concerning this example, we note two things. Firstly, the successful application of a statistical method (Kalman filtering has its roots in mathematical statistics and the theory of random processes) to estimate parameters in a deterministic system illustrates again a close interaction between deterministic and stochastic approaches to modelling discussed in Chap. 2 (see, e.g., Sect. 2.6). Secondly, the unscented Kalman filtering seems to be more efficient than the multiple shooting approach (Sect. 8.2.1) in many cases since the former technique does not require a continuous model orbit over the entire observation interval. In this respect, the unscented Kalman filtering is similar to the modified multiple shooting approach which allows several discontinuity points. However, the unscented Kalman filtering is easier in implementation, since it does not require to solve optimisation problem for the parameter estimation. On the other hand, the multiple shooting technique would certainly give more accurate parameter estimates for a deterministically chaotic system if one manages to find good starting guesses for the parameters and find the global minimum of the cost function (8.7).

The Kalman filtering-based approach also resembles the synchronisation-based approach: It also involves parameter updates depending on the current model state. However, the difference is that the Kalman filtering does not require any synchronisation between a model and the observed data and is appropriate when a dynamical noise is present in contrast to the synchronisation-based technique developed for deterministic systems.

A proper selection of starting guesses for the model parameters and the hidden variables is of importance for the unscented Kalman filtering as well, since an arbitrary starting guess does not assure convergence of the non-linear recursive procedure [Eqs. (8.16) and (8.13)] to the true parameter values. More detailed discussion of the unscented Kalman filtering can be found in Sitz et al. (2002, 2004); Voss et al. (2004).

Even more general techniques are based on the Bayesian approach (Bremer and Kaplan, 2001; Davies, 1994; Meyer and Christensen, 2000), where the state and parameter estimators are calculated based on the entire available set of observations rather than only on the previous ones. However, full Bayesian estimation is more difficult to use and implement, since it requires to solve a complicated numerical problem of sampling from multidimensional probability distributions.

8.2.2.1 Concluding Remark

Model validation for the considered "transparent box" problems is performed along two main lines. The first one is the analysis of the model residuals (Box and Jenkins, 1970), i.e. the check for their correspondence to the assumed noise properties (see Sect. 7.3). The second one is the computation of dynamical, geometrical and topological characteristics of a model attractor and their comparison with the respective properties of an original (Gouesbet et al., 2003b) (see Sect. 10.4).

8.3 What One Can Learn from Modelling Successes and Failures

Successful realisation of the above techniques (Sect. 8.2) promises an opportunity to obtain parameter estimates and time courses of hidden variables. It would allow several useful applications such as validation of model ideas, "measurement" of quantities inaccessible to measurement devices and restoration of lost or distorted segments of data. Let us comment it in more detail.

In studying of real-world objects, a researcher never meets a purely "transparent box" setting. He/she can only believe subjectively that a trial model structure is adequate to an original. Therefore, even with a perfect realisation of the procedures corresponding to the final modelling stages (Fig. 5.1), the result may appear negative, i.e. one may not get a valid model with a given structure. Then, a researcher should declare incorrectness of his/her ideas about the process under investigation and return to the stage of the model structure selection. If there are several alternative mathematical constructions, then modelling from time series may reveal the most adequate among them. Thus, the modelling procedure gives an opportunity to reject or confirm (possibly, to make more accurate) some substantial ideas about the object under investigation.

There are a number of practical examples of successful application of the approaches described above. Thus, in Horbelt et al. (2001) the authors confirm validity of their ideas about gas laser functioning and get directly immeasurable parameters of the transition rates between energetic levels depending on the pumping current. In Swameye et al. (2003) the authors are able to make substantial conclusions about the mechanism underlying a biochemical signalling process in cells which is described below.

8.3.1 An Example from Cell Biology

In many applications it is necessary to find out which cell properties determine an undesirable process in the cells most strongly and how one can purposefully affect those properties.[1] To answer such questions, it may be sufficient to get an adequate mathematical model as demonstrated in Swameye et al. (2003).

The authors investigate one of many intracellular signalling pathways, which provide a cell with an opportunity to produce necessary substances in response to variations in surroundings. In particular, such pathways provide reproduction, differentiation and survival of cells. The authors consider the so-called signalling pathway JAK-STAT, which transforms an external chemical signal into activation of a respective gene transcription in a cell nucleus (Fig. 8.5). One of the simplest mathematical models of the process can be written down based on the law of active mass (a usual approach in chemical kinetics) and reads as

$$
\begin{aligned}
&\mathrm{d}x_1/\mathrm{d}t = -k_1 x_1 E(t),\\
&\mathrm{d}x_2/\mathrm{d}t = -k_2 x_2^2 + k_1 x_1 E(t),\\
&\mathrm{d}x_3/\mathrm{d}t = -k_3 x_3 + k_2 x_2^2/2,\\
&\mathrm{d}x_4/\mathrm{d}t = k_3 x_3.
\end{aligned}
\tag{8.18}
$$

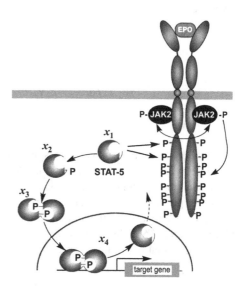

Fig. 8.5 A scheme of a biochemical signalling process in a cell (Swameye et al., 2003)

---

[1] Thus, growth of cancer cells is determined by the fact that they produce substances "inadequately" to the surrounding situation. A method of struggle against the disease, which is currently only hypothetical, could rely on the empirical modelling similar to that described in Swameye et al. (2003).

Here, k_i are reaction rates, $E(t)$ is the concentration of erythropoietin in the extracellular space (denoted Epo in Fig. 8.5), whose variations lead to the activation of respective receptors on a cell membrane. The receptors are bound to tyrosine kinase of the type JAK-2 existing in a cell cytoplasm. Tyrosine kinase reacts with molecules of the substance STAT5, whose concentration is denoted x_1. As a result of the reaction, the latter are phosphorylated. There arise monomeric tyrosine phosphorylated molecules STAT5, whose concentration is denoted x_2. This reaction leads to a decrease in x_1 [the first line in Eq. (8.18)] and an increase in x_2 [a positive term in the second line of Eq. (8.18)]. The monomeric molecules dimerise when they meet each other. Concentration of the dimeric molecules is denoted x_3. This reaction leads to a decrease in x_2 [a negative term in the second line of Eq. (8.18)] and an increase in x_3 [a positive term in the third line of Eq. (8.18)]. The dimeric molecules penetrate into the nucleus, where their concentration is denoted x_4. This process leads do a decrease in x_3 [a negative term in the third line of Eq. (8.18)] and an increase in x_4 [the fourth line in Eq. (8.18)]. The dimeric molecules activate the transcription of a target gene. As a result, a specific protein is produced. At that, the dimeric molecules dissociate into the monomeric ones, which *degrade inside the nucleus according to the hypothesis underlying the model* (8.18).

However, there is another hypothesis according to which the monomeric molecules STAT5 *are relocated from the cell nucleus to the cytoplasm after a certain delay time*. Under such an assumption, the mathematical model slightly changes and takes the form

$$
\begin{aligned}
dx_1(t)/dt &= -k_1 x_1(t) E(t) + 2k_4 x_3(t - \tau), \\
dx_2(t)/dt &= -k_2 x_2^2(t) + k_1 x_1(t) E(t), \\
dx_3(t)/dt &= -k_3 x_3(t) + k_2 x_2^2(t)\big/2, \\
dx_4(t)/dt &= -k_4 x_3(t - \tau) + k_3 x_3(t),
\end{aligned}
\tag{8.19}
$$

where additional time-delayed terms appear in the first and the fourth lines and τ is the delay time. Which of the two models (i.e. which of the two hypotheses) is valid is unknown. Opportunities of the observations are quite limited, an equipment is quite expensive and a measurement process is quite complicated (Swameye et al., 2003). In an experiment, the authors could measure only the total mass of the phosphorylated STAT5 in the cytoplasm η_1 and the total mass of STAT5 in the cytoplasm η_2 up to constant multipliers:

$$
\begin{aligned}
\eta_1 &= k_5(x_2 + 2x_3), \\
\eta_2 &= k_6(x_1 + x_2 + 2x_3),
\end{aligned}
\tag{8.20}
$$

where k_5, k_6 are unknown proportionality coefficients. Along with those two observables, the concentration $E(t)$ is measured up to a proportionality coefficient (Fig. 8.6a). We stress that all model dynamical variables x_k are hidden. There are

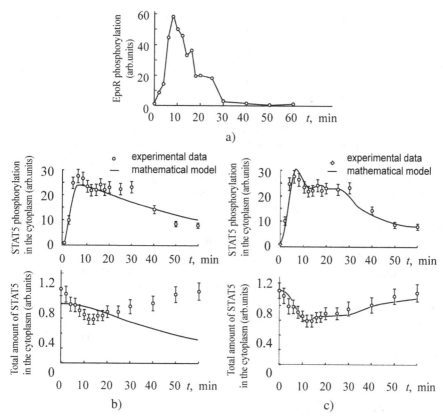

Fig. 8.6 Modelling of a biochemical cell signalling process (Swameye et al., 2003): (**a**) a time series of an external driving, variations in erythropoietin concentration; (**b**), (**c**) the results of an empirical model construction in the form (8.18) and (8.19), respectively

only two observables, which are related to the four dynamical variables in the known way (8.20).

The parameters in both models (8.18) and (8.19) are estimated in Swameye et al. (2003) from the described experimental data. The authors have shown invalidity of the former model and a good correspondence between the experiments and the latter model. Namely, the time series were measured in three independent experiments and one of them is shown in Fig. 8.6. Model parameters were estimated from all the available data via the initial value approach (Sect. 8.2.1). This approach was appropriate since the time series were rather short. They contained only 16 data points per experiment session and represented responses to "pulse" changes in the erythropoietin concentration (Fig. 8.6a). Thus, one should not expect problems with finding the global minimum of the cost function. The model (8.18) appeared incapable of reproducing the observed time series (Fig. 8.6b), while the model (8.19) was adequate in this respect (Fig. 8.6c). The estimate of the delay time was $\tau \approx 6$ min, which agreed by the order of magnitude with the results of other authors obtained with different techniques for similar objects.

Thus, only the modelling from time series allowed the authors to make a non-trivial conclusion that relocation of STAT5 molecules to the cytoplasm plays a significant role in the process under study. They found out some details of the process, which cannot be observed directly, for instance, the stay of the STAT5 molecules in the nucleus approximately for 6 min.

Moreover, by studying the model (8.19), one can predict what happens if some parameters of the process are varied. For example, the authors studied how the total mass of the protein produced by a cell (which is proportional to a total number of STAT5 molecules participating in the process) changes under variations in different model parameters. This quantity appeared to depend very weakly on k_1, k_2 and quite strongly on k_3, k_4, τ. In other words, the processes in the nucleus play a major role. According to the model (8.19), decreasing k_4 down to zero (inhibition of a nuclear export) leads to the decrease in the produced protein mass by 55%. In experiments with leptomycin B, the authors inhibited a nuclear export (an analogue to the parameter k_4) only by 60%. According to the model, it should lead to the reduction in the amount of activated STAT5 by 40%. In experiment, the reduction by 37% was observed. Thus, a model prediction was finely confirmed, which further increases one's confidence to the model and allows to use it for a more detailed study of the process and its control. Having such achievements, one could think about opportunities to use empirical modelling for medical purposes.

8.3.2 Concluding Remarks

Despite the successes mentioned above, the problem of modelling may often appear technically unsolvable even under the relatively simple "transparent box" setting and for an adequate model structure. To date, examples of successful modelling are observed when the difference between the model dimension and the number of observables is not greater than two or three and the number of estimated parameters is not greater than 3–5. These are only rough figures to give an impression about typical practical opportunities. In each concrete case, modelling results depend on specific non-linearities involved in the model equations. In any case, the greater the number of hidden variables and unknown model parameters, the weaker the chances for a successful modelling and the lower the accuracy of parameter estimators.

References

Baake, E., Baake, M., Bock, H.J., Briggs, K.M.: Fitting ordinary differential equations to chaotic data. Phys. Rev. A. **45**, 5524–5529 (1992)

Bar-Shalom, Y., Fortmann, T.E.: Tracking and Data Association. Academic Press, Orlando (1988)

Bezruchko, B.P., Smirnov, D.A., Sysoev, I.V.: Identification of chaotic systems with hidden variables (modified Bock's algorithm). Chaos, Solitons Fractals **29**, 82–90 (2006)

Bock, H.G.: Numerical treatment of inverse problems in chemical reaction kinetics. In: Ebert, K.H., Deuflhard, P., Jaeger, W., et al. (eds.) Modelling of Chemical Reaction Systems, pp. 102–125 Springer, New York (1981)

Box, G.E.P., Jenkins, G.M.: Time Series Analysis. Forecasting and Control. Holden-Day, San Francisco (1970)

Breeden, J.L., Hubler, A.: Reconstructing equations of motion from experimental data with unobserved variables. Phys. Rev. A. **42**, 5817–5826 (1990)

Bremer, C.L., Kaplan, D.T.: Markov chain Monte Carlo estimation of nonlinear dynamics from time series. Phys. D. **160**, 116–126 (2001)

Butkovsky, O.Ya., Kravtsov Yu.A., Logunov, M.Yu. Analysis of a nonlinear map parameter estimation error from noisy chaotic time series. Radiophys. Quantum Electron **45**(1), 55–66, (in Russian) (2002)

Chen, M., Kurths, J.: Chaos synchronization and parameter estimation from a scalar output signal. Phys. Rev. E. **76**, 027203 (2007)

Davies, M.E.: Noise reduction schemes for chaotic time series. Physica D. **79**, 174–192 (1994)

Dennis, J., Schnabel, R.: Numerical Methods for Unconstrained Optimization and Nonlinear Equations. Prentice-Hall, Engle Wood Cliffs, NJ (1983)

Freitas, U.S., Macau, E.E.N., Grebogi, C.: Using geometric control and chaotic synchronization to estimate an unknown model parameter. Phys. Rev. E. **71**, 047203 (2005)

Gouesbet, G., Meunier-Guttin-Cluzel, S., Ménard, O.: Global reconstructions of equations of motion from data series, and validation techniques, a review. In: Gouesbet, G., Meunier-Guttin-Cluzel, S., Ménard, O. (eds.) Chaos and Its Reconstructions, pp. 1–160. Nova Science Publishers, New York, (2003b)

Horbelt, W., Timmer, J., Bünner, M.J., et al. Identifying physical properties of a CO_2 laser by dynamical modeling of measured time series. Phys. Rev. E. **64**, 016222 (2001)

Horbelt, W., Timmer, J.: Asymptotic scaling laws for precision of parameter estimates in dynamical systems. Phys. Lett. A. **310**, 269–280 (2003)

Horbelt, W.: Maximum Likelihood Estimation in Dynamical Systems: PhD Thesis. University of Freiburg, Freiburg. Available at http://webber.physik.uni-freiburg.de/~horbelt/diss. (2001)

Horbelt, W.: Maximum Likelihood Estimation in Dynamical Systems: PhD thesis. University of Freiburg, Freiburg. Available at http://webber.physik.uni-freiburg.de/~horbelt/diss (2001)

Hu, M., Xu, Z., Zhang, R., Hu, A.: Parameters identification and adaptive full state hybrid projective synchronization of chaotic (hyper-chaotic) systems. Phys. Lett. A. **361**, 231–237 (2007)

Huang, D.: Synchronization-based estimation of all parameters of chaotic systems from time series. Phys. Rev. E. **69**, 067201 (2004)

Ibragimov, I.A., Has'minskii R.Z.: Asymptotic Theory of Estimation. Nauka, Moscow (1979). Translated into English Under the Title Statistical Estimation: Springer, New York (1981)

Jaeger, L., Kanrz, H.: Unbiased reconstruction of the dynamics underlying a noisy chaotic time series. Chaos. **6**, 440–450 (1996)

Judd, K.: Chaotic time series reconstruction by the Bayesian paradigm: Right results by wrong methods?. Phys. Rev. E. **67**, 026212 (2003)

Kalman, R.E., Bucy, R.S.: New results in linear filtering and prediction theory. J. Basic Eng. ASME. **83**, 95–108 (1961)

Konnur, R.: Synchronization-based approach for estimating all model parameters of chaotic systems. Phys. Rev. E. **67**, 027204 (2003)

Marino, I.P., Miguez, J.: Adaptive approximation method for joint parameter estimation and identical synchronization of chaotic systems. Phys. Rev. E. **72**. 057202 (2005)

Maybhate, A., Amritkar, R.E.: Use of synchronization and adaptive control in parameter estimation from a time series. Phys. Rev. E. **59**, 284–293 (1999)

McSharry, P.E., Smith, L.A.: Better nonlinear models from noisy data: attractors with maximum likelihood. Phys. Rev. Lett. **83**, 4285–4288 (1999)

Meyer, R., Christensen, N.: Bayesian reconstruction of chaotic dynamical systems. Phys. Rev. E. **62**, 3535–3542 (2000)

Parlitz, U., Junge, L., Kocarev, L.: Synchronization-based parameter estimation from time series. Phys. Rev. E. **54**, 6253–6259 (1996)

Parlitz, U.: Estimating model parameters from time series by auto-synchronization. Phys. Rev. Lett. **76**, 1232–1235 (1996)

Pisarenko, V.F., Sornette, D.: Statistical methods of parameter estimation for deterministically chaotic time series. Phys. Rev. E. **69**, 036122 (2004)

Sitz, A., Schwartz, U., Kurths, J., Voss, H.U.: Estimation of parameters and unobserved components for nonlinear systems from noisy time series. Phys. Rev. E. **66**, 016210 (2002)

Sitz, A., Schwarz, U., Kurths, J.: The unscented Kalman filter, a powerful tool for data analysis. Int. J. Bifurc. Chaos. **14**, 2093–2105 (2004)

Smirnov, D.A., Vlaskin, V.S., Ponomarenko, V.I.: Estimation of parameters in one-dimensional maps from noisy chaotic time series. Phys. Lett. A. **336**, 448–458 (2005)

Swameye, I., Muller, T.G., Timmer, J., et al.: Identification of nucleocytoplasmic cycling as a remote sensor in cellular signaling by data based modeling. Proc. Natl. Acad. Sci. USA. **100**, 1028–1033 (2003)

Tao, C., Zhang Yu., Du, G., Jiang, J.J.: Estimating model parameters by chaos synchronization. Phys. Rev. E. **69**, 036204 (2004)

Voss, H.U., Timmer, J., Kurths, J.: Nonlinear dynamical system identification from uncertain and indirect measurements. Int. J. Bif. Chaos. **14**, 1905–1933 (2004)

Yule, G.U.: On a method of investigating periodicities in disturbed series, with special reference to Wolfer's sunspot numbers. Phil. Trans. R. Soc. London A. **226**, 267–298 (1927)

Chapter 9
Model Equations: Restoration of Equivalent Characteristics

In the case called "grey box" in Sect. 5.2, a researcher has *partial* knowledge about the structure of model equations $\mathbf{x}_{n+1} = \mathbf{f}(\mathbf{x}_n, \mathbf{c})$ or $d\mathbf{x}/dt = \mathbf{f}(\mathbf{x}, \mathbf{c})$. More concretely, some *components* of the function $\mathbf{f}$ are unknown. Then, the problem gets more complicated, than just *parameter estimation* discussed in Chap. 8, and more interesting from a practical viewpoint.

To illustrate its peculiarities, let us consider modelling of a harmonically driven non-linear dissipative oscillator described by the equations

$$dx_1/dt = x_2,$$
$$dx_2/dt = -\gamma_0 x_2 + F(x_1) + A_0 \cos(\omega_0 t + \phi_0), \tag{9.1}$$

where $\gamma_0, A_0, \omega_0, \phi_0$ are parameters and F is a non-linear restoring force, whose *form is unknown*. Let the variable x_1 be an observable, i.e. $\eta = x_1$. Note that the function F is only a component of the entire dynamical system (9.1): the function F together with the other terms in the right-hand side specifies the phase velocity field. In practice, F is a characteristic of the object, which makes clear physical sense and can be of significant interest by itself. Its values may be unavailable for direct measurements due to experimental conditions, i.e. it may be impossible to get experimental data points directly on the plane (x_1, F). However, information about the function F is contained in the time series, since F influences the dynamics. One can "extract" the values of F indirectly, i.e. via construction of an empirical model whose structure includes a model function corresponding to F. Namely, one should construct a model in the form

$$dx_1/dt = x_2,$$
$$dx_2/dt = -\gamma x_2 + f(x_1, \mathbf{c}) + A \cos(\omega t + \phi), \tag{9.2}$$

where $f(x_1, \mathbf{c})$ should approximate F. Approximation of a *one-variable* function is a much more feasible task in practice, than a general problem of multivariable function approximation arising in "black box" reconstruction (Chap. 10). If one manages to get a "good" model, the ideas behind the model structure (9.2) are validated and the characteristic F is restored in the form $f(x_1, \hat{\mathbf{c}})$. We stress that it can be the only way to get the characteristic F.

B.P. Bezruchko, D.A. Smirnov, *Extracting Knowledge From Time Series*, Springer Series in Synergetics, DOI 10.1007/978-3-642-12601-7_9,
© Springer-Verlag Berlin Heidelberg 2010

Due to the importance of information about characteristics of non-linear elements inaccessible for direct measurements, we call the considered modelling problem "restoration of equivalent characteristics". Components of a model function **f** can make different physical sense: a restoring force, non-linear friction, etc. Opportunities to extract such information arise if physical sense is introduced into a model structure in advance. More often, this is achieved with differential equations, since many laws of nature are formulated in such a form.

As in Chap. 8, the models considered here determine either a dependency "a future state x_{n+1} versus a current state x_n" or "a phase velocity dx/dt versus a state **x**". The difference from Chap. 8 is that one must specify a functional form of the characteristics to be restored before the stage of parameter estimation (Sect. 9.1). Hence, it gets more important to optimise a model structure (Sect. 9.2) and even to select it in a specific way for a certain object (Sect. 9.3) or a class of objects (Sect. 9.4).

9.1 Restoration Procedure and Peculiarities of the Problem

9.1.1 Discrete Maps

Let an original be a one-dimensional map $x_{n+1} = F(x_n)$, where an observable is $\eta_n = x_n$. Let the dimension of the system be known and the form of the function F unknown.

A model is constructed as a one-dimensional map $x_{n+1} = f(x_n, \mathbf{c})$. For this simple example, data points on the plane (η_n, η_{n+1}) represent the plot of F. One should just select the form of $f(x, \mathbf{c})$ and find the values of **c** so as to approximate the data points in the best way (Fig. 9.1). Only the entire function $f(x, \hat{\mathbf{c}})$ makes physical sense, rather than each single parameter, that is typical under the "grey box" setting.

The problem is almost the same as in Chap. 7, see, e.g., Fig. 7.1b. The difference is that the quantities (η_n, η_{n+1}) are shown along the coordinate axes rather than the quantities (t, η). Therefore, one can use the techniques discussed in Sect. 7.2.1

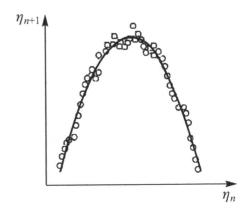

Fig. 9.1 Construction of a model map: finding a dependence of the next value of an observable on the previous one from experimental data (*circles*)

by replacing the pair of quantities (t, η) by (η_n, η_{n+1}). Thus, it is convenient to use a low-order algebraic polynomial or cubic splines (Sect. 7.2.4) to approximate a one-variable function shown in Fig. 9.1. Parameters can be estimated with the ordinary LS technique (8.3).

If a dynamical noise is present in the original map as $x_{n+1} = F(x_n) + \xi_n$, nothing changes in the model construction procedure. A measurement noise, i.e. $\eta_n = x_n + \zeta_n$, makes the parameter estimation more complicated (Sect. 8.1.2). If its level is low, the ordinary LS technique is still suitable. For higher levels of the measurement noise, it would be desirable to use more sophisticated techniques (Sects. 8.1.2 and 8.2.1), but under the "grey box" setting, a model typically contains many parameters to be estimated, which makes the use of those techniques much more troublesome.

9.1.2 Ordinary Differential Equations

To describe complex motions, including chaotic ones, one uses non-linear model ODEs $d\mathbf{x}/dt = \mathbf{f}(\mathbf{x},\mathbf{c})$ with at least three dynamical variables. Some components of the velocity field $\mathbf{f}$ can be unknown as in the examples (9.1) and (9.2). Those "equivalent characteristics" are built into the model structure and one can find them via the construction of the entire model. Let us consider some details.

The first case is when all the dynamical variables x_k are observed: $\eta_k(t_i) = x_k(t_i) + \zeta_k(t_i)$, $k = 1, \ldots, D$. To construct a model, one approximates a dependence of the derivative $dx_k(t_i)/dt$ on $\mathbf{x}(t_i)$ with a function $f_k(\mathbf{x},\mathbf{c}_k)$ for each k. The values of $dx_k(t_i)/dt$ are usually obtained from the observed data $\eta_k(t_i)$ via numerical differentiation (Sect. 7.4.2). Let us denote their estimates $d\hat{x}_k(t_i)/dt$. "Smoothed" values of the dynamical variables $\hat{x}_k(t_i)$ emerge as a by-product of the differentiation procedure. From the values $\hat{x}_k(t_i)$, $d\hat{x}_k(t_i)/dt$, one estimates model parameters $\mathbf{c}_k$ with the ordinary LS technique:

$$S(\mathbf{c}_k) = \sum_i \left(d\hat{x}_k(t_i)\big/dt - f_k(\hat{\mathbf{x}}(t_i), \mathbf{c}_k)\right)^2 \rightarrow \min, k = 1, \ldots, D. \qquad (9.3)$$

The functions $f_k(\mathbf{x},\mathbf{c}_k)$ contain, in particular, sought equivalent characteristics.

The second typical case is when one observes a single dynamical variable: $\eta(t_i) = x_1(t_i) + \zeta(t_i)$. The dimension D of the system is known. Successful modelling gets more probable if the dynamical equations for the original take the standard form (3.27), where the components of the state vector $\mathbf{x}$ are successive derivatives of the variable x_1. A model is constructed in the corresponding form

$$d^D x(t)\big/dt^D = f(x(t), dx(t)\big/dt, d^{D-1}x(t)\big/dt^{D-1}, \mathbf{c}). \qquad (9.4)$$

Firstly, one gets the time series of $\hat{x}, \hat{x}^{(1)}, \hat{x}^{(2)}, \ldots, \hat{x}^{(D-1)}$, where superscript denotes an order of the derivative, via numerical differentiation of the observable

$\eta(t_i)$. Parameters of the function f, which includes equivalent characteristics to be restored, are estimated with the ordinary LS technique:

$$S(\mathbf{c}) = \sum_i \left(\hat{x}^{(D)}(t_i) - f(\hat{x}(t_i), \hat{x}^{(1)}(t_i), \ldots, \hat{x}^{(D-1)}(t_i), \mathbf{c}) \right)^2 \to \min. \qquad (9.5)$$

The described techniques perform well under sufficiently low levels of the measurement noise. Moreover, the results are more reliable if the functions f_k in Eq. (9.3) and f in Eq. (9.5) depend on the parameters in a linear way, i.e. represent pseudo-linear models (Sect. 7.2.4). For higher noise levels, the modelling gets much more difficult, since numerical differentiation amplifies any noise, especially when $D > 1$ derivatives are computed. Then, the ordinary LS technique becomes unsuitable, while the use of more sophisticated techniques (Sect. 8.2.1) in the case of *multidimensional* models with *many* unknown parameters is also unrealistic.

9.1.3 Stochastic Differential Equations

A more general idea, which is seemingly advantageous in the case of a multiplicative dynamical noise, is suggested in Friedrich et al. (2000) and Siegert et al. (1998). It is based on the estimation of parameters in the Fokker – Planck equation for a considered non-linear system, i.e. the estimation of the drift and diffusion coefficients (Sects. 4.3 and 4.5). Thus, let an object of modelling be given by the Langevin equations

$$dx_k / dt = F_k(\mathbf{x}) + G_k(\mathbf{x})\xi_k(t), \, k = 1, \ldots, D,$$

where independent zero-mean white noises $\xi_k(t)$ have auto-covariance functions $\langle \xi_k(t)\xi_k(t') \rangle = \delta(t - t')$. One assumes that the dimension D and noise properties are known, and all D state variables are observed. Only concrete functional forms of $F_k(\mathbf{x})$, $G_k(\mathbf{x})$ are unknown so that these functions are to be determined from a time series.

Let us consider the case of $D = 1$ for simplicity of notations, i.e. the system $dx/dt = F(x) + G(x)\xi(t)$. Recall that the Fokker – Planck equation (4.8) is defined as

$$\frac{\partial p(x, t)}{\partial t} = -\frac{\partial}{\partial x}(c_1(x, t)p(x, t)) + \frac{1}{2}\frac{\partial^2}{\partial x^2}(c_2(x, t)p(x, t)),$$

where the drift coefficient is

$$c_1(x, t) = \lim_{\tau \to 0} \frac{1}{\tau} \int_{-\infty}^{\infty} (x' - x)p(x', t + \tau | x, t)dx'$$

and the diffusion coefficient is

$$c_2(x, t) = \lim_{\tau \to 0} \frac{1}{\tau} \int_{-\infty}^{\infty} (x' - x)^2 p(x', t + \tau | x, t) dx'$$

The functions F, G are related to these coefficients as

$$c_1(x) = F(x) + \frac{1}{2} \frac{dG(x)}{dx} G(x)$$

and $c_2(x) = G^2(x)$ (Sect. 4.3). If $c_1(x), c_2(x)$ are known, the functions F, G can be restored from them (e.g., if one requires positivity of $G(x)$). Moreover, to answer many questions, it is possible to use the Fokker – Planck equation directly, rather than the original stochastic DE with the functions F, G.

As one can see from the above definitions, $c_1(x)$ and $c_2(x)$ are directly related to the conditional mean and conditional variance of the next value of x, given the current value of x. The conditional mean and variance can be estimated from data just as the sample mean and sample variance (Sect. 2.2.1) over all observed states close to a given state x. Having the estimates of the conditional mean and variance for different intervals τ (the smallest possible value of τ being equal to the sampling interval Δt), one can estimate the limits $\tau \to 0$ by extrapolation (Friedrich et al., 2000). Thereby, the estimates $\hat{c}_1(x)$ and $\hat{c}_2(x)$ are obtained. They are reliable at a given state x if an observed orbit passes near this state many times. Thus, the estimates $\hat{c}_1(x)$ and $\hat{c}_2(x)$ are more accurate for the often visited regions in the state space. They are poorly defined for "rarely populated" regions.

The estimates $\hat{c}_1(x)$ and $\hat{c}_2(x)$ are obtained in a non-parametric form (just as tables of numerical values). However, one may approximate the obtained dependencies $\hat{c}_1(x)$ and $\hat{c}_2(x)$ with any smooth functions if necessary, e.g. with a polynomial (Friedrich et al., 2000). The obtained functions $\hat{c}_1(x)$ and $\hat{c}_2(x)$ can also be considered as (non-linear) characteristics of the system under study. Deriving the estimates of the functions F and G, entering the original stochastic equation, from the estimates $\hat{c}_1(x)$ and $\hat{c}_2(x)$ is possible under some conditions on the function G assuring uniqueness of the relationship. Getting the functions F and G is of a specific practical interest if they have a clearer physical interpretation compared to the coefficients $\hat{c}_1(x)$ and $\hat{c}_2(x)$.

Several examples of successful applications of the approach to numerically simulated time series, electronic experiments and physiological data, as well as a detailed discussion are given in Friedrich et al. (2000), Ragwitz and Kantz (2001); and Siegert et al. (1998). The approach directly applies if there is no measurement noise. Its generalisation to the case of measurement noise is presented in Siefert et al. (2003).

9.2 Model Structure Optimisation

Model structure selection is as important for the restoration of equivalent characteristics as for the problem considered in Sect. 7.2.1. To choose a model size, e.g.
a polynomial order, one can use the criteria described in Sect. 7.2.3. However, the
question remains: How to choose small subset of function terms from a large set of
basis functions to provide the best model of a given size?

Let us consider an efficient approach (Bezruchko et al., 2001a) with the example
of the reconstruction of equations for the van der Pol – Toda oscillator

$$\mathrm{d}x_1 / \mathrm{d}t = x_2,$$
$$\mathrm{d}x_2 / \mathrm{d}t = (1 - x_1^2)x_2 - 1 + \mathrm{e}^{-x_1}. \qquad (9.6)$$

A time series of the observable is $\eta = x_1$ is supposed to be available. The corresponding phase orbit, containing a transient process, is shown in Fig. 9.2. A model
is constructed in the form

$$\mathrm{d}x_1 / \mathrm{d}t = x_2,$$
$$\mathrm{d}x_2 / \mathrm{d}t = f(x_1, x_2, \mathbf{c}), \qquad (9.7)$$

where

$$f(x_1, x_2, \mathbf{c}) = \sum_{i,j=0}^{K} c_{i,j} x_1^i x_2^j, \quad i + j \le K.$$

Many terms in the polynomial are "superfluous" since they have no analogues in
Eq. (9.6), e.g. the terms $c_{0,0}$, $c_{1,1}x_1x_2$, $c_{0,2}x_2^2$ and others. Estimates of the coefficients corresponding to the *superfluous terms* can appear non-zero due to various
errors and fluctuations. This circumstance can strongly reduce model quality. Thus,
it is desirable to exclude the superfluous terms from the model equations.

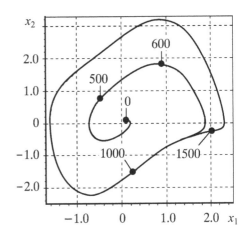

Fig. 9.2 A phase orbit of the
van der Pol – Toda oscillator
(9.6) which contains a
transient process. The
attractor is a limit cycle. The
numbers are temporal indices
of some data points in the
time series, which is recorded
with a sampling interval of
0.01. There are about 600
data points per basic period
of oscillations

Superfluous terms can be identified via the estimation of the model coefficients from different segments of the time series, i.e. from the data points lying in different domains of the phase space. The estimates of the "necessary" coefficients must not depend on a time series segment used. In contrast, the estimates of the coefficients corresponding to superfluous terms are expected to exhibit considerable variations. Such variations are stronger if a time series contains a transient process (Anishchenko et al., 1998; Bezruchko et al., 2001a), since the phase orbit then explores different domains in the phase space (Fig. 9.2).

We constructed the model (9.7) with a polynomial of a high-order K from subsequent time series segments of length W: $\{\eta_{(k-1)W+1}, \ldots, \eta_{(k-1)W+W}\}$, $k = 1, 2, \ldots, L$. Thus, we obtained a set of estimates for each coefficient $\hat{c}_{i,j}^{(k)}$ (Fig. 9.3a). A degree of stability of each coefficient estimate $\hat{c}_{i,j}$ is defined as $\left|\langle\hat{c}_{i,j}\rangle\right|/\sigma_{i,j}$, where

$$\langle\hat{c}_{i,j}\rangle = (1/L)\sum_{k=1}^{L}\hat{c}_{i,j}^{(k)}$$

and

$$\sigma_{i,j} = \sqrt{(1/L)\sum_{k=1}^{L}\left(\hat{c}_{i,j}^{(k)} - \langle c_{i,j}\rangle\right)^2}.$$

The term corresponding to the least stable coefficient was excluded. The entire procedure was repeated for the simplified model structure. By repeating the exclusion procedure many times, we sequentially removed the "unstable terms".

The procedure was stopped when the model quality did no longer improve. The criterion of quality was a minimum of the approximation error over a wide area V in the phase space (shown in Fig. 9.3b)

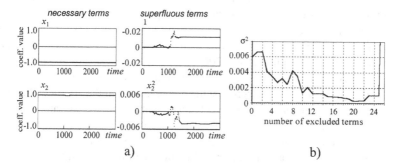

a)   b)

Fig. 9.3 Construction of a model for the van der Pol – Toda oscillator (9.6) starting with a full two-variable polynomial of order $K = 7$ in Eq. (9.7) from time series segments of length $W = 2000$ data points: (**a**) estimates of the coefficients, corresponding to the indicated terms, versus the starting time instant of a time series segment (in units of sampling interval), the *thick lines*; (**b**) the model error versus the number of excluded terms, it is minimal for 20 terms excluded

$$\sigma^2 = \iint\limits_V \left\{ f(x_1, x_2, \hat{\mathbf{c}}) - \left[(1 - x_1^2)x_2 - 1 + \mathrm{e}^{-x_1} \right] \right\}^2 \mathrm{d}x_1 \, \mathrm{d}x_2.$$

After excluding 20 terms from an initial polynomial with $K = 7$, the error was reduced by an order of magnitude in comparison with its starting value (Bezruchko et al., 2001a). At that, the final model reproduced much more accurately the dynamics of an object (including transient processes starting from different initial conditions) in the entire square region of the phase space shown in Fig. 9.2. This is because the described procedure of the model structure optimisation allows better generalisation of some essential dynamical features by reducing the danger of overfitting.

9.3 Equivalent Characteristics for Two Real-World Oscillators

In this section, we describe our results on the restoration of equivalent characteristics of oscillatory processes from the fields of physiology (Stoop et al., 2006) and electronics (Bezruchko et al., 1999a).

9.3.1 Physiological Oscillator

The cochlear amplifier is a fundamental, generally accepted concept in cochlear mechanics, having a large impact on our understanding of how hearing works. The concept, first brought forward by Gold in 1948 (Gold, 1948), posits that an active mechanical process improves the mechanical performance of the ear (Robles and Ruggero, 2001). Until recently, the study of this amplifying process has been restricted to the ears of vertebrates, where the high complexity and the limited accessibility of the auditory system complicate the in situ investigation of the mechanisms involved. This limitation has hampered the validation of cochlear models that have been devised (Dallos et al., 1996; Kern and Stoop, 2003). The hearing organs of certain insects have recently been shown to exhibit signal-processing characteristics similar to the mammalian cochlea by using active amplification (Goepfert and Robert, 2001; 2003; Goepfert et al., 2005); the ears of these insects are able to actively amplify incoming stimuli, display a pronounced compressive non-linearity, exhibit power gain and generate self-sustained oscillations in the absence of sound. In both vertebrates and insects, the mechanism that promotes this amplification resides in the motility of auditory mechanosensory cells, i.e. vertebrate hair cells and insect chordotonal neurons. Both types of cells are developmentally derived by homologous genes and share similar transduction machineries, pointing to a common evolutionary origin (Boekhoff-Falk, 2005). In line with such an evolutionary scenario, it seems possible that the fundamental mechanism of active amplification in the ears of insects and vertebrates is also evolutionarily conserved (Robert and Goepfert, 2002).

Since insect's hearing organs are located on the body surface, they are accessible to non-invasive examination. Moreover, because the external sound receiver is often directly coupled to the auditory sense cells, insect auditory systems can be expected to provide profound experimental and theoretical insights into the in situ mechanics of motile sense cells and their impact on the mechanical performance of the ear. Such information is technically relevant: providing natural examples of refined active sensors, the minuscule ears of insects promise inspiration for the design of nanoscale artificial analogues. Here, we present the results of modelling *self-sustained oscillations* of the antennal ear of the fruit fly *Drosophila melanogaster* with non-linear oscillator equation (9.7) and restoring its equivalent characteristics (Stoop et al., 2006).

In *Drosophila*, hearing is mediated by mechanosensory neurons that directly connect to an external sound receiver formed by the distal part of the antennas (Goepfert and Robert, 2000). These neurons actively modulate the receiver mechanics and, occasionally, give rise to self-sustained receiver oscillations (SOs). SOs occur spontaneously and are reliably induced by thoracic injection of dimethyl sulphoxide (DMSO), a local analgesic known to affect insect's auditory transduction. The precise action of DMSO on the auditory neurons remains unclear. However, as spontaneous and DMSO-induced SOs are both physiologically vulnerable and display similar temporal patterns, the latter can be used to probe the nature of the amplification mechanism in the fly's antennal ear (Goepfert and Robert, 2001). As revealed by measurements of the receiver vibrations (Fig. 9.4), about 20 min after the administration of DMSO, fully developed SOs are observed (Fig. 9.4b). They exhibit the temporal profile of *relaxation oscillations* with a characteristic frequency of about 100 Hz (Goepfert and Robert, 2003). About 10 min later, the SOs start to decrease in amplitude (Fig. 9.4c) and finally converge to a sinusoidal profile (Fig. 9.4d). The evoked SOs may last for up to 1–1.5 h.

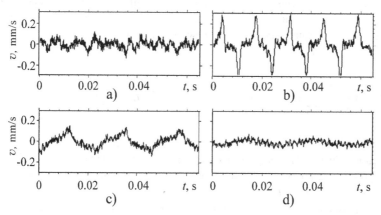

Fig. 9.4 Self-sustained oscillations of the *Drosophila* hearing sensor (velocity measurements): (**a**) 10 min, (**b**) 20 min, (**c**) 30 min, (**d**) 34 min after DMSO injection

The temporal profile of these oscillations is reminiscent of limit cycle oscillations generated by the van der Pol oscillator

$$
\begin{aligned}
dx_1/dt &= x_2, \\
dx_2/dt &= (\mu - x_1^2)x_2 - x_1,
\end{aligned}
\tag{9.8}
$$

where x_1 is identified with the receiver's vibrational position and the control parameter $\mu > 0$ is slowly decreased in order to account for the changes in the SO shape during time. It is well known that at $\mu = 0$, the van der Pol oscillator undergoes the *Andronov – Hopf bifurcation*; for $\mu > 0$, a stable limit cycle emerges that can be interpreted as undamping (i.e. amplification). A more detailed examination of the experimental data reveals a pronounced asymmetry (see Fig. 9.4b) by comparing the onsets and extents of the upward and downward excursions within one period, which requires a more general model for the SO generation than the standard van der Pol system.

In order to capture this asymmetry, we construct a model in the form of the generalised van der Pol oscillator (9.7) with $f(x_1, x_2) = f_1(x_1)x_2 - f_2(x_1)$, where $f_1(x_1)$ and $f_2(x_1)$ describe polynomials of the orders n and m, respectively. From the viewpoint of physics, $-f_1(x_1)$ describes a non-linear and possibly negative friction, whereas $-f_2(x_1)$ describes a non-linear restoring force. It is necessary to determine the orders n and m and polynomial coefficients that yield the optimal reproduction of the experimental data. One can expect that for a proper model, the polynomial orders n and m are unambiguously determined and only variations in the coefficients account for the observed changes in the SO temporal profile over time.

From the measurements with the sampling interval $\Delta t = 0.08$ ms, we are provided with a time series of the receiver's vibration velocities v which is described by the variable x_2 in the model. The values of the displacement and the acceleration are determined via numerical integration and differentiation, respectively. The latter is performed by applying the first-order Savitzky – Golay filter (Sect. 7.4.2). Quasi-stationary segments of the original data of lengths $N = 4000$ data points (i.e. the duration of 0.32 s) are used for the model construction. In order to determine the optimal polynomial orders n and m, we use the criterion of the training error saturation (Sect. 7.2.3):

$$
\hat{\varepsilon}^2 = \min_{\mathbf{c}_1, \mathbf{c}_2} \frac{1}{N} \sum_{i=1}^{N} \left(d\hat{x}_2(t_i)/dt - f_1(\hat{x}_1(t_i), \mathbf{c}_1)\hat{x}_2(t_i) + f_2(\hat{x}_1(t_i), \mathbf{c}_2) \right)^2.
\tag{9.9}
$$

The error $\hat{\varepsilon}$ saturates for $n = 2$ and $m = 5$ (Fig. 9.5). A further increase in n and m does not reduce $\hat{\varepsilon}$. The emergence of such a conspicuous saturation point is a rare case in practice and indicates that the model structure (9.7) faithfully reproduces the auditory data of *Drosophila*.

A comparison between realisations of the model and the measurements corroborates the validity of our modelling. For the fully developed SOs (after 20 min, Fig. 9.6), the comparison reveals that the measured velocities are faithfully

Fig. 9.5 Mean-squared error $\hat{\varepsilon}$ of the model fitting (9.9) showing a precipitous decay and saturation around the orders $n = 2$ and $m = 5$

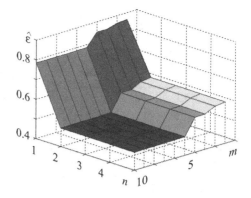

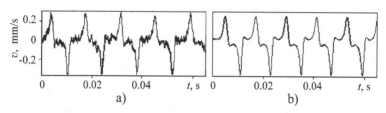

Fig. 9.6 Diagnostic check: (**a**) experimentally measured receiver's vibration velocity observed 20 min after DMSO injection, i.e. fully developed SOs; (**b**) a corresponding time series generated by the model (9.7) with $n = 2$ and $m = 5$

reproduced. This is further illustrated in Fig. 9.7, where the modelled and the measured data are compared on the phase plane (x_1, x_2). Similar observations take place for the time series recorded 10, 30 and 34 min after DMSO injection, respectively.

The shapes of the polynomials $f_1(x_1)$ and $f_2(x_1)$ reflect the asymmetry of the observed receiver oscillations, specifically when the SOs are fully developed (Fig. 9.4b). The asymmetry of $f_1(x_1)$ (Fig. 9.8a) and, in particular, $f_2(x_1)$ (Fig. 9.8b) becomes effective at large displacements and may have its origin in structural-mechanical properties of the antenna. An enlightening interpretation of

Fig. 9.7 Phase-space representation of the measured (*dots*) and model (the *solid line*) receiver vibrations in the case of fully developed SOs

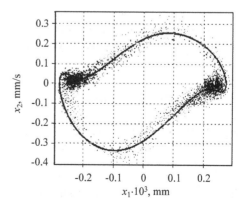

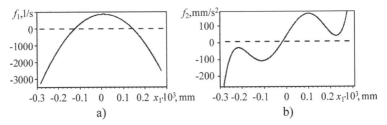

Fig. 9.8 Restored non-linear characteristics of the receiver for the fully developed SOs 20 min after DMSO injection: (**a**) the second-order polynomial $f_1(x_1)$, which means a non-linear friction with the opposite sign and shows the undamping $f_1(x_1) > 0$ (see the *dashed line*); (**b**) the fifth-order polynomial $f_2(x_1)$, which means a non-linear restoring force with the opposite sign and displays a noticeable asymmetry

the amplification dynamics can be given for the behaviour around zero displacement position $x_1 \approx 0$, where $f_1(x_1)$ attains positive values for small displacements x_1 (Fig. 9.8a). Since $-f_1(x_1)$ represents friction, the inequality $f_1(x_1) > 0$ implies that energy is injected into the system. This is a characteristic feature of an active amplification process. Around $x_1 = 0$, the non-linear restoring force $-f_2(x_1)$ and its first and second derivatives are relatively small. This implies that for small receiver displacements, virtually no restoring force is present. By means of the negative friction term, the system is thus easily driven out to relatively large amplitudes.

In the course of time, i.e. with decreasing DMSO concentration, the non-linear contributions to friction and restoring force decay. In particular, the range, where the friction is negative, gradually decreases and finally vanishes in agreement with the observed reduction in SO amplitude (see Fig. 9.4). When the SO starts to disappear, the restoring force function $f_2(x_1)$ gets approximately linear with a very small slope. At the same time, the friction term remains to be very small. As a consequence, weak stimuli will be sufficient to elicit considerable antennal vibrations. Although the amplifier has now returned into a stable state, where limit cycles do not occur, it remains very sensitive. Only small parameter variations are necessary in order to render the friction term negative and to lead to an amplification of incoming vibrations.

Thus, the model obtained captures several characteristics of the antennal ear oscillations, indicating that the empirical modelling may be useful for analysing the physics of the cochlear amplifier as well (Stoop et al., 2006).

9.3.2 Electronic Oscillator

An illustrative example from the field of electronics refers to the case when a chaotic motion of a non-linear system is successfully modelled under the "grey box" setting. An object is an *RLC* circuit with switched capacitors under an external sinusoidal driving with the amplitude U_0 and the angular frequency ω_0. Its scheme is presented in Fig. 9.9a, where K is an electronic key, a micro-scheme containing dozens of

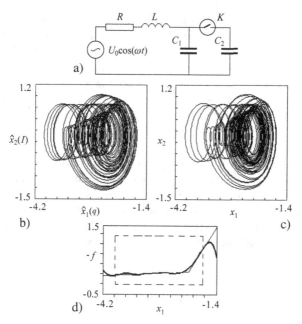

Fig. 9.9 Modelling of an electronic oscillator: (**a**) the scheme of an experimental set-up; (**b**) an experimental chaotic orbit on the plane "charge versus current" in dimensionless units (Bezruchko et al., 1999a); the values of I are measured with a 12-bit ADC at $\Delta t = 4\,\mu s$, $C_1 = 0.1\,\mu F$, $C_2 = 4.4\,\mu F$, $L = 20\,mH$, $R = 10\,\Omega$, $U_{thr} = -0.2\,V$, the driving period $T \approx 84.02\,\Delta t$, and $U_0 = 2.344\,V$; (**c**) an orbit of the reconstructed model (9.2) with the polynomial f of the ninth order; (**d**) a plot for the function $-f$ (*thick line*) and an expected piecewise linear dependence (*thin line*)

transistors and other passive elements, which is fed from a special source of direct voltage. Under small values of the voltage U on the capacity C_1, linear oscillations in the circuit RLC_1 take place, since the resistance of the key is very high. When U reaches a threshold value U_{thr}, resistance of the key reduces abruptly so that it closes the circuit and connects the capacity C_2. Reverse switching occurs at the value of U somewhat lower than U_{thr}, i.e. the key exhibits a hysteresis. It is the presence of non-linearity that leads to the possibility of chaotic oscillations in the circuit.

A model of this system derived from Kirchhoff's laws takes the form of the non-autonomous non-linear oscillator (9.2). The dimensionless variables read $t = t'/\sqrt{LC_1}$ and $x_1 = q/C_2|U_{thr}|$, where t' is the physical time and q is the total charge on the capacities C_1 and C_2. It is expected that the original function F is piecewise linear due to such voltage – capacity characteristic of the non-linear element represented by the switched capacitors.

Experimental measurements provide us with a chaotic time series of the current I through the resistor R, which corresponds to the quantity x_2 in Eq. (9.2). The time series of the variable x_1 is obtained via numerical integration of the observed signal and the time series of the variable dx_2/dt is obtained via numerical differentiation. We do not use information about the piecewise linear form of F, especially recalling that it is a theoretical approximation which ignores hysteresis of the key and other realistic features. Models are constructed in the form (9.2) with polynomials f of different orders K. Figure 9.9c shows the results for the best model with $K = 9$, which reproduces well the observed chaotic motion illustrated in Fig. 9.9b. The theoretical piecewise linear "restoring force" and the model polynomial f coincide with a good accuracy in the observed range of the x_1 values bounded by the dashed line in Fig. 9.9d. We note that without prior information about the model structure (9.2), it is impossible to get an adequate empirical model making physical sense (Bezruchko et al., 1999a).

This example illustrates restoring equivalent characteristics of a non-linear element via empirical modelling even in regimes of large amplitudes and chaos, where such characteristics may be inaccessible to direct measurements with ordinary tools. The empirical modelling has been successfully used to study dynamical characteristics of a ferroelectric capacitor (Hegger et al., 1998), semiconductor diodes (Sysoev et al., 2004) and optical fibre systems (Voss and Kurths, 1999).

9.4 Specific Choice of Model Structure

Uncertainty with respect to a model structure may not be so small as in the above examples. The "box" can be "dark grey" rather than "light grey" (Fig. 5.1) which makes empirical modelling much more difficult. However, in some cases, even small amount of a priori information along with a preliminary analysis of an observed time series can lead to a success if it is properly taken into account in a model structure. This is illustrated below with two wide classes of objects: systems under regular external driving and time-delay systems.

9.4.1 Systems Under Regular External Driving

If the presence of regular (periodic or quasi-periodic) driving is known a priori or assumed from a preliminary data analysis (e.g., strong discrete components in a power spectrum), then it is fruitful to include functions *explicitly* depending on time into model equations to describe such driving.

Thus, to describe an additive harmonic driving, one can reasonably use the model structure

$$\mathrm{d}^D x \big/ \mathrm{d}t^D = f(x, \mathrm{d}x/\mathrm{d}t, \dots, \mathrm{d}^{D-1}x \big/ \mathrm{d}t^{D-1}, \mathbf{c}) + a \cos \omega t + b \sin \omega t, \quad (9.10)$$

where x is an observable and f is an algebraic polynomial (Bezruchko and Smirnov, 2001; Bezruchko et al., 1999a). One may use smaller number of variables D in Eq. (9.10) than it would be necessary for the autonomous standard model (9.4). This circumstance determines advantages of the special model structure (9.10). The oscillator equation (9.2) is a particular case of Eq. (9.10) for $D = 2$ and an incomplete two-variable polynomial.

Along with the choice of the model structure, one should overcome a specific technical problem. It consists of the estimation of the driving frequency ω, which enters the model equations (9.10) in a non-linear way. As usual, one makes a starting guess and solves a minimisation problem for a cost function like Eq. (9.5) with an iterative technique. However, the right-hand side of Eq. (9.10) is very sensitive with respect to ω at large t analogous to example (7.19) in Sect. 7.1.2. Therefore, the cost function S of the type (9.5) is sensitive with respect to ω for a large time series length N. It implies that the variance of the resulting estimator of the frequency ω rapidly decreases with the time series length *if one manages* to find the global minimum of S. Namely, the variance scales as N^{-3} analogous to example (7.19). On the one hand, it gives an opportunity to determine ω to a very high accuracy. On the other hand, it is more difficult to find the *global minimum* since one needs a very lucky starting guess for ω. Taking it into account, one should carefully try multiple starting guesses for ω.

If ω is known a priori to a certain error, it is important to remember that for a very long time series, a small error in ω can lead to a bad description of the "true" driving with the corresponding terms in Eq. (9.10) due to the increase in "phase difference" between them over time (Bezruchko et al., 1999a). Then, the model structure (9.10) would get useless. Therefore, it is reasonable to consider the a priori known value as a starting guess for ω and determine the value of ω more accurately from the observation data. This discussion applies to other non-autonomous systems considered below.

For an arbitrary additive regular driving (complex periodic or quasi-periodic), a more appropriate model structure is

$$\mathrm{d}^D x \big/ \mathrm{d}t^D = f(x, \mathrm{d}x \big/ \mathrm{d}t, \ldots, \mathrm{d}^{D-1} x \big/ \mathrm{d}t^{D-1}, \mathbf{c}) + g(t, \mathbf{c}), \qquad (9.11)$$

where the function $g(t)$ describes the driving and can be represented as a trigonometric polynomial (Smirnov et al., 2003):

$$g(t) = \sum_{i=1}^{k} \sum_{j=1}^{K_i} c_{i,j} \cos(j\omega_i t + \phi_{i,j}). \qquad (9.12)$$

One can get good models with trigonometric polynomials of very high orders K_i, while approximation with a high-order algebraic polynomial typically leads to globally unstable model orbits.

To allow for multiplicative or even more complicated forms of driving, an explicit temporal dependence can be introduced into the coefficients of the polynomial f in Eq. (9.10) (Bezruchko and Smirnov, 2001). Thus, Fig. 9.10 shows an example of modelling of the non-autonomous Toda oscillator under combined harmonic driving

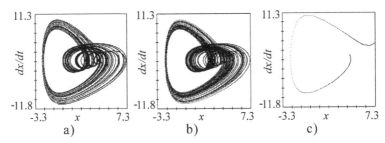

Fig. 9.10 Reconstruction of the equations for the non-autonomous Toda oscillator (9.13) from a time series of the variable x: (**a**) an original attractor; (**b**) an attractor of the non-autonomous polynomial model whose coefficients depend on time ($D = 2$, $K = 9$); (**c**) an orbit of the standard model (9.4) with $D = 4$, $K = 6$

$$\mathrm{d}^2 x \big/ \mathrm{d}t^2 = -0.45 \mathrm{d}x \big/ \mathrm{d}t + (5 + 4\cos t)(\mathrm{e}^{-x} - 1) + 7\sin t. \qquad (9.13)$$

A model is constructed in the form (9.4) with an explicit temporal dependence introduced into all the coefficients of the polynomial f, i.e. one replaces all c_k in the model structure with $c_k + a_k \cos \omega t + b_k \cos \omega t$. The best model is obtained for $D = 2$ and $K = 9$. Its phase orbit looks very similar to the original one (Fig. 9.10a,b). At that, the dimension of the standard model (9.4) without explicit temporal dependence should be not less than 3 to describe a chaotic regime. However, the standard model typically exhibits divergent orbits for $D > 2$ (Fig. 9.10c).

Efficiency of the special choice of a model structure is demonstrated in a similar way for the periodic pulse driving, periodic driving with suharmonics and quasi-periodic driving in Smirnov et al. (2003).

9.4.2 Time-Delay Systems

Modelling of *time-delay systems* has been actively considered in the last years (Bunner et al., 1996, 2000; Bezruchko et al, 2001b; Horbelt et al, 2002; Pono-marenko and Prokhorov, 2004; Ponomarenko et al., 2005; Prokhorov et al., 2005; Voss and Kurths, 1997, 1999). Despite such systems being infinite-dimensional, many of the above techniques are suitable to model them with some technical complications, e.g., the multiple shooting approach (Horbelt et al., 2002). Some principal differences (Bunner et al., 2000) are beyond the scope of our discussion.

Let us consider an example where modelling of a time-delay system corresponds to the "grey box" setting and can be performed with the techniques similar to those described above. We deal with the systems of the form

$$\varepsilon_0 \, \mathrm{d}x(t) \big/ \mathrm{d}t = -x(t) + F(x(t - \tau)), \qquad (9.14)$$

where an observable is $\eta = x$. Let us illustrate a modelling procedure with the reconstruction from a chaotic time realisation of the *Ikeda equation* (Fig. 9.11a):

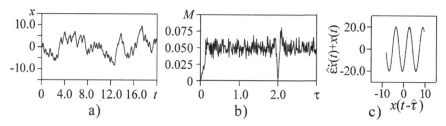

Fig. 9.11 Reconstruction of a time-delay system: (**a**) a time realisation of the Ikeda equation (9.15) with $x_0 = \pi/3$, $\varepsilon_0 = 1.0$, $\mu_0 = 20.0$, $\tau_0 = 2.0$; (**b**) the number of pairs of extrema $M(\tau)$ divided by the total number of extrema in the time series, $M_{\min}(\tau) = M(2.0)$; (**c**) a restored non-linear function. Numerical experiments with measurement noise show that modelling is successful for the ratio of the standard deviations of noise and signal up to 20%

$$\varepsilon_0 \, dx(t)/dt = -x(t) + \mu_0 \sin(x(t - \tau_0) - x_0), \qquad (9.15)$$

which describes the dynamics of a passive optical resonator.

Models are constructed in the form

$$\varepsilon \, dx(t)/dt = -x(t) + f(x(t - \tau), \mathbf{c}). \qquad (9.16)$$

Similar to the above examples, one may solve the minimisation problem $\sum_n (\varepsilon dx(t_n)/dt + x(t_n) - f(x(t_n - \tau), \mathbf{c}))^2 \to$ min, where the response constant ε and the *delay time* τ are considered as additional unknown parameters (Bunner et al., 2000). However, there is a special efficient approach (Bezruchko et al., 2001b), which is based on the statistical analysis of the time intervals separating extrema in a time series of the time-delay system (9.14). It appears that the number of pairs of extrema M separated by a given interval τ exhibits a clear minimum as a function of τ at τ equal to the true delay time of the system (9.14), Fig. 9.11b. This observation gives an opportunity to estimate the delay time and diagnose that a system under study belongs to the class of time-delay systems (9.14). Having an estimate $\hat{\tau} \approx \tau_0$, one can assess a response characteristic ε by checking different trial values of ε and selecting such value $\hat{\varepsilon}$ for which experimental data points on the plane $(x(t - \hat{\tau}), \hat{\varepsilon} dx(t)/dt + x(t))$ lie on a smooth one-dimensional curve. This curve is a plot of the sought function f, which is an approximation to F. Figure 9.11c illustrates such a restoration of the "true" function F for the system (9.15). Having such a plot, one can find an approximating function f using an expansion in a certain functional basis or a special formula.

The described approach to the determination of the delay time and reconstruction of the entire equation can be extended to the delay differential equations of higher orders and to the systems with several delay times. It is parsimonious with respect to the computation time and not highly sensitive to the measurement noise (Ponomarenko et al., 2005; Prokhorov et al., 2005).

Thus, as illustrated by several examples in this chapter, special selection of model structure based on the preliminary analysis of data and some (even rather general) a priori information about an object under study can essentially improve an empirical model quality and make possible meaningful interpretations of the modelling results.

References

Anishchenko, V.S., Janson, N.B., Pavlov, A.N.: Global reconstruction in the presence of a priori information. Chaos, Solitons Fractals. **9**(8), 1267–1278 (1998)

Bezruchko, B.P., Dikanev, T.V., Smirnov, D.A.: Role of transient processes for reconstruction of model equations from time series. Phys. Rev. E. **64**, 036210 (2001a)

Bezruchko, B.P., Karavaev, A.S., Ponomarenko, V.I., Prokhorov, M.D.: Reconstruction of time-delay systems from chaotic time series. Phys. Rev. E. **64**, 056216 (2001b)

Bezruchko, B.P., Seleznev, Ye.P., Smirnov, D.A.: Reconstructing equations of a non-autonomous nonlinear oscillator from time series: models and experiment. Izvestiya VUZ. Appl. Nonlinear Dynamics (ISSN 0869-6632). **7**(1), 49–67, (in Russian) (1999a)

Bezruchko, B.P., Smirnov, D.A.: Constructing nonautonomous differential equations from a time series. Phys. Rev. E. **63**, 016207, (2001)

Boekhoff-Falk, G.: Hearing in *Drosophila*: development of Johnston's organ and emerging parallels to vertebrate ear development. Dev. Dyn. **232**, 550–558 (2005)

Bünner, M.J., Ciofini, M., Giaquinta, A., et al. Reconstruction of systems with delayed feedback. Eur. Phys. J. D. **10**, 165–185 (2000)

Bünner, M.J., Popp, M., Meyer, Th., et al.: Tool to recover scalar time-delay systems from experimental time series. Phys. Rev. E. **54**, 3082–3085 (1996)

Dallos, P., Popper, A.N., Fay, R.R. (eds.): The Cochlea. Springer Handbook of Auditory Research. Springer, Berlin (1996)

Friedrich, R., Siegert, S., Peinke, J., Luck St., Siefert, M., Lindemann, M., Raethjen, J., Deuschl, G., Pfister, G.: Extracting model equations from experimental data. Phys. Lett. A. **271**, 217–222 (2000)

Goepfert, M.C., Humpfries, A.D.L., Albert, J.T., Robert, D., Hendrich, O.: Power gain exhibited by motile neurons in *Drosophila* ears. Proc. Natl. Acad. Sci. USA. **102**, 325–330 (2005)

Goepfert, M.C., Robert, D.: Active auditory mechanics in mosquitoes. Proc. R. Soc. Lond. B. **268**, 333–339 (2001)

Goepfert, M.C., Robert, D.: Motion generation by *Drosophila* mechanosensory neurons. Proc. Natl. Acad. Sci. USA. **100**, 5514–5519 (2003)

Goepfert, M.C., Robert, D.: Nanometer-range acoustic sensitivity in male and female mosquitoes. Proc. R. Soc. Lond. B. **267**, 453–457 (2000)

Gold, T.: Hearing. II. The physical basis of the action of the cochlea. Proc. R. Soc. Lond. B. **135**, 492–498 (1948)

Hegger, R., Kantz, H., Schmuser, F., et al. Dynamical properties of a ferroelectric capacitors observed through nonlinear time series analysis. Chaos. **8**, 727–754 (1998)

Horbelt, W., Timmer, J., Voss, H.U.: Parameter estimation in nonlinear delayed feedback systems from noisy data. Phys. Lett. A. **299**, 513–521 (2002)

Kern, A., Stoop, R.: Essential role of couplings between hearing nonlinearities. Phys. Rev. Lett. **91**, 128101 (2003)

Ponomarenko, V.I., Prokhorov, M.D., Karavaev, A.S., Bezruchko, B.P.: Recovery of parameters of delayed feedback systems from chaotic time series. J. Exp. Theor. Phys. **100**(3), 457–467 (2005)

Ponomarenko, V.I., Prokhorov, M.D.: Coding and recovery of information masked by the chaotic signal of a time-delay system. J. Commun. Technol. Electron. **49**(9), 1031–1037 (2004)

Prokhorov, M.D., Ponomarenko, V.I., Karavaev, A.S., Bezruchko, B.P.: Reconstruction of time-delayed feedback systems from time series. Phys. D. **203**, 209–223 (2005)

Ragwitz, M., Kantz, H.: Indispensable Finite time corrections for Fokker-Planck equations from time series data. Phys. Rev. Lett. **87**, 254501 (2001)

Robert, D., Goepfert, M.C.: Novel schemes for hearing and orientation in insects. Curr. Opin. Neurobiol. **12**, 715–720 (2002)

Robles, L., Ruggero, M.A.: Mechanics of the mammalian cochlea. Physiol. Rev. **81**, 1305–1352 (2001)

Siefert, M., Kittel, A., Friedrich, R., Peinke, J.: On a quantitative method to analyze dynamical and measurement noise. Europhys. Lett. **61**, 466–472 (2003)

Siegert, S., Friedrich, R., Peinke, J.: Analysis of data sets of stochastic systems. Phys. Lett. A. **243**, 275–280 (1998)

Smirnov, D.A., Sysoev, I.V., Seleznev Ye.P., Bezruchko, B.P.: Reconstructing nonautonomous system models with discrete spectrum of external action. Tech. Phys. Lett. **29**(10), 824–828 (2003)

Stoop, R., Kern, A., Goepfert, M.C., Smirnov, D.A., Dikanev, T.V., Bezrucko, B.P.: A generalization of the van-der-Pol oscillator underlies active signal amplification in Drosophila hearing. Eur. Biophys. J. **35**, 511–516 (2006)

Sysoev, I.V., Smirnov, D.A., Seleznev Ye.P., Bezruchko, B.P.: Reconstruction of nonlinear characteristics and equivalent parameters from experimental time series. Proc. 2nd IEEE Int. Conf. Circuits and Systems for Communications. Paper No. 140. Moscow (2004)

Voss, H.U., Kurths, J.: Reconstruction of non-linear time delay models from data by the use of optimal transformations. Phys. Lett. A. **234**, 336–344 (1997)

Voss, H.U., Kurths, J.: Reconstruction of nonlinear time delay models from optical data. Chaos, Solitons Fractals. **10**, 805–809 (1999)

Voss, H.U., Schwache, A., Kurths, J., Mitschke, F.: Equations of motion from chaotic data: A driven optical fiber ring resonator. Phys. Lett. A. **256**, 47–54 (1999)

Chapter 10
Model Equations: "Black Box" Reconstruction

Black box reconstruction is both the most difficult and the most tempting modelling problem when any prior information about an appropriate model structure is lacking. An intriguing thing is that a model capable of reproducing an observed behaviour or predicting further evolution should be obtained only from an observed time series, i.e. "from nothing" at first sight. Chances for a success are not large. Even more so, a "good" model would become a valuable tool to characterise an object and understand its dynamics. Lack of prior information causes one to utilise *universal model structures*, e.g. artificial neural networks, radial basis functions and algebraic polynomials are included in the right-hand sides of dynamical model equations. Such models are often multi-dimensional and involve quite many free parameters.

Since time series of all variables for such a model must be obtained from observed data, "restoration" of lacking variables gets extremely important. One often calls it "phase orbit reconstruction" or "state vector reconstruction". A theoretical justification is given by celebrated *Takens' theorems* (Sect. 10.1).

Not less important and difficult is the approximation stage, where one fits a dependence of the next state on the current one $\mathbf{x}_{n+1} = \mathbf{f}(\mathbf{x}_n, \mathbf{c})$ or of the phase velocity on the state vector $d\mathbf{x}/dt = \mathbf{f}(\mathbf{x}, \mathbf{c})$. In practice, one usually manages to get a valid model if it appears sufficient to use its moderate dimension, roughly, not greater than 5–6. To construct higher dimensional models, one needs huge amounts of data and deals with approximation of multivariable functions (Sect. 10.2) which is much more difficult than that of one-variable functions (Sects. 7.2, 9.1 and 9.3). Moreover, troubles quickly rise with the model dimension (Kantz and Schreiber, 1997). This is the so-called "curse of dimensionality", the main obstacle in the modelling of multitude of real-world processes.

Yet, successful results have sometimes been obtained for complex real-world objects even under the black box setting. Also, there are several nice theoretical results and many practical algorithms of reconstruction, which appear efficient for prediction and other modelling purposes.

B.P. Bezruchko, D.A. Smirnov, *Extracting Knowledge From Time Series*, Springer Series in Synergetics, DOI 10.1007/978-3-642-12601-7_10,

© Springer-Verlag Berlin Heidelberg 2010

10.1 Reconstruction of Phase Orbit

To get lacking model variables in modelling from a time series $\{\eta(t_1), \eta(t_2), \ldots, \eta(t_N)\}$, one can use subsequent values of η, i.e. a state vector $\mathbf{x}(t_i) = [\eta(t_i), \eta(t_i + \tau), \ldots, \eta(t_i + (D - 1)\tau)]$, where τ is the *time delay*, or successive derivatives, i.e. a state vector $\mathbf{x}(t_i) = [\eta(t_i), d\eta(t_i)/dt, \ldots, d^{D-1}\eta(t_i)/dt^{D-1}]$. These approaches have been applied for a long time without special justification (Sect. 6.1.2). Thus, the former one is, in fact, used since 1927 for the widely known *autoregression models* (4.12), where a future value of an observable is predicted based on several previous values (Yule, 1927). It seems just reasonable. If there is no other information besides a time series, then one can use only the previous values of an observable or their combinations to make a forecast.

At the beginning of the 1980s, relationships between both mentioned approaches and the theory of dynamical systems were revealed. It was proven that in reconstruction from a scalar time realisation of *a dynamical system* (under some conditions of smoothness), both time delays and successive derivatives assure an equivalent description of the original dynamics if the dimension of the restored vectors D is large enough. Namely, the condition $D > 2d$ should be fulfilled, where d is the dimension of a set M in the phase space of an original system, where a modelled motion occurs.[1] These statements constitute celebrated Takens' theorems (Takens, 1981) as discussed in Sect. 10.1.1. We note that the theorems are related to the case when an object is a *deterministic dynamical system* (Sect. 2.2.1).

In the modelling of real-world objects, one can use the above approaches without referring to the theorems, since it is impossible to check whether the conditions of the theorems are fulfilled and the dimension d is unknown (if one may speak about all that in respect of a real-world object at all). Yet, the value of the theoretical results obtained by Takens is high. Firstly, after their formulation it has become clear that both above approaches are suitable for the modelling of a sufficiently wide class of systems. Thus, the theorems "bless" practical application of the approaches, especially if one has any ideas confirming that the conditions of the theorems are fulfilled in a given situation. Secondly, based on the theory of dynamical systems, one has developed new fruitful approaches to the choice of the reconstruction parameters, such as the time delay τ, the model dimension D and others, as discussed in Sect. 10.1.2.

---

[1] The set M is a compact smooth manifold and the quantity d is its topological dimension (Sect. 10.1.1). There are generalisations of the theorem to the case of non-smooth sets M and fractal dimension d, which are beyond the scope of our discussion (Sauer et al., 1991). We note that the set M mentioned in the theorems *does not inevitably* correspond to the motion on an attractor. For instance, let an attractor be a limit cycle C "reeled" on a torus M. If one is interested only in the description of an established periodic motion on the cycle C, then it is sufficient to use $D > 2$ model variables for reconstruction according to Takens' theorems. If one needs to describe motions on the entire torus M, including transient processes, then it is necessary to use $D > 4$ variables. In practice, one often has a single realisation corresponding to established dynamics. Therefore, one usually speaks of the reconstruction of an attractor.

10.1.1 Takens' Theorems

We start with illustrating the theorems with a simple example and then give their mathematical formulations and discuss some details in a more strict way. Throughout this subsection, we denote the state vector of an original system **y** as distinct from the reconstructed vectors **x**. The notation d is related to the dimension of the set M in the phase space of an *original* system. It is not necessarily the dimension of the entire phase space, i.e. of the vector **y**. D is the dimension of the reconstructed vectors **x** and, hence, of a resulting model.

10.1.1.1 An Illustrative Example

Let an object be a continuous-time three-dimensional dynamical system. Its state vector is $\mathbf{y} = (y_1, y_2, y_3)$. Let a motion to occur on a limit cycle (Fig. 10.1a), i.e. on a set M of the dimension $d = 1$.

If all three variables y_1, y_2, y_3 were observed, one could proceed directly to the approximation of the dependence of $\mathbf{y}(t + \tau)$ on $\mathbf{y}(t)$, which is unique since **y** is a state vector. The latter means that whenever a certain value $\mathbf{y} = \mathbf{y}^*$ is observed, a unique future value follows it in a fixed time interval. The same present leads to the same future. If not all the components of the state vector are observed, then the situation is more complicated. One may pose a question: How many variables suffice for an equivalent description of an original dynamics? Which variables are suitable for that and which ones are not?

Since the set M, where the considered motion takes place, is one dimensional ($d = 1$), there should exist such a scalar dynamical variable which is sufficient to describe this motion. For instance, a closed curve M (Fig. 10.1a) can be mapped on

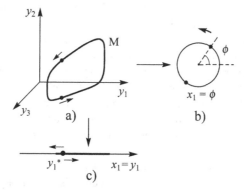

Fig. 10.1 One-dimensional representations of a limit cycle: (**a**) an original limit cycle; (**b**) its mapping on a circle; (**c**) its projection on a coordinate axis. The dimension of an original system is equal to three; the dimension of the set M, which is a closed curve, is $d = 1$; the dimension of the "reconstructed" vectors is $D = 1$ in both cases. Two different states [filled circles in the panel (**a**)] correspond to the two different points on the circle [the panel (**b**)] and to a single point on the axis [the panel (**c**)]. The mapping of a cycle on a circumference is one-to-one and its mapping on a line segment is not one-to-one

a circumference (Fig. 10.1b). It is important that the vectors $\mathbf{y}(t)$ on the cycle M can be related to the angle of rotation $\phi(t)$ of a point around the circumference in a one-to-one way. The variable $\phi(t)$ is the "wrapped" phase of oscillations (Sect. 6.4.3). Due to one-to-oneness, the variable ϕ completely determines the state of the system: The value of the phase ϕ^* corresponds to a unique simultaneous value of the vector $\mathbf{y}^*$. Having an observable ϕ, one can construct a one-dimensional *deterministic dynamical model* ($D = 1$) with $x_1 = \phi$.

However, not any variable is appropriate for a one-dimensional representation. Thus, if one observes just a single component of the vector $\mathbf{y}$, e.g. a coordinate y_1, then a closed curve is mapped on a line segment (a simple projection onto the y_1-axis). This mapping is not one-to-one. Almost any point $y_1^*(t)$ of the segment corresponds to two state vectors $\mathbf{y}(t)$ differing by the direction of the further motion (to the left or to the right along the y_1-axis, see Fig. 10.1a, c). Thus, y_1 does not uniquely determine the state of the system. If one observes some value $y_1 = y_1^*$, then one of the two possible future values can follow. Therefore, a deterministic one-dimensional description of the observed motion with the variable $x_1 = y_1$ is impossible.

In general, if one uses the model dimension D equal to the dimension of the observed motion d, the construction of a dynamical model may appear successful if one is "lucky". However, empirical modelling may fail as well. Both results are typical in the sense that the situation does not change under weak variations of an original system, an observable and parameters of the reconstruction algorithm.

What changes if one uses a two-dimensional description for the above example? The same two situations are typical as illustrated in Fig. 10.2. If the two components of the original state vector y_1 and y_2 are observables, i.e. the model state vector is $\mathbf{x} = (y_1, y_2)$, it corresponds to a projection of the closed curve onto the plane (y_1, y_2). In such a projection, one may get a curve either without self-intersections (Fig. 10.2a) or with them (Fig. 10.2b) depending on the shape of the original curve and its spatial orientation. The former case provides a one-to-one relationship between the original curve and its projection, i.e. the two-dimensional vector $\mathbf{x}$ completely determines the state of the system. The latter case differs, since the self-intersection point y_1^*, y_2^* on the plane (y_1, y_2) in Fig. 10.2b corresponds to two different states of the original system, i.e. the relationship between $\mathbf{x}$ and $\mathbf{y}$ is not one-to-one. Therefore, one cannot uniquely predict the future following the current values y_1^*, y_2^*. Hence, the vector $\mathbf{x}$ is not suitable as the state vector of a global deterministic model. It can be used only locally, far from the self-intersection point.

A similar situation takes place if one uses any two variables instead of y_1 and y_2. For instance, let $\eta = h(\mathbf{y})$ be an observable, where h is an arbitrary smooth function and let the components of $\mathbf{x}$ be the time-delayed values of η: $\mathbf{x}(t) = (\eta(t), \eta(t+\tau))$. Depending on h and τ, one may observe a situation either without self-intersections on the plane (x_1, x_2) as in Fig. 10.2a or with self-intersections as in Fig. 10.2b. Thus, even the number of model variables exceeding the dimension of an observed motion, $D = 2 > d = 1$, does not *assure* the possibility of a deterministic description.

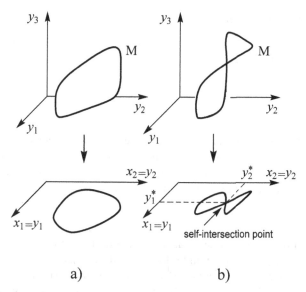

Fig. 10.2 Projections of a one-dimensional manifold from a three-dimensional space onto planes: (**a**) one-to-one mapping; (**b**) many-to-one mapping with a self-intersection point in the projection

Finally, let us consider a three-dimensional representation, e.g. when model state vectors are constructed as $\mathbf{x}(t) = (\eta(t), \eta(t+\tau), \eta(t+2\tau))$. An image of the original closed curve in the three-dimensional space (x_1, x_2, x_3) is also a closed curve, which typically does not exhibit self-intersections, i.e. there is a one-to-one correspondence between $\mathbf{x}$ and $\mathbf{y}$. An original motion on the limit cycle can be equivalently described with the vectors $\mathbf{x}$. In general, a self-intersection of an image curve may be observed in the space x_1, x_2, x_3 as a non-generic situation, i.e. it is eliminated by weak variations in an original system, an observable or reconstruction parameters. Intuitively, one easily agrees that self-intersections of a curve in a three-dimensional space are very unlikely.

Thus, in our example, an equivalent description of the dynamics is achieved *for sure* only if the state vectors are reconstructed in the space of the dimension $D > 2d$.[2] This is the main contents of Takens' theorems. We stress that this is a *sufficient* condition. Sometimes, an equivalent description is possible even for $D = d$ as illustrated above. In practical modelling, Takens' theorems serve just as a psychological support, because they state that there is a finite model dimension D at which deterministic modelling should be appropriate. Technically, one tries different values of D, starting from small ones, and aims at obtaining a "good" model with as low dimension as possible to avoid difficulties related to the above-mentioned "curse of dimensionality".

---

[2] An equivalent description of a motion on a limit cycle is assured for $D = 3$ even if the cycle "lives" in an infinite-dimensional phase space.

10.1.1.2 Mathematical Details

To formulate the theorems in a more rigorous way, let us introduce some notations. Let an object be a dynamical system

$$\mathbf{y}(t_0 + t) = \Phi_t(\mathbf{y}(t_0)),$$
$$\boldsymbol{\eta}(t) = \mathbf{h}(\mathbf{y}(t)), \tag{10.1}$$

where $\mathbf{y}$ is a state vector, Φ_t is an evolution operator and $\mathbf{h}$ is a measurement function.[3] The vector of observables is finite-dimensional: $\boldsymbol{\eta} \in R^m$. We discuss further only the case of a scalar time series $\eta(t_i)$, i.e. $m = 1$. It is the most widespread situation, which is also the most difficult for modelling.

Manifold

Let the motion of the system occur at some manifold M of the finite dimension d that can be observed even for infinite-dimensional systems. *Manifold* is a generalisation of the concept of a smooth surface in the Euclidean space (Gliklikh, 1998; Makarenko, 2002; Malinetsky and Potapov, 2000; Sauer et al., 1991). Roughly speaking, a d-dimensional manifold M is a surface which can be locally parameterised with d Euclidean coordinates in the vicinity of any of its points. In other words, any point $p \in M$ together with its local neighbourhood $U(p)$ can be mapped on a d-dimensional fragment (e.g. a ball) of the space R^d in a one-to-one and continuous way. A corresponding image $\Psi : U \to \Psi(U)$ is called a *chart* of the neighbourhood. The continuous map Ψ is called a *homeomorphism*. Examples of two-dimensional manifolds in a three-dimensional Euclidean space are a sphere, a torus, a bottle with a handle, etc. (Fig. 10.3a), but not a (double) cone (Fig. 10.3b).

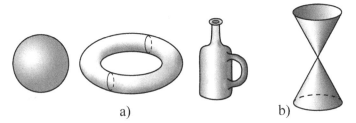

Fig. 10.3 Examples of the sets which (**a**) are manifolds and (**b**) is not a manifold

---

[3] If an object is a map $\mathbf{y}(t_{n+1}) = \mathbf{F}(\mathbf{y}(t_n))$, then an evolution operator $\Phi_{\Delta t}(\mathbf{y}(t_0))$ is just the function $\mathbf{F}$. If an object is a set of ODEs $d\mathbf{y}/dt = \mathbf{F}(\mathbf{y}(t))$, then the function $\Phi_t(\mathbf{y}(t_0))$ is the result of the integration of the ODEs over a time interval of length t. If an original system is given by a partial differential equation $\partial \mathbf{y}/\partial t = \mathbf{F}(\mathbf{y}, \partial \mathbf{y}/\partial \mathbf{r}, \partial^2 \mathbf{y}/\partial \mathbf{r}^2, \ldots)$, where $\mathbf{r}$ is a spatial coordinate, then $\mathbf{y}$ is a vector belonging to an *infinite-dimensional* space of functions $\mathbf{y}(\mathbf{r})$ and Φ_t is an operator acting in that space.

If Ψ is an n times differentiable mapping with the n times differentiable inverse, then one says that M belongs to the class C^n. If $n \geq 1$, the mapping Ψ is called a *diffeomorphism*. If the manifold M is mapped on a manifold $S \in R^D$, $D \geq d$, via a diffeomorphism, then M and S are called *diffeomorphic* to each other. One says that S is an *embedding* of the manifold M into the Euclidean space R^D. Below, we speak of a bounded and closed M. Boundedness means that M can be included into a ball of a finite radius. Closedness means that all limit points of M belong to M. Such a manifold in a finite-dimensional space is called *compact*.

The Question and Notations

Each phase orbit of the system (10.1) $y(t)$, $0 \leq t < \infty$, on a manifold M corresponds to a time realisation of an observable η: $\eta(t) = h(y(t))$, $0 \leq t < \infty$. The vector $y(t_0)$ determines the entire future behaviour of the system (10.1), in particular, the entire realisation $\eta(t)$, $t \geq t_0$. Is it possible to determine a state on the manifold M at a time instant t_0 and, hence, the entire future evolution from a segment of the realisation $\eta(t)$ around t_0? In other words, can one "restore" a state of the system from the values of $\eta(t)$ on a finite-time interval? This is a key question and Takens' theorems give a positive answer under some conditions.

Let us introduce some notations necessary to formulate rigorously the time-delay embedding theorem. A vector $y(t)$ corresponds to a D-dimensional vector $x(t) = [\eta(t), \eta(t+\tau), \ldots, \eta(t+(D-1)\tau)]$. Dependence of x on a simultaneous value of y is given by a *unique* mapping $\Psi : M \to R^D$ expressed via the evolution operator $\Phi_t : M \to M$ and the measurement function $h : M \to R$ as

$$x(t) = \Psi(y(t)) \equiv \begin{bmatrix} \Psi_1(y(t)) \\ \Psi_2(y(t)) \\ \cdots \\ \Psi_D(y(t)) \end{bmatrix} = \begin{bmatrix} h(y(t)) \\ h(\Phi_\tau(y(t))) \\ \cdots \\ h(\Phi_{(D-1)\tau}(y(t))) \end{bmatrix}. \tag{10.2}$$

Smoothness of Ψ (continuity, differentiability, existence and continuity of the higher order derivatives) is determined by the smoothness of Φ_τ and h. An image of the manifold M under the mapping Ψ is a certain set $S \subset R^D$.

The above question can now be formulated as follows: Is Ψ a diffeomorphism? If yes, then S is an embedding of M and each vector x on S corresponds to a single vector y on M.[4] Then, $x(t)$ can be used as a state vector to describe the dynamics on M and Eq. (10.1) can be rewritten as

$$x(t_0 + t) = \varphi_t(x(t_0)), \tag{10.3}$$

---

[4] It means that for the same segments $[\eta(t), \eta(t+\tau), \ldots, \eta(t+(D-1)\tau)]$ encountered at different time instants t, one observes the same continuation (i.e. the same future). It gives a justification to the predictive *method of analogues* applied already by E. Lorenz. The method is based on the search of the time series segments, which "resemble" a current segment, in the past and subsequent usage of a combination of their "futures" as a forecast. In a modern formulation, it is realised with local models (Sect. 10.2.1).

where a new evolution operator is $\varphi_t(\mathbf{x}) = \Psi(\Phi_t(\Psi^{-1}(\mathbf{x})))$. Due to the diffeomorphism, local properties of the dynamics such as stability and types of fixed points and others are preserved. Each phase orbit $\mathbf{y}(t)$ on M corresponds to an orbit $\mathbf{x}(t)$ on S in a one-to-one way. If a system (10.1) has an attractor in M, then a system (10.3) has an attractor in S. Such characteristics as fractal dimension and Lyapunov exponents coincide for both attractors. In other words, the system (10.3) on the manifold S and the system (10.1) on the manifold M can be considered as two representations of the same dynamical system.

Obviously, the mapping Ψ (10.2) is not always a diffeomorphism. Thus, Fig. 10.2b gives an example where a smooth mapping Ψ has a non-unique inverse Ψ^{-1}. Another undesirable situation is encountered if Ψ^{-1} is unique but *non-differentiable* (Fig. 10.4). The latter property takes place at the *return point* on the set S. In its neighbourhood, the two-dimensional vector (y_1, y_2) cannot be used to describe the dynamics with a set of ODEs, since the return point would be a fixed point so that S could not be a limit cycle. Here, the differentiability properties of M and S differ due to non-differentiability of Ψ^{-1}.

Formulation of the Time-Delay Embedding Theorem

Coming back to the system (10.1) and the mapping (10.2), one can say that any one of the above mentioned situations can be encountered for some Φ, M, h, d, D and τ. Sets of self-intersections and return points on $S = \Psi(M)$ can be vast, which is very undesirable. However, one can also meet a "good" situation of embedding (Fig. 10.2a). The result formulated below was first obtained rigorously by Dutch mathematician Floris Takens (1981) and then generalised in Sauer et al. (1991). It shows under what conditions an embedding of an original compact d-dimensional manifold M in the space R^D is assured with the mapping (10.2). Takens' theorem

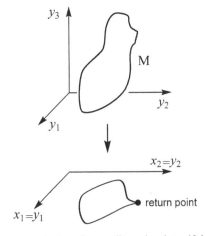

Fig. 10.4 The situation when a projection of a one-dimensional manifold M exhibits a return point on a plane

is related to the Whitney's embedding theorem (from the courses of differential geometry), which concerns arbitrary mappings. Takens' statement differs in that it concerns the special case of the mappings (10.2) determined by an evolution operator of a dynamical system.

Theorem 1 *Let M be a compact d-dimensional C^2 manifold. For almost any pair of functions Φ_t and h, which are twice continuously differentiable on M, the mapping $\Psi : M \to R^D$ given by the formula (10.2) is a diffeomorphism for almost any $\tau > 0$ and $D > 2d$.*

Comments

Diffeomorphism implies that an image of M under the mapping (10.2) is its embedding. The space R^D containing the image $S = \Psi(M)$ is called *embedding space*. The term "almost any pair" is understood by Takens in the sense of *genericity*. For instance, if for some Φ_t the mapping (10.2) does not provide an embedding, then there exists such an arbitrarily weak variation $\Phi_t + \delta\Phi_t$ that an embedding is achieved. More rigorously, generic properties are fulfilled on an intersection of open and everywhere dense sets. A metric analogue to genericity is *prevalence* (Sauer et al., 1991). "Almost any τ" should be understood in a similar way. In particular, if a limit cycle exists within M, the value of τ should not be equal to the period of that cycle, see Sauer et al. (1991) for more detail.

Discussion

Thus, if the dimension of the time-delay vector $\mathbf{x}$ (10.2) is high enough, one typically gets an embedding of the manifold M and can use $\mathbf{x}$ as a state vector of a deterministic model. It is possible to interpret the condition $D > 2d$ vividly as follows (Malinetsky and Potapov, 2000). To establish possible non-uniqueness of the mapping Ψ^{-1}, one must find such vectors $\mathbf{y}_1$ and $\mathbf{y}_2$ on M so that $\Psi(\mathbf{y}_1) = \Psi(\mathbf{y}_2)$. The latter equality is a set of D equations with $2d$ variables (d components for the two vectors $\mathbf{y}_1$ and $\mathbf{y}_2$ specifying their location on M). Roughly speaking, this set of equations has *typically* no solutions if the number of equations is greater than the number of variables, i.e. if $D > 2d$. This is the contents of Takens' theorem.

We stress again that the condition $D > 2d$ is sufficient, but not necessary. If it is fulfilled, a diffeomorphism is assured. However, if one is "lucky", a good reconstruction can be obtained for lower D as in Fig. 10.1a, b, where an embedding of a one-dimensional manifold M is achieved at $D = 1$ and is not a degenerate case.

What are those non-generic cases when the theorem is invalid? Let us indicate two examples (Malinetsky and Potapov, 2000):

(1) A measurement function is constant: $h(\mathbf{y}) = a$. This is a smooth function, but it maps the entire dynamics to a single point. This situation is almost surely eliminated via a weak variation in the measurement function, i.e. via adding an almost arbitrary "small" function of $\mathbf{y}$ to a.

(2) A system consisting of two unidirectionally coupled subsystems $d\mathbf{y}_1/dt = \mathbf{F}(\mathbf{y}_1, \mathbf{y}_2)$, $d\mathbf{y}_2/dt = \mathbf{G}(\mathbf{y}_2)$ when only the driving subsystem is observed,

i.e. $\eta = h(\mathbf{y}_2)$. In a non-synchronous regime, such an observable does not carry complete information about the driven subsystem $\mathbf{y}_1$. Therefore, an embedding of the original dynamics is not achieved. This situation is eliminated almost surely if an arbitrarily weak dependence on $\mathbf{y}_1$ is introduced into η.

Similar Theorems

A more general version of the theorem 1 is proven in Sauer et al. (1991). It concerns the *filtered embedding*, where coordinates of $\mathbf{x}$ are not just subsequent values of an observable but their linear combinations, which can be considered as outputs of a linear non-recursive filter.

Moreover, Takens proved a similar theorem for successive derivatives used as components of a state vector:

$$\mathbf{x}(t) = \begin{bmatrix} \eta(t) \\ \mathrm{d}\eta(t)/\mathrm{d}t \\ \dots \\ \mathrm{d}^{D-1}\eta(t)/\mathrm{d}t^{D-1} \end{bmatrix}, \tag{10.4}$$

where $D > 2d$. The theorem is formulated in the same way as theorem 1, but with stricter requirements to the smoothness of Φ_t and h. Namely, one demands continuous derivatives of the Dth order for each of these functions to assure the existence of the derivatives entering Eq. (10.4). If the latter derivatives are approximated with finite differences, then the relationship (10.4) becomes a particular case of the filtered embedding (Gibson et al., 1992).

In practice, one must always cope with noises. Takens' theorems are not directly related to such a case, although there are some generalisations (Casdagli et al., 1991; Stark et al., 1997). Nevertheless, the theorems are of significant value for practical modelling as discussed at the beginning of Sect. 10.1.

10.1.2 Practical Reconstruction Algorithms

10.1.2.1 Time-Delay Technique

This is the most popular reconstruction technique. One gets the vectors $\{\mathbf{x}_i = (\eta_i, \eta_{i+l}, \dots, \eta_{i+(D-1)l})\}_{i=1}^{N-(D-1)l}$ from an observed scalar time series $\{\eta_i = \eta(t_i)\}_{i=1}^{N}$, $t_i = i\Delta t$. Theoretically, the value of the time delay $\tau = l\Delta t$ can be almost arbitrary, but in practice one avoids both too small l, giving strongly correlated components[5] of the state vector, and too large l, introducing considerable complications into the geometrical structure of the reconstructed attractor.

---

[5] For $l = 1$ and a small sampling interval Δt, a reconstructed phase orbit stretches along the main diagonal, since it appears that $x_1 \approx x_2 \approx \dots \approx x_D$.

Therefore, it was suggested to choose the value of τ equal to the first zero of the autocorrelation function (Gibson et al., 1992), first minimum of the mutual information function (Fraser and Swinney, 1986) and so on (Liebert and Schuster, 1989). One also uses a *non-uniform embedding*, where time intervals between subsequent components of **x** are not the same, which is relevant for the dynamics with several characteristic timescales (Eckmann and Ruelle, 1985; Judd and Mees, 1998). For the dynamics representing alternating intervals of almost periodic and very complicated behaviour, one has developed the *variable embedding*, where a set of time delays depends on the location of **x** in the state space (Judd and Mees, 1998). Each of the ideas is appropriate for a specific kind of systems and does not assure successful results in general (Malinetsky and Potapov, 2000).

How to choose the model dimension D based on the analysis of an observed time series? There are different approaches including the false nearest neighbour technique (Kennel et al., 1992), the principal component analysis (Broomhead and King, 1986), the Grassberger and Procaccia method (Grassberger and Procaccia, 1983) and the "well-suited basis" approach (Landa and Rosenblum, 1989). Moreover, one should often try different values of D and construct model equations for each trial value until a "good" model is obtained. Then, the selection of D and even of the time delays can be a part of a united modelling procedure, rather than an isolated first stage.

10.1.2.2 False Nearest Neighbour Technique

It gives an integer-valued estimate of the attractor dimension. It is based on checking the property that a phase orbit reconstructed in the space of the sufficient dimension must not exhibit self-intersections. Let us illustrate the technique with a simple example of reconstruction from a time realisation of a sinusoid $\eta(t) = \sin t$, Fig. 10.5a.

At $D = 1$, i.e. $\mathbf{x}(t) = \eta(t)$, the reconstructed set lies on a line segment, Fig. 10.5b. Then, a data point at the instant t_k has the data points at the instants t_s and t_l as its close neighbours. However, the latter two states of an original system differ by the sign of the derivative of $\eta(t)$. In a two-dimensional space with $\mathbf{x}(t) = [\eta(t), \eta(t + \tau)]$, all the points go away from each other. However, the points at the instants t_k and t_l get *weakly* more distant, while the points at the instants t_k and t_s become very far from each other, Fig. 10.5c. Accordingly, one calls the neighbours at t_k and t_l "true" and the neighbours at t_k and t_s "false".

One of the version of the algorithm is as follows. At a trial dimension D, one finds a single nearest neighbour for each vector $\mathbf{x}_k$. After increasing D by 1, one determines which neighbours appear false and which ones are true. Then, one computes the ratio of the number of the false neighbours to the total number of the reconstructed vectors. This ratio is plotted versus D as in Fig. 10.5d. If this relative number of self-intersections reduces to zero at some value $D = D^*$, the latter is the dimension of the space, where an embedding of the original phase orbit is achieved. In practice, the number of the false neighbours becomes sufficiently small, starting from some "correct" value D^*, but does not decrease to zero due to noises and other

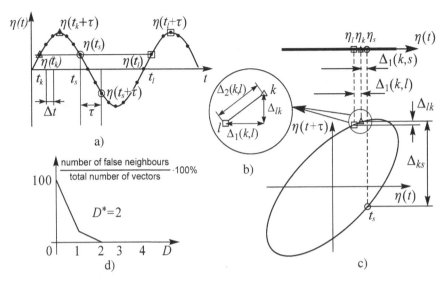

Fig. 10.5 An illustration to the false nearest neighbour technique: (**a**) a time realisation $\eta(t)$, where symbols indicate the data points $\eta(t_k)$, $\eta(t_s)$, $\eta(t_l)$, and the close values of η together with the points shifted by $\tau = 3\Delta t$; (**b**) an orbit reconstructed in a one-dimensional space; (**c**) an orbit reconstructed in a two-dimensional space; (**d**) the number of the false nearest neighbours divided by the total number of the reconstructed vectors in a time series versus the trial dimension of the reconstructed vectors D

factors. Then, D^* can be taken as a trial model dimension. It equals 2 for the example illustrated in Fig. 10.5d (see, e.g., Malinetsky and Potapov, 2000 for details).

10.1.2.3 Principal Component Analysis

It can be used both for the dimension estimation and for the reconstruction of state vectors. The technique is used in different fields and has many names. Its application to the reconstruction was suggested in Broomhead and King (1986). The idea is to rotate coordinate axes in a multidimensional space and choose a small subset of directions, along which the motion mainly develops.

For simplicity of notations, let the mean value of η be zero. The vectors $\mathbf{w}(t_i) = (\eta_i, \eta_{i+1}, \ldots, \eta_{i+k-1})$ of a sufficiently high dimension k are constructed. Components of these vectors are strongly correlated if the sampling interval is small. Figure 10.6 illustrates the case of a sinusoidal signal and the reconstruction of the phase orbit in a three-dimensional space ($k = 3$).

One performs a rotation in this space so that the directions of new axes (e.g. $\{\mathbf{s}_1, \mathbf{s}_2, \mathbf{s}_3\}$ in Fig. 10.6b, c) correspond to the directions of the most intensive motions in the descending order. Quantitatively, the characteristic directions and the extensions of an orbit along them are determined from the covariance matrix Θ of the vector $\mathbf{w}$, which is a square matrix of the order k:

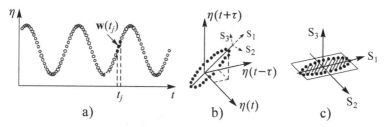

Fig. 10.6 Noise-free harmonic oscillations: (**a**) reconstruction of the time-delay vectors **w** of the dimension $k = 3$ from a scalar time series; (**b**) a reconstructed orbit is an ellipse stretched along the main diagonal of the space R^k; (**c**) a reconstructed orbit in a new coordinate system (after the rotation), where the component of the reconstructed vectors along the direction s_3 is zero

$$\Theta_{i,j} = \sum_{n=0}^{N-k} \eta_{i+n}\eta_{j+n}, \ i, j = 1, \ldots, k$$

It is symmetric, real valued and positive definite. Hence, its eigenvectors constitute a complete orthonormal basis of the space R^k. Its eigenvalues are non-negative. Let us denote them as $\sigma_1^2, \sigma_2^2, \ldots, \sigma_k^2$ in the non-ascending order and the corresponding eigenvectors as $s_1, s_2, \ldots, s_k$. The transformation to the basis $s_1, s_2, \ldots, s_k$ can be performed via the coordinate change $x'(t_i) = S^T \cdot w(t_i)$, where S is a matrix with the columns $s_1, s_2, \ldots, s_k$ and T means transposition. This is known in the theory of information as the Karhunen and Loeve transform. One can easily show that the covariance matrix of the components of the vector x' is diagonal:

$$\Theta' = S^T \Theta S = \begin{bmatrix} \sigma_1^2 & 0 & \ldots & 0 \\ 0 & \sigma_2^2 & \ldots & 0 \\ \ldots & \ldots & \ldots & \ldots \\ 0 & 0 & \ldots & \sigma_k^2 \end{bmatrix}$$

i.e. the components of x' are uncorrelated, which is a sign of a "good" reconstruction. Each diagonal element σ_i^2 is the mean-squared value of the projection of $w(t_i)$ onto the coordinate axis s_i. The values σ_i^2 determine the extensions of the orbit along the respective directions. Rank of the matrix Θ equals the number of non-zero eigenvalues (these are σ_1^2 and σ_2^2 for the situation shown in Fig. 10.6b, c) and the dimension of the subspace, where the motion occurs.

If a measurement noise is present, then all σ_i^2 are non-zero, since noise contributes to the directions, which are not explored by the deterministic component of an orbit. In such a case, the dimension can be estimated as the number D of considerable eigenvalues as illustrated in Fig. 10.7. Projections of $w(t_i)$ onto the corresponding directions (i.e. the first D components of the vector x') are called its *principal components*. The remaining eigenvalues constitute the so-called noise floor and the respective components can be ignored. Thus, one gets D-dimensional vectors $x(t_i)$ with coordinates $x_k(t_i) = s_k \cdot w(t_i)$, $k = 1, \ldots, D$.

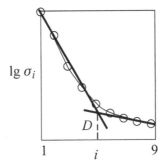

Fig. 10.7 Eigenvalues of the covariance matrix versus their order number: a qualitative illustration for $k = 9$. The "break point" D at the plot is an estimate of the dimension of an observed motion

If there is no characteristic break at the plot, then one increases a trial dimension k until the break emerges. The dimension estimate D is more reliable if the break is observed at the same value of D under the increase in k.

The principal component analysis is a particular case of the filtered embedding. It is very useful in the case of a considerable measurement noise, since it allows to filter the noise out to a significant extent: a realisation of $x_1(t)$ is "smoother" than that of the observable $\eta(t)$.

10.1.2.4 *Successive Derivatives* **and Other Techniques**

The usage of the reconstructed vectors (10.4) is attractive due to the clear physical meaning of their components. Many processes are described with a higher order model ODE (9.4), which involves successive derivatives of a single variable (Sect. 9.1). Some ODEs can be rewritten in such a form *analytically*, e.g. the *Roessler* system (see Sect. 10.2.2). However, an essential shortcoming in exploiting the vectors (10.4) is high sensitivity of the approach to the measurement noise, since the derivatives must be computed numerically (Sect. 7.4.2).

To summarise, there are many techniques to reconstruct a phase orbit. Having only a scalar time series, one can use successive derivatives or time delays. At that, several parameters can be selected in different ways, e.g. a time delay and a numerical differentiation scheme. Besides, one can use weighted summation (Brown et al., 1994; Sauer et al., 1991); and integration (Janson et al., 1998), which is advantageous for strongly non-uniform signals. One often exploits principal components, empirical modes, conjugated signal and phase (Sect. 6.4.3). It is possible to use combinations of all the techniques, e.g. to get some components via time delays, additional ones via integration and the rest via differentiation (Brown et al., 1994). In the case of a vector observable, one can restore variables from each of its components with any combination of the above techniques. Hence, the number of possible variants strongly increases (Cao et al., 1998; Celucci et al., 2003).

10.1.2.5 Choice of Dynamical Variables

Which of the state vector versions should be preferred? This question is important and attracts considerable attention (Letellier and Aguirre, 2002; Letellier et al., 1998b; Small and Tse, 2004). Trying all possible variants in turn and approximating a dependence $d\mathbf{x}/dt = \mathbf{f}(\mathbf{x,c})$ or $\mathbf{x}_{n+1} = \mathbf{f}(\mathbf{x}_n, \mathbf{c})$ for each of them is unfeasible, since solving the approximation problem often requires significant computational efforts and special approaches. Therefore, one should select a small number of reasonable sets of dynamical variables in advance. It can be done based on the preliminary analysis of experimental dependencies to be approximated (Rulkov et al., 1995; Smirnov et al., 2002). The respective procedures exploit an obvious circumstance that one needs such set of variables which would provide uniqueness and continuity of the dependencies $d\mathbf{x}/dt(\mathbf{x})$ or $\mathbf{x}_{n+1}(\mathbf{x}_n)$, where components of $\mathbf{x}$ are either observed or computed from the observed data.

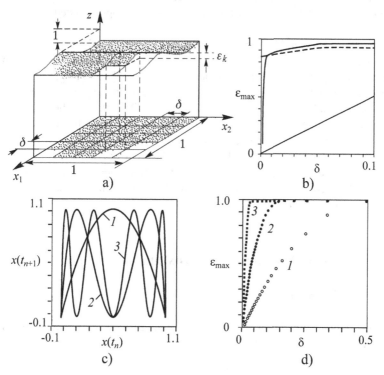

Fig. 10.8 Checking uniqueness and continuity of a dependence $z(\mathbf{x})$: (**a**) an illustration for $D = 2$; (**b**) typical plots $\varepsilon_{max}(\delta)$ for different choices of variables; the *straight line* is the best case, the *dashed line* corresponds to non-uniqueness or discontinuity of $z(\mathbf{x})$, the *broken line* corresponds to a complicated dependence $z(\mathbf{x})$ with the domains of fast and slow variations; (**c**) the plots of the first, the second and the third iterates of a quadratic map; (**d**) the plots $\varepsilon_{max}(\delta)$ for the dependence of $x(t_{n+1})$ on $x(t_n)$ in the three cases shown in panel (**c**)

Let us denote the left-hand side of model equations as $\mathbf{z}$: $\mathbf{z}(t) = d\mathbf{x}(t)/dt$ for a set of ODEs $d\mathbf{x}(t)/dt = \mathbf{f}(\mathbf{x}(t))$ and $\mathbf{z}(t_n) = \mathbf{x}(t_{n+1})$ for a map $\mathbf{x}(t_{n+1}) = \mathbf{f}(\mathbf{x}(t_n), \mathbf{c})$. After the reconstruction of the vectors $\mathbf{x}$ from an observable η, one should get a time series $\{\mathbf{z}(t_i)\}$. It is achieved via the numerical differentiation of the series $\{\mathbf{x}(t_i)\}$ for a set of ODEs and via the time shift of $\{\mathbf{x}(t_i)\}$ for a map. Further, it is necessary to check whether close vectors $\mathbf{x}(t_1)$ and $\mathbf{x}(t_2)$ correspond to close simultaneous vectors $\mathbf{z}(t_1)$ and $\mathbf{z}(t_2)$. A possible procedure is as follows (Smirnov et al., 2002).

A domain V containing the set of vectors $\{\mathbf{x}(t_i)\}$ is divided into equal hypercubic cells with the side δ (Fig. 10.8a). One selects all cells $s_1, \ldots, s_M$ such that each s_k contains more than one vector $\mathbf{x}(t_i)$. Thus, the cell s_k corresponds also to more than one vector $\mathbf{z}(t_i)$. The difference between the maximal and the minimal value of z (one of the components of the vector $\mathbf{z}$) over the cell s_k is called *local scattering* ε_k. Suitability of the quantities $\mathbf{x}$ and z for the global modelling is assessed from the maximal local scattering $\varepsilon_{\max} = \max\limits_{1 \le k \le M} \varepsilon_k$ and the plot $\varepsilon_{\max}(\delta)$. To construct a global model, one should choose variables such that the plot $\varepsilon_{\max}(\delta)$ gradually tend to the origin (Fig. 10.8b, straight line) for each of the approximated dependencies $z_k(\mathbf{x})$, $k = 1, \ldots, D$.

Moreover, it is desirable to provide the least slope of the plot $\varepsilon_{\max}(\delta)$, since one needs then a simpler approximating function, e.g. a low-order polynomial. This is illustrated in Fig. 10.8c, d, where the next value of an observable is shown versus the previous one and an observable is generated by the first, the second or the third iterate of the quadratic map $x(t_{n+1}) = \lambda - x^2(t_n)$. The plot for the first iterate is the "least oscillating" and, therefore, the slope of $\varepsilon_{\max}(\delta)$ is the smallest. In this case, one can get a "good" model most easily, since it requires the usage of only the second-order polynomial. At that, the eighth-order polynomial is necessary to describe the third iterate of the map. These three cases are even more different in respect of the reconstruction difficulties in the presence of noise. Additional details are given in Smirnov et al. (2002).

10.2 Multivariable Function Approximation

10.2.1 Model Maps

The time-delay embedding is typically used to construct multidimensional model maps

$$x_n = f(x_{n-D}, x_{n-D+1}, \ldots, x_{n-1}, \mathbf{c}), \tag{10.5}$$

where the variable x corresponds to an observable and the time delay is set equal to $l = 1$ for the simplicity of notations. Various choices of the function f in Eq. (10.5) are possible. One says that the function f, which is specified in a closed form (Sect. 3.5.1) in the *entire* phase space, provides a *global approximation*. Then, one also speaks of a *global model* and a *global reconstruction*. Alternatively, one can use a *local approximation*, i.e. the function f with its own set of parameter values for each small domain of the phase space. Then, one speaks of a *local model*.

In practice, a global approximation with algebraic polynomials often performs badly already for two-variable functions (Bezruchko and Smirnov, 2001; Casdagli, 1989; Judd and Mees, 1995; Pavlov et al., 1997. A pronounced feature is that the number of model parameters and the model prediction errors rise *quickly* with the model dimension D. The techniques with such a property are characterised as *weak approximation*. They also include trigonometric polynomials and wavelets. In practical black box modelling, one often has to use D at least as large as 5–6. Therefore, algebraic polynomials are not widely used.

Much efforts of researchers have been spent to *strong approximation* approaches, i.e. the approaches which are relatively insensitive to the rise in D. They include local techniques with low-order polynomials (Casdagli, 1989; Abarbanel et al., 1989; Farmer and Sidorowich, 1987; Kugiumtzis et al., 1998; Sauer, 1993; Schroer et al., 1998), radial, cylindrical, and elliptical basis functions (Giona et al., 1991; Judd and Mees, 1995, 1998; Judd and Small, 2000; Small and Judd, 1998; Small et al., 2002; Smith, 1992) and artificial neural networks (Broomhead and Lowe, 1988; Makarenko, 2003; Wan, 1993). All these functions usually contain many parameters so that a careful selection of the model structure and the model size is especially important to avoid overfitting (see Sects. 7.2.3 and 9.2).

10.2.1.1 A Generalised Polynomial

To construct a global model (10.5), one selects the form of f and estimates its parameters via the ordinary LS technique:

$$S(\mathbf{c}) = \sum_{i=D+1}^{N} (\eta_i - f(\eta_{i-D}, \eta_{i-D+1}, \ldots, \eta_{i-1}, \mathbf{c}))^2 \to \min. \qquad (10.6)$$

To simplify computations, it is desirable to select the function f, which is linear in its parameters $\mathbf{c}$. This is the case for a function

$$f(\mathbf{x}) = \sum_{k=1}^{P} c_k f_k(\mathbf{x}) \qquad (10.7)$$

which is called a *generalised polynomial* with respect to a *set of basis functions* $f_1, f_2, \ldots, f_P$. Then, the problem (10.6) is linear so that the local minima problem is avoided. A particular case of such an approach is represented by an algebraic polynomial. A trial polynomial order is increased until an appropriate model is obtained or another condition is fulfilled as discussed in Sect. 7.2.3.

10.2.1.2 Radial Basis Functions

These are functions $\phi_k(\mathbf{x}) = \phi(\|\mathbf{x} - \mathbf{a}_k\| / r_k)$, where $\| \cdot \|$ denotes a vector norm, a "mother" function ϕ is usually represented by a well-localised function,

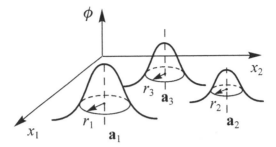

Fig. 10.9 The plots of two-variable radial basis functions (qualitative outlook): three "Gaussian hills"

e.g., $\phi(y) = \exp(-y^2/2)$, the quantities $\mathbf{a}_k$ are called "centres" and r_k are "radii". The model function f is taken to be a generalised polynomial with respect to the set of functions ϕ_k : $f(\mathbf{x}, \mathbf{c}) = \sum_k c_k \phi_k(\mathbf{x})$. Each term essentially differs from zero only within the distance about r_k from the centre $\mathbf{a}_k$ (Fig. 10.9). Intuitively, one can see that such a superposition can approximate a very complicated smooth relief. Radial basis functions possess many attractive properties and are often used in the approximation practice. However, we stop their discussion here and describe in more detail two approaches, which are even more widespread.

10.2.1.3 Artificial Neural Networks

Models with ANNs (Sect. 3.8) are successfully used to solve many tasks. Their right-hand side is represented by a *composition* of basis functions, rather than by their sum. In contrast to the generalised polynomial (10.7), ANNs are almost always non-linear with respect to the estimated parameters. This is the "most universal" way of the multivariable function approximation in the sense that along with a firm theoretical justification, it successfully performs in practice.

Let us introduce an ANN formally (in addition to the discussion of Sect. 3.8) with an example of a multilayer perceptron. Let $\mathbf{x} = (x_1, \ldots, x_D)$ be an argument of a multivariable function f. Let us consider the set of functions $f_j^{(1)}(\mathbf{x})$:

$$f_j^{(1)}(\mathbf{x}) = \phi \left(\sum_{i=1}^{D} w_{j,i}^{(0)} \cdot x_i - v_j^{(0)} \right), \qquad (10.8)$$

where $j = 1, \ldots, K_1$, the constants $w_{j,i}^{(0)}$ are called weights, $v_j^{(0)}$ are thresholds, ϕ is an activation function. The function ϕ is usually non-linear and has a step-like plot. One often uses the *classical sigmoid*: $\phi(x) = 1/(1 - e^{-x})$. Let us say that each function $f_j^{(1)}$ represents an output of *a standard formal neuron* with an order number j, whose input is the vector $\mathbf{x}$. Indeed, a living neuron sums up external stimuli and reacts to them in a threshold way that determines the properties of the

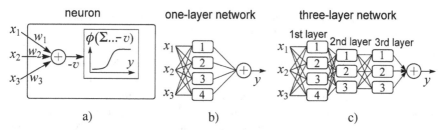

Fig. 10.10 Illustrations to artificial neural networks: (**a**) a standard formal neuron; (**b**) a scheme for a one-layer ANN with a single output; a single rectangle denotes a single neuron; (**c**) a scheme for a *multi-layer ANN* with a single output

function ϕ (Fig. 10.10a). The set of functions $f_1^{(1)}, \ldots, f_{K_1}^{(1)}$ is called the set of first-layer neurons (Fig. 10.10b). The values of $f_j^{(1)}$ are the outputs of the first-layer neurons. Let us denote them as vector $\mathbf{y}^{(1)}$ with components $y_j^{(1)} = f_j^{(1)}(\mathbf{x})$.

By defining the function f as a linear combination of $f_j^{(1)}$, one gets a *one-layer ANN* model

$$f(\mathbf{x}) = \sum_{j=1}^{K_1} w_j^{(1)} y_j^{(1)} - v^{(1)} \equiv \sum_{j=1}^{K_1} w_j^{(1)} \phi \left(\sum_{i=1}^{D} w_{j,i}^{(0)} x_i - v_i^{(0)} \right) - v^{(1)}, \quad (10.9)$$

where $w_j^{(1)}$, $v^{(1)}$ are additional weights and a threshold, respectively. The number of free parameters is $P = K_1(D+1)+1$. This representation resembles the generalised polynomial (10.7), but the ANN depends on $w_{j,i}^{(0)}$ and $v_j^{(0)}$ in a non-linear way.

By induction, let us consider a set of K_1-variable functions $f_k^{(2)}$, $k = 1, \ldots, K_2$, of the form (10.9). These are second-layer neurons, whose input is the output $\mathbf{y}^{(1)}$ of the first-layer neurons (Fig. 10.10c). Let us denote their output values as a vector $\mathbf{y}^{(2)}$ of the dimension K_2 and define the function f as a linear combination of the output values of the second-layer neurons:

$$f(\mathbf{x}) = \sum_{j_2=1}^{K_2} w_{j_2}^{(2)} \phi \left(\sum_{j_1=1}^{K_1} w_{j_2,j_1}^{(1)} \phi \left(\sum_{i=1}^{D} w_{j_1,i}^{(0)} x_i - v_{j_1}^{(0)} \right) - v_{j_2}^{(1)} \right) - v^{(2)}. \quad (10.10)$$

This is a *two-layer ANN* which involves compositions of functions. The latter circumstance makes it essentially different from the *pseudo-linear model* (10.7). Increasing the number of layers is straightforward.

To solve the approximation problems, one most often uses two-layer ANNs (10.10) and sometimes three-layer ones (Malinetsky and Potapov, 2000). The increase in the number of layers does not lead to a significant improvement. Improvements can be more often achieved via the increase in the number of neurons in each layer K_1, K_2. A theoretical base underlying the usage of the ANNs is the *generalised approximation theorem* (*Weierstrass' theorems* are its partial cases), which states that any continuous function can be arbitrarily accurately uniformly approximated with an ANN. A rigorous exposition is given, e.g., in Gorban' (1998).

The procedure for the estimation of parameters in an ANN via the minimisation (10.6) is called *learning* of an ANN. This is a problem of multidimensional non-linear optimisation. There are special "technologies" for its solution including backward error propagation algorithm, scheduled learning, learning with noise, stochastic learning (genetic algorithms and simulated annealing), etc. An ANN may contain many superfluous elements so that it is very desirable to make the structure of such a model (i.e. a network architecture) "more compact". For that, one excludes from a network those neurons whose weights and thresholds remain almost unchanged during the learning process.

If several alternative ANNs with different architectures are obtained from a training time series, then the best of them is usually selected according to the least test error (Sect. 7.2.3). To get an "honest" indicator of its predictive ability, one uses one more data set (not the training one and not the test one, since both of them are used to get the model), which is called a validation time series.

An advantage of an ANN over other constructions in empirical modelling is not easy to understand (Malinetsky and Potapov, 2000). If one gets an ANN, which performs well, it is usually unclear why this model is so good. It is the problem of the "network transparency"; a model of a black box is also a black box in a certain sense. Yet, even such a model can be investigated numerically and used to generate predictions.

10.2.1.4 Local Models

Local models are constructed so as to minimise the sum of squares like Eq. (10.6) over a local domain of the phase space. Thus, to predict the value η_{i+D}, which follows a current state $\mathbf{x}_i = [\eta_i, \eta_{i+1}, \ldots, \eta_{i+D-1}]$, one uses the following procedure. One finds k *nearest neighbours* of the vector $\mathbf{x}_i$ among all the vectors in the training time series (in the past). These are vectors with time indices n_j, whose distance to $\mathbf{x}_i$ are smallest:

$$\left\| \mathbf{x}_{n_j} - \mathbf{x}_i \right\| \le \left\| \mathbf{x}_l - \mathbf{x}_i \right\|, j = 1, \ldots, k, l \ne i, l \ne n_j. \tag{10.11}$$

They are also called the *analogues* of $\mathbf{x}_i$, see Figs. 10.11 and 10.12.

The values of an observable, which followed the neighbours $\mathbf{x}_{n_j}$ in the past, are known. Hence, one can construct the model (10.5) from those data. For that, one typically uses a simple function $f(\mathbf{x}, \mathbf{c})$, whose parameters are found with the ordinary LS technique (Sect. 8.1.1), although more sophisticated estimation techniques are available (Kugiumtzis et al., 1998). An obtained function $f(\mathbf{x}, \hat{\mathbf{c}}_i)$ is used to generate a prediction of the value η_{i+D} according to the formula $\hat{\eta}_{i+D} = f(\mathbf{x}_i, \hat{\mathbf{c}}_i)$, see Fig. 10.12. The vector $\hat{\mathbf{c}}_i$ has a subscript i, since it corresponds only to the vicinity of the vector $\mathbf{x}_i$. According to the so-called *iterative forecast* (Sect. 10.3), one predicts the next value η_{i+D+1} by repeating the same procedure of the neighbour search and

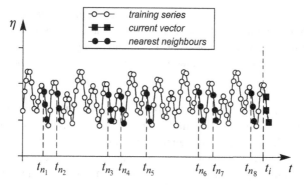

Fig. 10.11 Illustration for a three-dimensional local model: nearest neighbours (*filled circle*) of a vector x_i (*filled squares*) found in a training time series

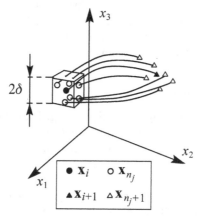

Fig. 10.12 Nearest neighbours (*open circles*) of a vector x_i (*filled circles*) and the vectors following them in time (*open triangles*). The latter are used to predict the vector x_{i+1} (*filled triangle*)

parameter estimation for the model state vector $\hat{x}_{i+1} = (\eta_{i+1}, \ldots, \eta_{i+D-1}, \hat{\eta}_{i+D})$. Thereby, one gets a new forecast $\hat{\eta}_{i+D+1} = f(\hat{x}_{i+1}, \hat{c}_{i+1})$ and so on.

Relying on the Taylor polynomial expansion theorem, one uses such approximating functions as the constant $f(\mathbf{x}, \mathbf{c}) = c_1$, the linear function $f(\mathbf{x}, \mathbf{c}) = c_1 + \sum_{j=1}^{D} c_{j+1} x_j$ and the polynomials of a higher order K. On the one hand, an approximation error is smaller if the neighbours are closer to the current vector. Therefore, it should decrease with an increasing time series length, since closer returns to the vicinity of each vectors would occur. On the other hand, one should use a greater number of neighbours k to reduce the noise influence. Thus, a trade-off is necessary: one cannot use *too distant* "neighbours" to keep an error of approximation with a low-order polynomial small, but one cannot take *too a small number* of the nearest neighbours as well.

Local constant models are less requiring to the amount of data and more robust to noise since they contain a single free parameter for each small domain. *Local linear models* are superior for weak noise and sufficiently long time series: The concrete values depend on the necessary value of D. To construct a local linear model, one must use at least $k = D+1$ neighbours, since a model contains $D+1$ free parameters for each "cell". Its approximation error scales as λ^2 for a very long time series and "clean" data, where λ is a characteristic distance between the nearest neighbours in the time series. Local models with higher order polynomials are rarely used.

For the above local models, the function f is usually discontinuous, since different "pieces" of local approximation are not matched with each other. Sometimes, it leads to undesirable peculiarities of the model dynamics, which are not observed for the original system. One can avoid the discontinuity via triangulation (Small and Judd, 1998). At that, a model acquires some properties of the global approximation (f becomes continuous) and is described as a global-local model. However, the triangulation step makes the modelling procedure much more complicated.

Local models are often exploited for practical predictions. There are various algorithms taking into account delicate details. In essence, this is a contemporary version of the predictive "method of analogues" (Fig. 10.11).

10.2.1.5 Nearest Neighbour Search

It can take much time if a training time series is long. Thus, if one naively computes distances from a current vector to each vector in the time series and selects the least ones, the number of operations scales as N^2. Below, an efficient search algorithm based on the preliminary partition of the training set into cells (Kantz and Schreiber, 1997) is described.

The above local models are characterised by *fixed number of neighbours*. Let us consider another (but similar) version: local models with *fixed neighbourhood size*. The difference is that one looks for the neighbours of a vector $\mathbf{x}_i$, which are separated from $\mathbf{x}_i$ by a distance not greater than δ (Fig. 10.12):

$$\left\| \mathbf{x}_{n_j} - \mathbf{x}_i \right\| \le \delta. \tag{10.12}$$

The number of neighbours may differ for different $\mathbf{x}_i$, but it should not be less than $D+1$. If there are too a few neighbours, one should increase the neighbourhood size δ. Under a fixed time series length, an optimal neighbourhood size rises with the noise level and the model dimension. An optimal δ is selected via trials and errors. One can use any norm of a vector in Eq. (10.11) or (10.12). The most convenient one is $\|\mathbf{x}\| = \max\{|x_1|, |x_2|, \ldots, |x_D|\}$, since it is quickly computed. In such a case, the neighbourhood (10.12) is a cube with the side of length 2δ.

Computation of the distances from a reference vector $\mathbf{x}_i$ to all the vectors in the training time series would require a lot of time. It is desirable to skip the vectors, which deliberately cannot be close neighbours of $\mathbf{x}_i$. For that, one preliminarily *sorts* all the vectors based on the first and the last of their D coordinates. Let $\eta_{\min}$ and $\eta_{\max}$ be the minimal and maximal values, respectively, of an observable over

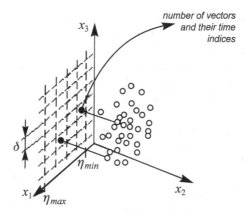

Fig. 10.13 Vectors of a training time series are sorted based on their first and last coordinates: one creates a square array whose elements contain information about the number of vectors in the respective cell and their time indices

the training series. Then, the corresponding orbit on the plane (x_1, x_D) lies within the square with the sides belonging to the straight lines defined by the equations $x_1 = \eta_{min}, x_1 = \eta_{max}, x_D = \eta_{min}, x_D = \eta_{max}$ (Fig. 10.13). The square is divided into square cells of size δ. One determines to which cell each vector falls and creates an array, whose elements correspond to the cells. Each element contains time indices of the vectors falling into the respective cell. To find the nearest neighbours of a vector **x**, one checks into which cell it falls and computes distances from **x** to the vectors belonging to the same cell or the cells having a common vertex with it. In total, one must check at most nine cells. This algorithm speeds up the process of neighbour search and requires the order of N operations if there are no too densely and too rarely "populated" domains in the reconstructed phase space.

10.2.1.6 A Real-World Example

Chaotic dynamics of a laser (Fig. 10.14) was suggested as a test data set for the *competition in time series prediction* at the conference in Santa-Fe in 1993 (Gerschenfeld and Weigend, 1993). Competitors had to provide a continuation of the time series, namely to predict the next 100 data points based on 1000 given data points. A winner was Eric Wan, who used a feed-forward ANN-based model of the form (10.5) (Wan, 1993).

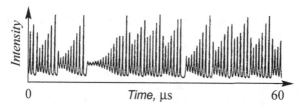

Fig. 10.14 Data from a ring laser in a chaotic regime (Hubner et al., 1993), $\Delta t = 40\,\text{ns}$

Figure 10.15a shows the observed time series (thin lines) and predictions (thick lines) with the ANN-based model for different starting instants (Wan, 1993). Similar prediction accuracy is provided by a local linear model (Sauer, 1993), see Fig. 10.15b. Accuracy of predictions for different intervals depends on how accurately the models predict an instant of switching from the high-amplitude oscillations to the low-amplitude ones. Thus, the local linear model performs worse than the ANN for the starting instants 1000 and 2180 and better for the three others. Local linear models appear to reproduce better some dynamical characteristics of the process and its long-term behaviour (Gerschenfeld and Weigend, 1993); the top panel in Fig. 10.15b shows an iterative forecast over 400 steps, which agrees well with the experimental data. The long-term behaviour is described a bit worse with the ANN (Sauer, 1993; Wan, 1993).

It is interesting to note that Eric Wan used ANN with 1105 free parameters trained on only 1000 data points. The number of parameters was even greater than the number of data points that usually leads to an ill-posed problem in statistics.

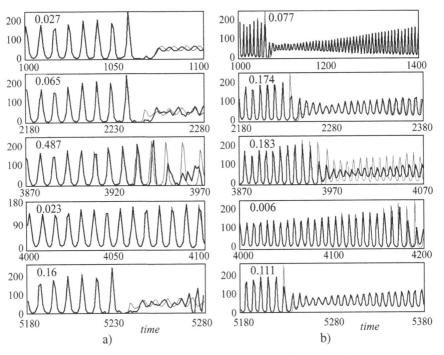

Fig. 10.15 Forecast of the laser intensity: (**a**) an ANN-based model (Wan, 1993); (**b**) a local linear model (Sauer, 1993). Laser intensity is shown along the vertical axis in arbitrary units. Time is shown along the horizontal axis in the units of sampling interval. The *thin lines* are the observed values and the *thick lines* are the predictions. Different panels correspond to predictions starting from different time instants: 1000, 2,180, 3,870, 4,000, and 5,180. The number at the top left corner of each panel is the normalised mean-squared prediction error over the first 100 data points of the respective data segment. The *top panels* show the segment proposed for the competition in Santa-Fe: the data points 1001–1100 were to be predicted

However, an ANN is a highly non-linear model function so that the number of "effective degrees of freedom" ("effectively free" parameters) is not equal to the full number of estimated parameters. There are constraints imposed on the possible model behaviour by the topology of the ANN. The author performed cross-validation (Sect. 7.2.3) by using only 900 data points as the training time series and 100 data points as the test one. He stated that there were no signs of overfitting when the size of the network was changed. Still, he noted an indirect sign of overfitting: after a good short-term forecast over several dozens of time steps, the ANN-based model exhibited "noisier" long-term behaviour than it is observed in the original data (Fig. 10.14). This can also be the reason why the local linear model of Tim Sauer appeared superior for the description of the long-term behaviour. An overall judgement seems to be that an ANN and a local linear model are approximately equally good in the example considered.

A number of applications of local models to predictions can be found, e.g., in Farmer and Sidorowich (1987); Kantz and Schreiber (1997); Kugiumtzis et al. (1998). ANN-based models are used probably more often, since they are less demanding with respect to the time series length and noise level. There are examples of their successful applications even to geophysical and financial predictions (Makarenko, 2003). Forecasts with other models of the form (10.5) are described, e.g., in Judd and Small (2000); Small and Judd (1998).

10.2.2 Model Differential Equations

In construction of model ODEs from a scalar time series, one often gets state vectors with successive derivatives $[\eta, d\eta/dt, \ldots, d^{D-1}\eta/dt^{D-1}]$ and uses the *standard form of model equations* (Sect. 3.5.3):

$$d^D x \big/ dt^D = f(x, dx/dt, \ldots, d^{D-1}x \big/ dt^{D-1}, \mathbf{c}) \qquad (10.13)$$

where $\eta = x$. The approximating function f is selected in the same way as described above for the models (10.5). Here, one more often observes "smoother" dependencies to be approximated and uses algebraic polynomials

$$f(x_1, x_2, \ldots, x_D, \mathbf{c}) = \sum_{l_1, l_2, \ldots, l_D = 0}^{K} c_{l_1, l_2, \ldots, l_D} \prod_{j=1}^{D} x_j^{l_j}, \quad \sum_{j=1}^{D} l_j \leq K. \qquad (10.14)$$

Model ODEs (10.13) with ANNs and other functions are rarely used (Small et al., 2002).

Some systems can be rewritten in the standard form (10.13) analytically. Thus, the *Roessler system,* which is a paradigmatic chaotic system, reads as

$$\begin{aligned} \mathrm{d}x/\mathrm{d}t &= -y - z, \\ \mathrm{d}y/\mathrm{d}t &= x + C_1 y, \\ \mathrm{d}z/\mathrm{d}t &= C_2 - C_3 z + xz. \end{aligned} \qquad (10.15)$$

One can show that it can be reduced to a three-dimensional system with successive derivatives of y and the second-order polynomial in the right-hand side:

$$\begin{aligned} \mathrm{d}x_1/\mathrm{d}t &= x_2, \\ \mathrm{d}x_2/\mathrm{d}t &= x_3, \\ \mathrm{d}x_3/\mathrm{d}t &= -C_2 - C_3 x_1 + (C_1 C_3 - 1)x_2 + (C_1 - C_3)x_3 - C_1 x_1^2 + \\ &\quad + (C_1^2 + 1)x_1 x_2 - C_1 x_1 x_3 - C_1 x_2^2 + x_2 x_3, \end{aligned} \qquad (10.16)$$

where $x_1 = y$. With successive derivatives of x and z, one gets the equations similar to Eq. (10.16), but with rational functions in the right-hand side (Gouesbet and Letellier, 1994).

The standard models are used in practice (see, e.g., Gouesbet, 1991; Gouesbet and Letellier, 1994; Gouesbet et al., 2003b Gribkov et al., 1994a, 1994b; Letellier et al., 1995, 1997, 1998a; Pavlov et al., 1999), but successful results are quite rare. The usage of the model structure (10.13) with an algebraic polynomial (10.14) often leads to quite cumbersome equations.

In construction of ODEs from a vector time series, everything is similar, but one looks for D scalar-valued functions rather than for a single one (analogous to the "grey box" case described in Sect. 9.1).

10.3 Forecast with Various Models

The novel techniques developed within the non-linear dynamics framework and discussed above are often the most efficient ones to predict complex real-world processes. Especially, this is the case when it appears sufficient to use a low-dimensional model. Multiple non-linear dynamics techniques can be distinguished according to different aspects: iterative forecast versus direct forecast; model maps of different kinds, e.g., global models versus local models; model maps versus model ODEs. Below, advantages and disadvantages of different approaches are briefly discussed and compared. To run ahead, the most efficient tool is usually a model map (global or local one depending on the amount of available data and the necessary model dimension) with an iterative, direct or combined prediction technique (depending on the required advance time). However, let us start the comparison with "older" approaches.

10.3.1 Techniques Which Are not Based on Non-linear Dynamics Ideas

For very simple signals, good predictions can be achieved even with explicit functions of time (Chap. 7). For stationary irregular signals without signs of non-linearity, the most appropriate tool is linear ARMA models (Sects. 4.4 and 8.1) despite their possibilities being quite limited. Thus, one can show that prediction with a linear ARMA model can be reasonably accurate only over an interval of the order of the process correlation time τ_{cor} (Anosov et al., 1995), i.e. a characteristic time of the autocorrelation function decay (see Sect. 2.3, Fig. 2.8).

For a chaotic time series, τ_{cor} can be quite small making possible only quite short-term predictions with an ARMA model. Although a chaotic process cannot be accurately predicted far in the future in principle, the prediction time can be much greater than τ_{cor} for non-linear models. It can be roughly estimated with the formula (2.34): $\tau_{\text{pred}} = (1/2\Lambda_1) \ln \left(\sigma_x^2 / (\sigma_\nu^2 + \sigma_\mu^2 + \sigma_{\Delta M}^2)\right)$. If noises and model errors are not large, then τ_{pred} can strongly exceed the correlation time roughly estimated as $\tau_{\text{cor}} \sim 1/\Lambda_1$ (see Sect. 2.4 and Kravtsov, 1989; Smith, 1997).

10.3.2 Iterative, Direct and Combined Predictors

One can predict the values of an observable following the last value in a time series η_N with a model (10.5) via the above-mentioned (Sect. 10.2.1) *iterative* way:

(i) One-step-ahead prediction is generated as

$$\hat{\eta}_{N+1} = f(\mathbf{x}_{N-D+1}, \hat{\mathbf{c}}) = f(\eta_{N-D+1}, \eta_{N-D+2}, \ldots, \eta_N, \hat{\mathbf{c}}). \qquad (10.17)$$

(ii) The predicted value $\hat{x}_{N+1}$ is considered as the last coordinate of the new state vector $\hat{\mathbf{x}}_{N-D+2} = (\eta_{N-D+2}, \eta_{N-D+3}, \ldots, \eta_N, \hat{\eta}_{N+1})$;

(iii) The vector $\hat{\mathbf{x}}_{N-D+2}$ is used as an argument of the function f to generate a new forecast $\hat{\eta}_{N+2} = f(\hat{\mathbf{x}}_{N-D+2}, \hat{\mathbf{c}})$ and so on.

Thus, one gets a forecast $\hat{\eta}_{N+l}$ over any number of steps l ahead.

Alternatively, one can get an *l*-step-ahead forecast by constructing a model which *directly* approximates dependence of η_{i+l} on $(\eta_{i-D+1}, \eta_{i-D+2}, \ldots, \eta_i)$ instead of making l iterations of the map (10.5). The form of such a dependence can be rather complicated for the chaotic dynamics and large l due to high sensitivity of the future behaviour η_{i+l} to the initial conditions $\mathbf{x}_{i-D+1}$. As a result, for a very large l, one gets an approximately constant model function $f \approx \langle \eta \rangle$ and, hence, a low prediction accuracy. However, the direct approach can be advantageous for moderate values of l.

How does the prediction time for both approaches depend on the time series length N and other factors? This question can be answered theoretically for a local model with a polynomial of an order K. According to the estimates of Casdagli (1989) and Farmer and Sidorowich (1987), the prediction error grows with l as $\sigma_{\Delta M} \cdot e^{\Lambda_1 l \Delta t}$ for the iterative technique and as $\sigma_{\Delta M} \cdot e^{(K+1)Hl\Delta t}$ for the direct technique, where H is the sum of the positive Lyapunov exponents. Thus, the error grows faster for the direct approach. The reason is mentioned above: it is difficult to approximate the dependence of the *far future* on the present state. However, this superiority of the iterative technique takes place only if the model (10.5) gives very accurate one-step-ahead predictions, which are achieved only for a very long training time series and a very low noise level. Otherwise, the direct technique can give more accurate l-step-ahead predictions for l greater than 1 but less than a characteristic time of the divergence of initially nearby orbits (Judd and Small, 2000). This is because a one-step-ahead predictor (10.5) can exhibit systematic errors (due to an inappropriate approximating function or an insufficient model dimension), whose accumulation over iterations can be more substantial than the approximation error of the direct approach.

To improve predictions for a moderate advance time l, a combined "predictor – corrector" approach is developed in Judd and Small (2000), which is as follows. One generates a forecast via a direct or an iterative way with an existing model. Let us call it a "base predictor". Then, one "corrects" the predictions with an additional model map (the so-called "corrector") which is also constructed from a training time series and describes the dependence of an l-step-ahead prediction error of the base predictor on the prediction itself. A functional form of the corrector is taken much simpler than that for the base predictor. The combination *"predictor – corrector"* can give essentially more accurate forecasts in comparison with the "pure" direct and iterative approaches.

Finally, we note an important conceptual distinction between the dynamical models (10.5) and explicit functions of time (Sect. 7.4.1). In contrast to explicit *extrapolation* of a temporal dependence, the model (10.5) relies on *interpolation in the phase space* and, therefore, appears much more efficient. Indeed, a current value of the state vector $\mathbf{x}_i$ used to generate a prediction typically lies "between" many vectors of the training time series, which are used to construct the model (Fig. 10.12). If the model state vector $\mathbf{x}$ under iterations of the model map goes out of the domain populated by the vectors of the training time series, the usage of the model to generate further predictions is tantamount to *extrapolation in the phase space*. Then, the forecasts get much less reliable and a model orbit may behave quite differently from the observed process, e.g. diverge to infinity. The latter often occurs if a model involves an algebraic polynomial, which usually quite badly extrapolates.

10.3.3 Different Kinds of Model Maps

Let us compare predictive abilities of the models (10.5) for different forms of the function f.

Algebraic polynomials of a moderate order K are quite efficient to approximate gradually varying one-variable functions without discontinuities and breaks. Only cubic or higher order splines are even better in this case (Johnson and Riess, 1982; Kalitkin, 1978; Press et al., 1988; Samarsky, 1982; Stoer and Bulirsch, 1980). The greater the necessary model dimension D and the polynomial order K, the less the probability of successful modelling results with algebraic polynomials.

Rational functions are efficient under the same conditions but can better describe dependencies with the domains of fast variations. Trigonometric polynomials and wavelets (Sect. 6.4.2) are also kinds of weak approximation. They perform well for dependencies with specific properties described in Sect. 7.2.4.

Radial basis functions (Judd and Mees, 1995) are much superior to the mentioned approaches in the case of a large model dimension (roughly speaking, the dimension greater than 3). Artificial neural networks exhibit similar performance. According to some indications (Casdagli, 1989; Wan, 1993), the ANNs approximate complicated dependencies even better. Requirements to the amount of data and noise level are not very strict for all the listed models, since all of them are global.

Local linear models are very efficient for moderate model dimensions (less than a moderate number depending on the length of a training time series), long time series (giving a considerable number of the close neighbours for each state vector) and low levels of the measurement noise. Requirements to the amount of data and noise level are quite strict. Local constant models are better than local linear ones for higher noise levels and shorter time series.

In any case, all the approaches suffer from the curse of dimensionality. In practice, very high-dimensional motions (roughly, with the dimensions about 10 or greater) typically are not successfully described with the above empirical models.

10.3.4 Model Maps Versus Model ODEs

In general, model maps give better short-term forecasts than do ODEs (Small et al., 2002). It can be understood in analogy with the situation, where the iterative predictions are less accurate than the direct ones due to significant errors of the one-step-ahead predictor (10.5). Model ODEs are constructed so as to approximate a dependence of the phase velocity dx/dt on x (9.3), i.e. to provide better forecasts over short time intervals: $x(t_{i+1}) \approx x(t_i) + \left(dx(t_i)/dt\right)\Delta t$. Then, the integration of ODEs to predict distant future values resembles an iterative forecast. It can be less precise if a systematic error is present in the model ODEs.

Empirical model maps are often superior even for long-term description of the observed dynamics (Small et al., 2002). Besides, their construction and exploitation are simpler: One does not need numerical differentiation of the signals and numerical integration of the equations.

Model ODEs are a good choice if they are "relatives" to an object, i.e. if an original dynamics yields almost exactly to a set of ODEs with some known structure. Such a case is more typical for the "transparent box" or the "grey box" settings and almost improbable without detailed prior information.

Yet, many authors deal with the construction of model ODEs even under the "black box" setting. It is related in part to the problem of the model "transparency". If one gets a "good" model, it is desirable to understand how it "works" and to interpret its variables and parameters from the physical viewpoint. One may hope for such physical interpretations when model ODEs with algebraic polynomials are used, since asymptotic models of many real-world processes take such a form, e.g., in chemical kinetics and laser physics. For the same reason, one can use model ODEs with the successive derivatives (10.4) rather than with the time-delay vectors:

derivatives can be easily interpreted as velocity, acceleration, etc. However, the hope for physical interpretations does not usually prove to be correct: If one does not include physical ideas into the structure of model ODEs in advance (Bezruchko and Smirnov, 2001; Smirnov and Bezruchko, 2006), it gets almost impossible to extract physical sense from an algebraic polynomial (10.14) or a similar universal construction a posteriori.

10.4 Model Validation

Though it is relevant to perform the *residual error analysis* (see Sect. 7.3), if a dynamical noise is assumed to influence an original dynamics, one typically computes model characteristics popular in the theory of dynamical systems and compares them to experimental estimates to validate a dynamical model.

(i) For a deterministic model, *predictability time* can be theoretically estimated as $\tau_{pred} = (1/2\Lambda_1) \ln (\sigma_x^2/(\sigma_v^2 + \sigma_\mu^2 + \sigma_{\Delta M}^2))$. For an adequate model, this estimate must coincide with the corresponding empirical estimate.

(ii) *Qualitative similarity of the phase orbits projected onto different planes.* This subjective criterion seems important. It is directed to the assessment of the similarity between *essential features* of an object and model dynamics. Its quantitative formulations lead to several ways of model validation, which are mentioned below following the review (Gouesbet et al., 2003).

(iii) *Comparison of invariant measures* (probability distribution densities for a state vector) or their projections (marginal probability distributions). The approach applies to stochastic models as well.

(iv) *Comparison of the largest Lyapunov exponent* of a model attractor with its estimate obtained from an observed time series.

(v) *Comparison of fractal dimensions and entropies* of a model attractor with their estimates obtained from an observed time series.

(vi) *Comparison of topological properties.* This delicate approach is based on the search and analysis of unstable periodic orbits embedded into an attractor and the determination of their mutual location in the phase space. Strictly speaking, it applies only to deterministic systems, whose dimensionality is not greater than three, and represents a very strict test for a model. If a model reproduces a major part of unstable orbits found from an observed time series, it is a strong evidence in favour of its validity.

(vii) *Comparison of the Poincare maps.* It is easily achieved for one-dimensional Poincare maps. As a rule, one analyses a dependence of the next local maximum value of an observable on its previous local maximum. The approach relates to the analysis of the topological properties of attractors and is often used in combination with the latter.

(viii) *Synchronisation of model dynamics by an original signal.* A model is regarded valid if its dynamics synchronise up to a given accuracy by an observed time series, used as a driving, under a moderate driving intensity (Brown et al., 1994). This approach applies if model parameters were not estimated via synchronisation-based technique (Sect. 8.2.1) from the same time series.

(ix) It has been suggested to check whether a model has the same number of attractors of the same type as an object; whether the attractors of a model and an object are located in the same domains of the phase space and whether their basins of attraction coincide. These are very strict tests and in practice no empirical model can usually pass them.

Finally, we note that modelling of spatially extended systems with partial differential equations and other tools is actively studied for the last years (Bar et al., 1999; Parlitz and Mayer-Kress, 1995; Parlitz and Merkwirth, 2000; Sitz et al., 2003; Voss et al., 1998) which is not discussed here. Also, we have only briefly touched upon stochastic model equations (Sitz et al., 2002; Timmer, 2000; Tong, 1990). Diverse useful information on those and adjacent subjects can be found, in particular, at the websites mentioned in the Preface.

References

Abarbanel, H.D.I., Brown, R., Kadtke, J.B.: Prediction and system identification in chaotic nonlinear systems: time series with broadband spectra. Phys. Lett. A. **138**, 401–408 (1989)

Anosov, O.L., Butkovsky, O.Ya., Kravtsov, Yu.A.: Limits of predictability for linear autoregression models. J. Communications Technol. Electron. **40**(12), 1866–1873, (in Russian) (1995)

Bär, M., Hegger, R., Kantz, H.: Fitting partial differential equations to space-time dynamics. Phys. Rev. E. **59**, 337–343 (1999)

Bezruchko, B.P., Smirnov, D.A.: Constructing nonautonomous differential equations from a time series. Phys. Rev. E. **63**, 016207, (2001)

Broomhead, D.S., King, G.P.: Extracting qualitative dynamics from experimental data. Physica D. **20**, 217–236 (1986)

Broomhead, D.S., Lowe, D.: Multivariable functional interpolation and adaptive networks. Complex Syst. **2**, 321–355 (1988)

Brown, R., Rulkov, N.F., Tracy, E.R.: Modeling and synchronizing chaotic systems from experimental data. Phys. Lett. A. **194**, 71–76 (1994)

Cao, L., Mees, A.I., Judd, K.: Dynamics from multivariate time series. Phys. D. **121**, 75–88 (1998)

Casdagli, M.: Nonlinear prediction of chaotic time series. Phys. D. **35**, 335–356 (1989)

Casdagli, M., Eubank, S., Farmer, J.D., Gibson, J.: State space reconstruction in presence of noise. Phys. D. **51**, 52–98 (1991)

Cellucci, C.J., Albano, A.M., Rapp, P.E.: Comparative study of embedding methods. Phys. Rev. E. **67**, 066210 (2003)

Eckmann, J.P., Ruelle, D.: Ergodic theory of chaos and strange attractors. Rev. Mod. Phys. **57**, 617–656 (1985)

Farmer, J.D., Sidorowich, J.J.: Predicting chaotic time series. Phys. Rev. Lett. **59**, 845–848 (1987)

Fraser, A.M., Swinney, H.L.: Independent coordinates for strange attractors from mutual information. Phys. Rev. A. **33**, 1131–1140 (1986)

Gerschenfeld, N.A., Weigend, A.S. (eds.): Time Series Prediction: Forecasting the Future and Understanding the Past. SFI Studies in the Science of Complexity, Proc. V. XV. Addison-Wesley, New York (1993)

Gibson, J.F., Farmer, J.D., Casdagli, M., Eubank, S.: An analytic approach to practical state space reconstruction. Phys. D. **57**, 1–30 (1992)

Giona, M., Lentini, F., Cimagalli, V.: Functional reconstruction and local prediction of chaotic time series. Phys. Rev. E. **44**, 3496–3502 (1991)

Gliklikh Yu.E.: What a smooth manifold is. Soros Educ. J. **11**, 155–159, (in Russian) (1998)

Gorban' A.N.: Multivariable functions and neural networks. Soros Educational J. No. 12, 105–112, (in Russian) (1998)

Gouesbet, G.: Reconstruction of the vector fields of continuous dynamical systems from scalar time series. Phys. Rev. A. **43**, 5321–5331 (1991)

Gouesbet, G., Letellier, C.: Global vector-field approximation by using a multivariate polynomial L_2 approximation on nets. Phys. Rev. E. **49**, 4955–4972 (1994)

Gouesbet, G., Meunier-Guttin-Cluzel, S., Ménard, O.: Global reconstructions of equations of motion from data series, and validation techniques, a review. In: Gouesbet, G., Meunier-Guttin-Cluzel, S., Ménard, O. (eds.) Chaos and Its Reconstructions, pp. 1–160. Nova Science Publishers, New York, (2003)

Grassberger, P., Procaccia, I.: Measuring the strangeness of strange attractors. Phys. D. **9**, 189–208 (1983)

Gribkov, D.A., Gribkova, V.V., Kravtsov Yu.A., et al. Reconstructing structure of dynamical system from time series. J. Commun. Technol. Electron. **39**(2), 269–277, (in Russian) (1994)

Hubner, U., Weiss, C.-O., Abraham, N.B., Tang, D.: Lorenz-like chaos in $NH_3 - FIR$ lasers (data set A). In: Gerschenfeld, N.A., Weigend, A.S. (eds.) Time Series Prediction: Forecasting the Future and Understanding the Past. SFI Studies in the Science of Complexity, Proc. V. XV, pp. 73–104. Addison-Wesley, Reading, MA (1993)

Janson, N.B., Pavlov, A.N., Anishchenko, V.S.: One method for restoring inhomogeneous attractors. Int. J. Bif. Chaos. **8**, 825–833 (1998)

Johnson, L.W., Riess, R.D.: Numerical Analysis, 2nd edn. Addison-Wesley, Reading, MA (1982)

Judd, K., Mees, A.I. On selecting models for nonlinear time series. Phys. D. **82**, 426–444 (1995)

Judd, K., Mees, A.I.: Embedding as a modeling problem. Phys. D. **120**, 273–286 (1998)

Judd, K., Small, M.: Towards long-term prediction. Phys. D. **136**, 31–44 (2000)

Kalitkin, N.N.: Numerical Methods. Nauka, Moscow, (in Russian) (1978)

Kantz, H., Schreiber, T.: Nonlinear Time Series Analysis. Cambridge University Press, Cambridge (1997)

Kartsev, V.P.: Adventures of Great Equations. Znaniye, Moscow, (in Russian) (1976)

Kennel, M.B., Brown, R., Abarbanel, H.D.I.: Determining embedding dimension for phase-space reconstruction using a geometrical construction. Phys. Rev. A. **45**, 3403–3411 (1992)

Kravtsov Yu.A.: Randomness, determinacy, predictability. Phys. Uspekhi. **158**(1), 93–115, (in Russian) (1989)

Kugiumtzis, D., Lingjaerde, O.C., Christophersen, N.: Regularized local linear prediction of chaotic time series. Phys. D. **112**, 344–360 (1998)

Landa, P.S., Rosenblum, M.G.: Comparison of methods for construction of phase space and determination of attractor dimension from experimental data. Tech. Phys. **59**(11), 1–8, (in Russian) (1989)

Letellier, C., Aguirre, L.A.: Investigating nonlinear dynamics from time series: The influence of symmetries and the choice of observables. Chaos. **12**, 549–558 (2002)

Letellier, C., Le Sceller, L., Gouesbet, G., et al.: Recovering deterministic behavior from experimental time series in mixing reactor. AIChE J. **43**(9), 2194–2202 (1997)

Letellier, C., Le Sceller, L., Maréchal, E., et al.: Global vector field reconstruction from a chaotic experimental signal in copper electrodissolution. Phys. Rev. E. **51**, 4262–4266 (1995)

Letellier, C., Maquet, J., Labro, H., et al. Analyzing chaotic behavior in a Belousov-Zhabotinskyi reaction by using a global vector field reconstruction. J. Phys. Chem. **102**, 10265–10273 (1998a)

Letellier, C., Macquet, J., Le Sceller, L., et al. On the non-equivalence of observables in phase space reconstructions from recorded time series. J. Phys. A: Math. Gen. **31**, 7913–7927 (1998b)

Liebert, W., Schuster, H.G.: Proper choice the of time delay for the analysis of chaotic time series. Phys. Lett. A. **142**, 107–111 (1989)

Makarenko, N.G.: Fractals, attractors, neural networks and so forth. Procs. IV All-Russian Conf. "Neuroinformatics-2002". Part 2, pp. 121–169. Moscow, (in Russian) (2002)

Makarenko, N.G.: Embedology and Neuro-Prediction. Procs. V All-Russian Conf. "Neuroinformatics-2003". Part 1, pp. 86–148. Moscow, (in Russian) (2003)

Malinetsky, G.G., Potapov, A.B.: Contemporary Problems of Nonlinear Dynamics. Editorial URSS, Moscow, (in Russian) (2000)

Parlitz, U., Mayer-Kress, G.: Predicting low-dimensional spatiotemporal dynamics using discrete wavelet transforms. Phys. Rev. E. **51**, R2709–R271 (1995)

Parlitz, U., Merkwirth, C.: Prediction of spatiotemporal time series based on reconstructed local states. Phys. Rev. Lett. **84**, 1890–1893 (2000)

Pavlov, A.N., Janson, N.B., Anishchenko, V.S.: Application of statistical methods to solve the problem of global reconstruction. Tech. Phys. Lett. **23**(4), 297–299 (1997)

Pavlov, A.N., Janson, N.B., Anishchenko, V.S.: Reconstruction of dynamical systems. J. Commun. Technol. Electron. **44**(9), 999–1014 (1999)

Press, W.H., Flannery, B.P., Teukolsky, S.A., Vetterling, W.T.: Numerical Recipes in C. Cambridge University Press, Cambridge (1988)

Rulkov, N.F., Sushchik, M.M., Tsimring, L.S., Abarbanel, H.D.I.: Generalized synchronization of chaos in directionally coupled chaotic systems. Phys. Rev. E. **51**, 980–994 (1995)

Samarsky, A.A.: Introduction to Numerical Methods. Nauka, Moscow, (in Russian) (1982)

Sauer, T., Yorke, J.A., Casdagli, M.: Embedology. J. Stat. Phys. **65**(3–4), 579–616 (1991)

Sauer, T.: Time series prediction by using delay coordinate embedding. in Time Series Prediction: Forecasting the Future and Understanding the Past. Eds. N.A. Gerschenfeld, A.S. Weigend. SFI Studies in the Science of Complexity, Proc. V. XV, pp. 175–193. Addison-Wesley (1993)

Schroer, C., Sauer, T., Ott, E., Yorke, J.: Predicting chaos most of the time from embeddings with self-intersections. Phys. Rev. Lett. **80**, 1410–1413 (1998)

Sitz, A., Kurths, J., Voss, H.U.: Identification of nonlinear spatiotemporal systems via partitioned filtering. Phys. Rev. E. **68**, 016202 (2003)

Sitz, A., Schwartz, U., Kurths, J., Voss, H.U.: Estimation of parameters and unobserved components for nonlinear systems from noisy time series. Phys. Rev. E. **66**, 016210 (2002)

Small, M., Judd, K., Mees, A.: Modelling continuous processes from data. Phys. Rev. E. **65**, 046704 (2002)

Small, M., Judd, K.: Comparisons of new nonlinear modeling techniques with applications to infant respiration. Physica, D. **117**, 283–298 (1998)

Small, M., Tse, C.K.: Optimal embedding: A modelling paradigm. Phys. D. **194**, 283–296 (2004)

Smirnov, D.A., Bezruchko, B.P., Seleznev Ye.P.: Choice of dynamical variables for global reconstruction of model equations from time series. Phys. Rev. E. **65**, 026205 (2002)

Smirnov, D.A., Bezruchko, B.P.: Nonlinear dynamical models from chaotic time series: methods and applications. In: Winterhalder, M., Schelter, B., Timmer, J. (eds.) Handbook of Time Series Analysis, pp. 181–212. Wiley-VCH, Berlin (2006)

Smith, L.A.: Identification and prediction of low-dimensional dynamics. Phys. D. **58**, 50–76 (1992)

Smith, L.A.: Maintenance of uncertainty. Proc. Int. School of Physics "Enrico Fermi", Course CXXXIII, pp. 177–246. Italian Physical Society, Bologna, (1997). Available at http://www.maths.ox.ac.uk/~lenny

Stark, J., Broomhead, D.S., Davies, M., Huke, J.: Takens embedding theorem for forced and stochastic systems. Nonlinear Analysis. Theory, Methods, and Applications. Proc. 2nd Congress on Nonlinear Analysis. Elsevier Science Ltd., **30**(8), 5303–5314 (1997)

Stoer, J., Bulirsch, R.: Introduction to Numerical Analysis. Springer, New York (1980)

Takens, F.: Detecting strange attractors in turbulence. Lect. Notes Math. **898**, 366–381 (1981)

Timmer, J.: Parameter estimation in nonlinear stochastic differential equations. Chaos, Solitons Fractals **11**, 2571–2578 (2000)

Tong, H.: Nonlinear Time Series Analysis: a Dynamical System Approach. Oxford University Press, Oxford (1990)

Voss, H.U., Bünner, M., Abel, M.: Identification of continuous, spatiotemporal systems. Phys. Rev. E. **57**, 2820–2823 (1998)

Wan, E.A.: Time series prediction by using a connectionist network with internal delay lines. In: Gerschenfeld, N.A., Weigend, A.S. (eds.) Time Series Prediction: Forecasting the Future and Understanding the Past. SFI Studies in the Science of Complexity, Proc. V. XV, pp. 195–217. Addison-Wesley (1993)

Yule, G.U.: On a method of investigating periodicities in disturbed series, with special reference to Wolfer's sunspot numbers. Phil. Trans. R. Soc. Lond. A. **226**, 267–298 (1927)

Chapter 11
Practical Applications of Empirical Modelling

It is difficult even to list all fields of knowledge and practice where *modelling from data series* is applied. One can say that they range from astrophysics to medicine. Purposes of modelling are diverse as well. Therefore, we confine ourselves with several examples demonstrating practical usefulness of empirical modelling.

The most well-known application is, of course, a *forecast* (see Sects. 7.4.1, 10.2.1 and 10.3). Indeed, it seems the most intriguing problem of the data analysis (Casdagli and Eubank, 1992; Gerschenfeld and Weigend, 1993; Kravtsov, 1997; Soofi and Cao, 2002): one often hears about weather and climate forecasts, prediction of earthquakes and financial indices, etc. Still, empirical models obtained as described above accurately predict such complicated processes only rarely. The main obstacles are the curse of dimensionality, deficit of experimental data, considerable noise levels and non-stationarity. However, the chances for accurate predictions rise in simpler and more definite situations.

Another useful possibility is *validation of physical ideas* about an object under study. Modelling from time series allows to improve one's understanding of the "mechanisms" of an object functioning. High quality of a model evidences validity of the underlying substantial ideas (see Sects. 8.3 and 9.3). Such a good model can be further used to reach practical purposes.

Less known are applications of empirical modelling to non-stationary data analysis. *Non-stationarity* is ubiquitous in nature and technology. In empirical modelling, it is important to cope with non-stationary signals properly. Quite often, non-stationarity is just a harmful obstacle in modelling. However, there are many situations where this is not the case. One the one hand, a natural non-stationarity of a real-world signal can be of interest by itself (Sect. 11.1). One the other hand, artificially introduced non-stationarity of a technical signal can be a way of information transmission (Sect. 11.2).

A very promising and widely required application is characterisation of directional (causal) couplings between observed processes. It is discussed in more detail, than the other mentioned problems, in Chap. 12. Several other opportunities such as prediction of bifurcations and signal classification are briefly described in Sect. 11.3.

B.P. Bezruchko, D.A. Smirnov, *Extracting Knowledge From Time Series*, Springer Series in Synergetics, DOI 10.1007/978-3-642-12601-7_11,
© Springer-Verlag Berlin Heidelberg 2010

11.1 Segmentation of Non-stationary Time Series

In the theory of random processes, *non-stationarity* of a process implies temporal
changes in its multidimensional distribution functions. Majority of real-world processes, especially in biology, geophysics or economics, look non-stationary since
their characteristics are not constant over an observation interval. Non-stationarity
leads to significant difficulties in modelling which are comparable with the curse
of dimensionality. However, character of non-stationarity may carry an important
information about properties of an object under study.

The term "dynamical non-stationarity" refers to a situation where an original
process can be described with differential or difference equations whose *parameters vary* in time (Schreiber, 1997, 1999). Proper characteristics of dynamical non-
stationarity are useful, e.g., to detect a time instant of a parameter change more accurately, than it can be done with characteristics of probability distribution functions.
Indeed, if a dynamical regime looses its stability after a parameter change, a phase
orbit may remain in the same area of the phase space for a certain time interval.
Hence, statistical properties of an observed time series do not change rapidly. Yet,
another regime will be established in the future and it may be important to anticipate
upcoming changes as early as possible.

According to a basic procedure for the investigation of a possibly non-stationary
process, one divides an original time series into M segments of length $L \leq N_{st}$,
over which the process is regarded stationary. Then, one performs reconstruction
of model equations or computes some other statistics (sample moments, power
spectrum, etc.) for each segment N_j separately. After that, the segments are compared to each other according to the closeness of their model characteristics or other
statistics. Namely, one introduces the distance d between segments and gets matrix
of distances $d_{i,j} = d(N_i, N_j)$. The values of the distances allow to judge about
stationarity of the process. The following concrete techniques are used.

(i) Comparison of the sample probability distributions according to the χ^2 criterion
(Hively et al., 1999). One divides the range of the observable values into H bins
and counts the number of data points from each segment falling into each bin.
A distance between two segments is computed as

$$d_{i,j}^2 = \sum_{k=1}^{H} \frac{(n_{k,i} - n_{k,j})^2}{n_{k,i} + n_{k,j}},$$

where $n_{k,i}$ and $n_{k,j}$ are the numbers of data points in the kth bin from the ith
and jth segment, respectively. This quantity can reveal non-stationarity in the
sense of a transient process for a system with a constant evolution operator.

(ii) Comparison of empirical models. One constructs global models and gets esti-
mates of their parameter vectors $\hat{\mathbf{c}}_i, \hat{\mathbf{c}}_j$ for the ith and jth segments, respectively. A distance between the segments can be defined, e.g., as a Euclidean
distance between those estimates: $d_{i,j}^2 = \sum_{k=1}^{P}(\hat{c}_{k,i} - \hat{c}_{k,j})^2$ (Gribkov and
Gribkova, 2000).

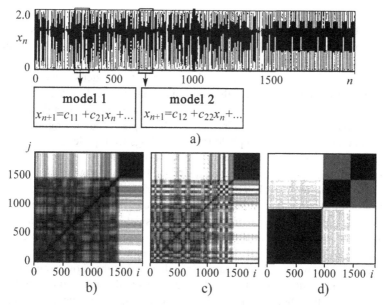

Fig. 11.1 Non-stationary data analysis: (**a**) a time realisation of the system (11.1); (**b**) a recurrence diagram where the distances between time series segments are computed based on the χ^2 criterion for the sample probability densities; (**c**), (**d**) the distances are the Euclidean distances between parameter vectors of the models $x_{n+1} = f(x_n, \mathbf{c})$ with a polynomial f of the order $K = 2$ (**c**) and $K = 6$ (**d**)

The results of such an analysis of an exemplary time series (Fig. 11.1a) are conveniently presented in a kind of "recurrence diagram" (Fig. 11.1b), where starting time instants of the ith and jth segment are shown along the axes. The distances between the segments are indicated in greyscale: white colour corresponds to strongly different segments (large distances) and black one to "almost the same" segments (zero distances). To illustrate possibilities of the approach, let us consider an exemplary one-dimensional map

$$x_{n+1} = c_0(n)\cos x_n \qquad (11.1)$$

with an observable $\eta = x$ and a time series of the length of 2000 data points.

The parameter c changes its value at a time instant $n = 1000$ from $c_0 = 2.1$ corresponding to a chaotic regime to $c_0 = 2.11735$ corresponding to a period-7 cycle. However, the latter regime is clearly established after a sufficiently long transient process (about 400 iterations) so that majority of statistical properties such as the sample mean and the sample variance in a relatively short moving window get clearly different only at a time instant of about $n = 1400$ (Fig. 11.1a). Figure 11.1b–d show the recurrence diagrams for the time series segments of the length of 200 data points. A big dark square on a diagonal corresponds to a quasi-stationary segment, while a boundary between dark areas is a time instant when the process properties reflected by the distance between the segments change. Figure 11.1b shows a diagram based on the χ^2 criterion. Figure 11.1c, d is obtained via the construction of the models $x_{n+1} = f(x_n, \mathbf{c})$ with algebraic polynomials of the orders $K = 2$ (a less adequate model, Fig. 11.1c) and $K = 6$ (a more adequate model, Fig. 11.1d). One can see that the statistic χ^2 detects the changes in the system much later (Fig. 11.1b) than does a good dynamical model (Fig. 11.1d). Figure 11.1c, d also demonstrates that dynamical non-stationarity is detected correctly if empirical models are of high quality. Examples of an analogous segmentation of intracranial electroencephalogram recordings from patients with temporal lobe epilepsy are described in Dikanev et al. (2005).

The selection of quasi-stationary segments can be of importance for obtaining a valid empirical model as well. Generally, in empirical modelling it is desirable to use an entire time series to get more accurate parameter estimates. However, a model with constant parameters is inappropriate in the case of dynamical non-stationarity. One should fit such a model to a *quasi-stationary segment*, which should be as long as possible. To learn a maximal length of a continuous quasi-stationary segment, one should find the widest black square on the diagonal of the above recurrence diagram. Then, one should add similar shorter segments to the first segment and repeat model fitting for the obtained maximally long quasi-stationary piece of data. This approach is realised and applied to the electroencephalogram analysis in Gribkov and Gribkova (2000). A similar problem concerning localisation of time instants of abrupt changes is considered in Anosov et al. (1997).

11.2 Confidential Information Transmission

Development of communication systems which use a chaotic carrier is a topical problem in communication technology (Dmitriev and Panas, 2002). In relation to this field, we note an interesting possibility to use non-stationary time series modelling for multichannel confidential information transmission (Anishchenko and Pavlov, 1998). Let us consider a non-linear dynamical system $d\mathbf{x}/dt = \mathbf{f}(\mathbf{x}, \mathbf{c}_0)$, whose parameter $\mathbf{c}_0$ slowly changes in time as $\mathbf{c}_0 = \mathbf{c}_0(t)$. Let a chaotic time realisation of this system, e.g., $\eta(t) = x_1(t)$, be a transmitted signal and parameter variations $\mathbf{c}_0(t)$ be information signals which are not directly transmitted. The conditions necessary to extract information signals from the chaotic observed signal are as follows:

(i) One completely knows the structure of the dynamical system used as a transmitter, i.e. as a generator of a chaotic time series with changing parameters;
(ii) A characteristic time of variation in its parameters $\mathbf{c}_0$ is much greater than its characteristic oscillation time.

Then, one can restore the time realisations $\mathbf{c}_0(t)$ from a single observable $\eta(t)$ by estimating the respective parameters in the model $d\mathbf{x}/dt = \mathbf{f}(\mathbf{x}, \mathbf{c})$ from subsequent quasi-stationary segments of the observed time series $\eta(t)$ with some of the techniques described in Chaps. 8, 9 and 10.

A numerical toy example of transmitting a greyscale graphical image is considered in Anishchenko and Pavlov (1998), where a system producing a chaotic signal is a modified generator with inertial non-linearity (also called the Anishchenko–Astakhov oscillator) given by the equations

$$
\begin{aligned}
\mathrm{d}x/\mathrm{d}t &= mx + y - xz, \\
\mathrm{d}y/\mathrm{d}t &= -x, \\
\mathrm{d}z/\mathrm{d}t &= -gz + 0.5g\,(x + |x|)\,x.
\end{aligned}
\tag{11.2}
$$

An information signal represents intensity of the greyscale (256 possible values) for the subsequent pixels on the portrait of Einstein (Fig. 11.2a). This signal modulates the values of the parameter g in the interval [0.15, 0.25] at $m = 1.5$. A transmitted signal is $\eta(t) = y(t)$ with a superimposed weak observational noise. A signal in the communication channel looks as "pure" noise (Fig. 11.2b).

If the structure of the generating dynamical system is unknown, one cannot restore a transmitted information (at least, it is very difficult). The result of extracting the information signal in a "receiver" via the estimation of the parameter g in Eq. (11.2) is shown in Fig. 11.2c. A single scalar information signal $g(t)$ is transmitted in this example, though one can change several parameters of the generator simultaneously. In practice, the number of transmitted signals $c_{0,k}(t)$ which can be successfully restored from an observed realisation $\eta(t)$ is limited by the intensity of noise in a communication channel (Anishchenko and Pavlov, 1998). Further developments in this research direction are given in Ponomarenko and Prokhorov (2004, 2002).

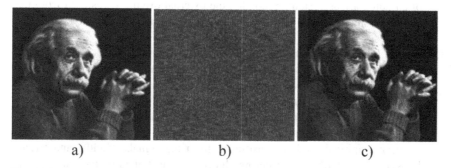

Fig. 11.2 Information transmission with the use of the reconstruction from a time series (Anishchenko and Pavlov, 1998): (**a**) an original image (500 × 464 pixels); (**b**) a signal y in a communication channel; (**c**) a restored image

11.3 Other Applications

Several more specific applications of empirical modelling are listed below and commented very briefly:

(1) *Prediction of bifurcations in weakly non-autonomous systems* (Casdagli, 1989; Feigin et al., 1998, 2001; Mol'kov and Feigin, 2003). When an object parameter changes slowly, one fits an autonomous model with the same structure to subsequent time series segments. Estimates of its parameters vary between the segments. Thus, one obtains a time dependence of the parameter estimates: $\hat{\mathbf{c}}_j$, $j = 1, \ldots, M$, where j is the number of a time series segment. From such a new time series, one can construct a model *predicting future parameter changes*, e.g. a model in the form of an explicit function of time (Sect. 7.4.1). Thereby, one gets predicted parameter values for each future time instant j and checks to what dynamical regime of the autonomous model they correspond. Thus, one can predict a change in the observed dynamical regime, i.e. a bifurcation in the autonomous system, which occurs when the value of $\mathbf{c}$ crosses a boundary of the domain where the current regime is stable. Such bifurcation forecast is possible under the strict conditions that the autonomous model adequately describes an object dynamics in a wide area of the parameter space including a "post-bifurcation" dynamical regime.

(2) *Restoration of external driving* from a chaotic time realisation of a single dynamical variable of a non-autonomous system. This is useful if the driving signal carrying important information cannot be measured directly and one observes only a result of its influence on some non-linear system. Principal possibility of such restoration is illustrated in Gribkov et al. (1995) even in cases where an external driving is not slowly varying. The conditions necessary for the restoration are as follows: (i) the structure of the driven system and the way how the driving enters the equations are known a priori; (ii) a time series from an autonomous system is also available. Then, one first estimates parameters of an autonomous system. Secondly, the external driving is restored from the non-autonomous system realisation taking into account the obtained parameter estimates.

(3) *Signal classification* is another important task where one divides a set of objects into groups (classes) of similar objects based on the experimental data analysis. A general approach involves definition of the distance between two signals, computation of the distances for each pair of signals and division of the signals into groups (clusters) with any of the existing clustering approaches. The distance between signals can be defined, in particular, by estimating a difference between empirical models constructed from the signals (Kadtke and Kremliovsky, 1997). Different clustering algorithms are studied by the cluster analysis (Kendall and Stuart, 1979). We note that the problem of quasi-stationary segment selection (Sect. 11.1) can be formulated in the same terms: different segments of a signal can be considered as different signals, which are united into clusters (quasi-stationary intervals) based on the kind of recurrence diagram (Fig. 11.1c, d).

(4) *Automatic control* of a technical object is realised via regulation of the available parameters. The search for optimal parameter values can be realised in practice as follows: (i) one performs measurements at different values of object parameters; (ii) one constructs an empirical model with a given structure from each set of data; (iii) one reveals the model parameters, whose variations correspond to variations in the governing parameters of an object; (iv) via investigation of the model, one reveals the model parameters which strongly influence the character of the model dynamics; (v) via investigation of the model, one finds such values of its most influential parameters, which correspond to "the best regime of an object functioning"; (vi) finally, one specifies the values of object parameters corresponding to the model parameters found at the previous step. Such an approach is suggested in Gribkov et al. (1994b) and partly realised in relation to a system of resonance frequency and temperature stabilisation in a section of a linear electron accelerator.

(5) *Estimation of an attractor characteristics from a short time series* (Pavlov et al., 1997). One of the important problems in non-linear time series analysis is the computation of such characteristics of an attractor as Lyapunov exponents and fractal dimensions (see, e.g., Eckmann and Ruelle, 1985, 1992; Ershov and Potapov, 1998; Kantz, 1995; Theiler, 1990; Wolf et al., 1985). However, reliable estimates of them can be obtained directly only from a time series, which is very long (so that an orbit could return to the vicinity of any of its points many times) and sufficiently "clean" (noise free). Such data are typically unavailable in practice. A global dynamical model can often be constructed from a much shorter time series if it involves a small number of free parameters. After getting a model, one can compute Lyapunov exponents and fractal dimensions of its attractor from a numerically simulated arbitrarily long time realisation (Sect. 2.1.4) and take them as the sought estimates.

(6) *Testing for non-linearity and determinism* (Gerschenfeld and Weigend, 1993; Small et al., 2001). In investigation of a complex object dynamics, one often does not manage to get a valid model, to reliably estimate dimension, etc. Then, one poses "more modest" questions, which can also be quite substantial. One of them is whether the dynamics of an object is non-linear? One can get an answer with the use of empirical models. A possible technique is as follows.

One constructs local linear models with different numbers of neighbours k [Eq. (10.11) in Sect. 10.2.1]. Root-mean-squared one-step-ahead prediction error ε computed from a test time series or via the cross-validation technique (Sect. 7.2.3) is plotted versus k. The value of k close to the length of a training time series corresponds to a global linear model. At small values of k, one gets different sets of model parameter values for different small domains of the phase space. If the original process is linear, then the prediction error ε monotonously decreases with k, since the parameter estimates become more accurate (due to the use of greater amount of data) without violation of the model validity. If the process is non-linear, then the error ε reaches its minimum at some intermediate value of k, which is sufficiently big to reduce the noise influence and sufficiently small to provide a good accuracy of the local linear approximation. Thereby, one can infer the presence of non-linearity and estimate the scale, where it manifests itself, from the plot $\varepsilon(k)$ (Gerschenfeld and Weigend, 1993).

(7) *Adaptive noise reduction* (Davies, 1994; Farmer and Sidorowich, 1991; Kostelich and Schreiber, 1993). An observed signal $\eta(t)$ often represents a sum of a "useful" signal $X(t)$ and a "harmful" signal $\xi(t)$, which is called "noise":

$$\eta(t) = X(t) + \xi(t). \tag{11.3}$$

Then, the problem is to extract the signal $X(t)$ from $\eta(t)$, i.e. to get a signal $\hat{X}(t)$, which differs from $X(t)$ less strongly than the observed signal $\eta(t)$ (Farmer and Sidorowich, 1991). The expression (11.3) describes a measurement noise, which is especially harmful in numerical differentiation where it is usually reduced with a Savitzky–Golay filter (Sect. 7.4.2). The latter is a kind of linear filtering (Hamming, 1983; Rabiner and Gold, 1975). However, all linear filters are based on the assumption that an "interesting" dynamics $X(t)$ and the noise $\xi(t)$ have different characteristic timescales, i.e. their powers are concentrated in different frequency bands (Sect. 6.4.2). As a rule, the noise is assumed to be a very high-frequency signal (this is implied when the Savitzky–Golay filter is used in differentiation) or a very low-frequency signal (slow drifts of the mean and so forth).

However, it may appear that the noise has the same timescale as the useful signal. Then, linear filtering cannot help, but non-linear noise reduction based on the construction of non-linear empirical models may appear efficient. The principal idea is to fit a non-linear model to an observed time series taking into account measurement noise (see, e.g., Sects. 8.1.2 and 8.2). Residual errors $\varepsilon(t)$ of the model fitting are considered as the estimates of the noise realisation $\xi(t)$. Then, one estimates the useful signal via subtraction of $\varepsilon(t)$ from the observed signal $\eta(t)$:

$$\hat{X}(t) = \eta(t) - \varepsilon(t). \tag{11.4}$$

There are many implementations of the approach. One of the first works (Farmer and Sidorowich, 1991) exploited local linear models. In any case, the model size selection and the choice of an approximating function form are important, since an inadequate model would give biased estimates $\hat{X}(t)$ which may strongly differ from the true signal $X(t)$, i.e. a further distortion of the signal would occur rather than a noise reduction.

(8) Finally, let us mention again such a promising possibility as *restoration of equivalent characteristics* of non-linear elements in electric circuits and other systems (Sect. 9.3). Such characteristics can be reconstructed via the empirical modelling even in the regimes of large oscillation amplitudes and chaos, where ordinary tools may be inapplicable. This approach has been successfully used to investigate dynamical characteristics of a ferroelectric capacitor (Hegger et al., 1998), semiconductor elements (Sysoev et al., 2004; Timmer et al., 2000) and fibre optical systems (Voss et al., 1999).

References

Anishchenko, V.S., Pavlov, A.N.: Global reconstruction in application to multichannel communication. Phys. Rev. E. **57**, 2455–2457 (1998)

Anosov, O.L., Butkovsky, O.Ya., Kravtsov, Yu.A.: Minimax Procedure for Identifying Chaotic Systems from the Observed Time Sequence. J. Commun. Technol. Electron. **42**(3), 288–293, (1997)

Casdagli, M., Eubank, S. (eds.): Nonlinear modeling and forecasting. SFI Studies in the Sciences of Complexity, vol XII. Addison-Wesley, New York, (1992)

Casdagli, M.: Nonlinear prediction of chaotic time series. Phys. D. **35**, 335–356 (1989)

Davies, M.E.: Noise reduction schemes for chaotic time series. Phys. D. **79**, 174–192 (1994)

Dikanev, T., Smirnov, D., Wennberg, R., Perez Velazquez, J.L., Bezruchko, B. EEG nonstationarity during intracranially recorded seizures: statistical and dynamical analysis. Clin. Neurophysiol. **116**, 1796–1807 (2005)

Dmitriev, A.S., Panas, A.I.: Dynamical chaos: Novel type of information carrier for communication systems. Fizmatlit, Moscow, (in Russian) (2002)

Eckmann, J.P., Ruelle, D.: Ergodic theory of chaos and strange attractors. Rev. Mod. Phys. **57**, 617–656 (1985)

Eckmann, J.P., Ruelle, D.: Fundamental limitations for estimating dimensions and Lyapunov exponents in dynamical systems. Phys. D. **56**, 185–187 (1992)

Ershov, S.V., Potapov, A.B.: On the concept of stationary Lyapunov basis. Phys. D. **118**, 167–198 (1998)

Farmer, J.D., Sidorowich, J.J.: Optimal shadowing and noise reduction. Phys. D. **47**, 373–392 (1991)

Feigin, A.M., Konovalov, I.B., Molkov, Y.I.: Toward an understanding of the nonlinear nature of atmospheric photochemistry: essential dynamic model of the mesospheric photochemical system. J. Geophys. Res. **103**(D19), 25447–25460 (1998)

Feigin, A.M., Mol'kov Ya.I., Mukhin, D.N., Loskutov, E.M.: Forecast pof qualitative behaviour of dynamical system from chaotic time series. Radiophys. Quantum Electron. **44**(5–6), 376–399, (in Russian) (2001)

Gerschenfeld, N.A., Weigend, A.S. (eds.): Time series prediction: forecasting the future and understanding the past. SFI Studies in the Science of Complexity, Proc. V. XV. Addison-Wesley, New York (1993)

Gribkov, D., Gribkova, V.: Learning dynamics from nonstationary time series: analysis of electroencephalograms. Phys. Rev. E. **61**, 6538–6545 (2000)

Gribkov, D.A., Gribkova, V.V., Kravtsov Yu.A., et al.: Construction of a model for systems of stabilisation of resonance frequency and temperature in a section of linear electron accelerator from experimental data. *Mosc*ow Univ. Bull. Ser. 3. **35**(1), 96–98, (in Russian) (1994)

Gribkov, D.A., Gribkova, V.V., Kuznetsov Yu.I.: Reconstruction of external driving from realisation of a single variable of self-stochastic system. Moscow Univ. Bull. Ser. 3. **36**(1): 76–78, (in Russian) (1995)

Hamming, R.W.: Digital Filters. 2nd edn. Prentice-Hall, Englewood Cliffs, NJ (1983)

Hegger, R., Kantz, H., Schmuser, F., et al. Dynamical properties of a ferroelectric capacitors observed through nonlinear time series analysis. Chaos. **8**, 727–754 (1998)

Hively, L.M., Gaily, P.C., Protopopescu, V.A.: Detecting dynamical change in nonlinear time series. Phys. Lett. A. **258**, 103–114 (1999)

Kadtke, J., Kremliovsky, M.: Estimating statistics for detecting determinism using global dynamical models. Phys. Lett. A. **229**, 97–106 (1997)

Kantz, H. A robust method to estimate the maximal Lyapunov exponent of a time series. Phys. Lett. A. **185**, 77 (1995)

Kendall, M.G., Stuart, A.: The Advanced Theory of Statistics, vol. 2 and 3. Charles Griffin, London (1979)

Kostelich, E.J., Schreiber, T.: Noise reduction in chaotic time series data: a survey of common methods. Phys. Rev. E. **48**, 1752–1763 (1993)

Kravtsov, Yu.A. (ed.): Limits of Predictability. TsentrCom, Moscow, (in Russian) (1997a)

Mol'kov Ya.I., Feigin, A.M.: Forecast of qualitative behaviour of dynamical system from chaotic time series. In: Gaponov-Grekhov, A.V., Nekorkin, V.I. (eds.) Nonlinear Waves – 2002, pp. 34–53. Institute of Applied Physics RAS, Nizhny Novgorod, (in Russian) (2003)

Pavlov, A.N., Janson, N.B., Anishchenko, V.S.: Application of statistical methods to solve the problem of global reconstruction. Tech. Phys. Lett. **23**(4), 297–299 (1997)

Ponomarenko, V.I., Prokhorov, M.D.: Coding and recovery of information masked by the chaotic signal of a time-delay system. J. Commun. Technol. Electron. **49**(9), 1031–1037 (2004)

Ponomarenko, V.I., Prokhorov, M.D.: Extracting information masked by the chaotic signal of a time-delay system. Phys. Rev. E. **66**, 026215 (2002)

Rabiner, L.R., Gold, B.: Theory and Applications of Digital Signal Processing. Prentice Hall, New York (1975)

Schreiber, T.: Detecting and Analyzing Nonstationarity in a time series using nonlinear cross predictions. Phys. Rev. Lett. **78**, 843–846 (1997)

Schreiber, T.: Interdisciplinary application of nonlinear time series methods. Phys. Rep. **308**, 3082–3145 (1999)

Small, M., Judd, K., Mees, A.I.: Testing time series for nonlinearity. Statistics Comput. **11**, 257–268 (2001)

Soofi, A.S., Cao, L. (eds.): Modeling and Forecasting Financial Data: Techniques of Nonlinear Dynamics. Kluwer, Dordrecht (2002)

Sysoev, I.V., Smirnov, D.A., Seleznev Ye.P., Bezruchko, B.P.: Reconstruction of nonlinear characteristics and equivalent parameters from experimental time series. Proc. 2nd IEEE Int. Conf. Circuits and Systems for Communications. Paper No. 140. Moscow (2004)

Theiler, J.: Estimating fractal dimension. J. Opt. Soc. Am. **7**, 1055 (1990)

Timmer, J., Rust, H., Horbelt, W., Voss, H.U.: Parametric, nonparametric and parametric modelling of a chaotic circuit time series. Phys. Lett. A. **274**, 123–130 (2000)

Voss, H.U., Schwache, A., Kurths, J., Mitschke, F.: Equations of motion from chaotic data: A driven optical fiber ring resonator. Phys. Lett. A. **256**, 47–54 (1999)

Wolf, A., Swift, J.B., Swinney, H.L., Vastano, J.A.: Determining Lyapunov exponents from a time series. Phys. D **16**, 285–317 (1985)

Chapter 12
Identification of Directional Couplings

An important piece of information, which can be extracted from parameters of empirical models, is quantitative characteristics of couplings between processes under study. The problem of *coupling detection* is encountered in multiple fields including physics (Bezruchko et al., 2003), geophysics (Maraun and Kurths, 2005; Mokhov and Smirnov, 2006, 2008; Mosedale et al., 2006; Palus and Novotna, 2006; Verdes, 2005; Wang et al., 2004), cardiology (Rosenblum et al., 2002; Palus and Stefanovska, 2003) and neurophysiology (Arnhold et al., 1999; Brea et al., 2006; Faes et al., 2008; Friston et al., 2003; Kreuz et al., 2007; Kiemel et al., 2003; Le Van Quyen et al., 1999; Mormann et al., 2000; Osterhage et al., 2007; Pereda et al., 2005; Prusseit and Lehnertz, 2008; Smirnov et al., 2005; Romano et al., 2007; Schelter et al., 2006; Schiff et al., 1996; Sitnikova et al., 2008; Smirnov et al., 2008, Staniek and Lehnertz, 2008; Tass, 1999; Tass et al., 2003). Numerous investigations are devoted to *synchronisation*, which is an effect of interaction between non-linear oscillatory systems (see, e.g., Balanov et al., 2008; Boccaletti et al., 2002; Hramov and Koronovskii, 2004; Kreuz et al., 2007; Maraun and Kurths, 2005; Mormann et al., 2000; Mosekilde et al., 2002; Osipov et al., 2007; Palus and Novotna, 2006; Pikovsky et al., 2001; Prokhorov et al., 2003; Tass et al., 2003). In the last decade, more careful attention is paid to *directional coupling* analysis. Such characteristics might help, e.g., to localise an epileptic focus (a pathologic area) in the brain from electroencephalogram (EEG) or magnetoencephalogram (MEG) recordings: hypothetically, an increasing influence of an epileptic focus on adjacent areas leads to the seizure onset for some kinds of epilepsy.

The most appropriate and direct approaches to the detection of *causal influences* are based on the construction of empirical models. These approaches include *Granger causality* (Sect. 12.1) and *phase dynamics modelling* (Sect. 12.2). Below, we present our results showing their fruitful applications to the problems of neurophysiology (Sects. 12.3 and 12.4) and climatology (Sects. 12.5 and 12.6).

12.1 Granger Causality

The problem is formally posed as follows. There are time series from M processes $\{x_k(t)\}_{t=1}^N$, $k = 1, \ldots, M$. One needs to detect and characterise couplings between

B.P. Bezruchko, D.A. Smirnov, *Extracting Knowledge From Time Series*, Springer Series in Synergetics, DOI 10.1007/978-3-642-12601-7_12,
© Springer-Verlag Berlin Heidelberg 2010

them, i.e. to find out how the processes influence each other. In the case of two linear processes (Granger, 1969; Pereda et al., 2005), one first constructs univariate autoregression models (Sect. 4.4)

$$x_k(t) = A_{k,0} + \sum_{i=1}^{d} A_{k,i} x_k(t-i) + \xi_k(t),$$ (12.1)

where $k = 1, 2$, d is a model order, ξ_k are Gaussian white noises with variances $\sigma_{\xi_k}^2$. Let us denote the vector of coefficients $\{A_{k,i}, i = 0, \ldots, d\}$ as $\mathbf{A}_k$, the sum of squared residual errors as

$$\Sigma_k^2 = \sum_{t=d+1}^{N} \left(x_k(t) - A_{k,0} - \sum_{i=1}^{d} A_{k,i} x_k(t-i) \right)^2,$$

and its minimal value as $s_k^2 = \min_{\mathbf{A}_k} \Sigma_k^2$. The model coefficients are estimated via the *ordinary least-squares technique* (Sect. 7.1.1), i.e. one gets $\hat{\mathbf{A}}_k = \arg\min_{\mathbf{A}_k} \Sigma_k^2$. An unbiased estimator for $\sigma_{\xi_k}^2$ would represent the mean-squared prediction error of the univariate model. Such an estimator is given by

$$\hat{\sigma}_k^2 = \frac{s_k^2}{N - d - (d+1)},$$

where $d + 1$ is the number of estimated coefficients in Eq. (12.1). The model order d is selected large enough to provide delta correlatedness of the residual errors. For automatic choice of d, one often uses criteria of Akaike (1974) or Schwarz (1978).

Then, one similarly constructs a bivariate AR model:

$$
\begin{aligned}
x_1(t) &= a_{1,0} + \sum_{i=1}^{d} a_{1,i} x_1(t-i) + \sum_{i=1}^{d} b_{1,i} x_2(t-i) + \eta_1(t), \\
x_2(t) &= a_{2,0} + \sum_{i=1}^{d} a_{2,i} x_2(t-i) + \sum_{i=1}^{d} b_{2,i} x_1(t-i) + \eta_2(t).
\end{aligned}
$$ (12.2)

where η_k are Gaussian white noises. Minimal values of the sums of squared residual errors are denoted $s_{1|2}^2$ and $s_{2|1}^2$ for the first and second processes, respectively. Unbiased estimators for the residual error variances are denoted $\hat{\sigma}_{1|2}^2$ and $\hat{\sigma}_{2|1}^2$. *Prediction improvement* for x_k, i.e. the quantity $\mathrm{PI}_{j \to k} = \hat{\sigma}_k^2 - \hat{\sigma}_{k|j}^2$, characterises the influence of the process x_j on x_k (denoted further as $j \to k$).

Note that $\mathrm{PI}_{j \to k}$ is an *estimate* obtained from a time series. To define a theoretical (true) prediction improvement $\mathrm{PI}_{j \to k}^{\mathrm{true}} = \sigma_k^2 - \sigma_{k|j}^2$, one should minimise the *expectations* of the squared prediction errors instead of the *empirical* sums Σ_k^2 and $\Sigma_{k|j}^2$ to get model coefficients, i.e. one should use an ensemble averaging or an averaging over an infinitely long-time realisation instead of the averaging over a finite time series. For uncoupled processes, one has $PI_{j \to k}^{true} = 0$, but the estimator $PI_{j \to k}$ can take positive values due to random fluctuations. Therefore, one needs a criterion to decide whether an obtained positive value of $PI_{j \to k}$ implies the presence of the influence $j \to k$. It can be shown that the quantity

$$F_{j \to k} = \frac{(N - 3d - 1)\left(s_k^2 - s_{k|j}^2\right)}{s_{k|j}^2 d}$$

is distributed according to *Fisher's F-law* with $(d, N - 3d - 1)$ degrees of freedom. Hence, one can conclude that $PI_{j \to k}^{true} > 0$ and the influence $j \to k$ exists at the *significance level p* (i.e. with the probability of random error not greater than p) if the value of $F_{j \to k}$ exceeds $(1 - p)$ quantile of the respective F-distribution. This is called *F-test* or *Granger – Sargent test* (see, e.g., Hlavackova-Schindler et al., 2007).

If a time series is short, it is problematic to use high values of d, since the number of the estimated coefficients is then large, which often leads to insignificant conclusions even in cases of really existing couplings. The difficulty can be overcome in part if one constructs a bivariate model in the form

$$x_k(t) = a_{k,0} + \sum_{i=1}^{d_k} a_{k,i} x_k(t - i) + \sum_{i=1}^{d_{j \to k}} b_{k,i} x_j(t - i - \Delta_{j \to k}), \qquad (12.3)$$

where $j, k = 1, 2$, $j \neq k$, and selects a separate univariate model order d_k for each process instead of the common d in Eq. (12.2), a separate value of $d_{j \to k}$ and a separate trial *delay time* $\Delta_{j \to k}$. If at least some of the values d_k and $d_{j \to k}$ can be made small, then the number of the estimated coefficients is reduced.

If one needs non-linear models, the difficulty gets even harder due to the curse of dimensionality. In a non-linear case, the procedure of coupling estimation remains the same, but the AR models must involve non-linear functions. Thus, univariate AR models take the form

$$x_k(t) = f_k(x_k(t - 1), x_k(t - 2), \ldots, x_k(t - d_k), \mathbf{A}_k) + \xi_k(t), \qquad (12.4)$$

where it is important to choose properly the kind of the non-linear functions f_k. Algebraic polynomials (Mokhov and Smirnov, 2006), radial basis functions (Ancona et al., 2004) and locally constant predictors (Feldmann and Bhattacharya, 2004) have been used. For relatively short time series, it is reasonable to use polynomials f_k of low orders P_k. Bivariate models are then constructed in the form (12.2), where the linear functions are replaced with polynomials of the order P_k. Yet, there is no regular procedure assuring an appropriate choice of the non-linear functions.

If the number of processes $M > 2$, then estimation of the influence $j \to k$ can be performed in two ways:

(i) Bivariate analysis of x_j and x_k results in an estimator, which reflects both a "direct" influence $j \to k$ and that mediated by other observed processes.

(ii) Multivariate analysis takes into account all the M processes and allows to distinguish between the influences $j \to k$ from different processes x_j. Namely, one computes a squared prediction error for x_k when a multivariate AR model containing all the processes except for x_j is used. Then, one computes such an error for a multivariate AR model containing all the M processes including x_j. If the predictions are more accurate in the latter case, one infers the presence of the direct influence $j \to k$.

To express prediction improvements in relative units, one normalises $\mathrm{PI}_{j \to k}$ by the variance $\mathrm{var}[x_k]$ of the process x_k or by the variance $\hat{\sigma}_k^2$ of the prediction error of the univariate model (12.1). The quantity $\mathrm{PI}_{j \to k}/\hat{\sigma}_k^2$ is used more often than $\mathrm{PI}_{j \to k}/\mathrm{var}[x_k]$. Both quantities are not greater than one and one may hope to give them a vivid interpretation. Thus, $\mathrm{PI}_{j \to k}/\hat{\sigma}_k^2$ is close to unity if the influence $j \to k$ describes almost all "external factors" ξ_k unexplained by the univariate model of x_k. $\mathrm{PI}_{j \to k}/\mathrm{var}[x_k]$ is close to unity if in addition the univariate model (12.1) explains a negligible part of the variance $\mathrm{var}[x_k]$, i.e. $\hat{\sigma}_k^2 \approx \mathrm{var}[x_k]$. These interpretations are often appropriate, even though they may appear insufficient to characterise an importance of the influence $j \to k$ from the viewpoint of long-term changes in the dynamics.

12.2 Phase Dynamics Modelling

A general idea of the approach is that such a characteristic as "intensity of coupling" between two oscillatory processes shows how strongly a future evolution of an oscillator phase depends on the current value of the other oscillator phase (Rosenblum and Pikovsky, 2001). In fact, it is similar to the Granger causality, since bivariate models for the phases are constructed to characterise couplings. It makes sense to model such variables as phases, since they are often especially sensitive to weak perturbations as known from the synchronisation theory (see, e.g., Pikovsky et al., 2001).

Phase dynamics of weakly coupled (deterministic) limit-cycle oscillators with close natural frequencies can be to a good approximation described with a set of ordinary differential equations (Kuramoto, 1984):

$$d\phi_1/dt = \omega_1 + H_1(\phi_2 - \phi_1),$$
$$d\phi_2/dt = \omega_2 + H_2(\phi_1 - \phi_2),$$
(12.5)

where ϕ_k are phases of the oscillators, ω_k are their natural frequencies and H_k are coupling functions. Model (12.5) does not apply if the phase dynamics of the oscillators is perturbed by noise (a typical situation in practice) or coupling functions depend on phases in a more complicated manner rather than only via phase difference due to strong non-linearities in the systems and their interactions. Yet, if noise level is low, the model can be generalised in a straightforward manner. One comes to stochastic differential equations (Kuramoto, 1984; Rosenblum et al., 2001)

$$
\begin{aligned}
d\phi_1/dt &= \omega_1 + G_1(\phi_1, \phi_2) + \xi_1(t), \\
d\phi_2/dt &= \omega_2 + G_2(\phi_2, \phi_1) + \xi_2(t),
\end{aligned}
\tag{12.6}
$$

where ω_k are not necessarily close to each other, ξ_k are independent zero-mean white noises with autocorrelation functions $\langle \xi_k(t)\xi_k(t') \rangle = \sigma_{\xi_k}^2 \delta(t - t')$, δ is the Dirac's delta function and $\sigma_{\xi_k}^2$ characterise noise intensities. The functions G_k are 2π periodic with respect to both arguments and describe both couplings between the oscillators and their individual phase non-linearity.

Let $\sigma_{\xi_k}^2$ and $|G_k|$ be reasonably small so that the contribution of the respective terms in Eq. (12.6) to the phase increment $\phi_k(t + \tau) - \phi_k(t)$ is small in comparison with the "linear increment" $\omega_k\tau$, where the finite time interval τ is of the order of the basic oscillation period. Then, by integrating Eq. (12.6) over the interval τ, one converts to difference equations and gets

$$
\phi_k(t + \tau) - \phi_k(t) = F_k(\phi_k(t), \phi_j(t), \mathbf{a}_k) + \varepsilon_k(t),
\tag{12.7}
$$

where $k, j = 1, 2$, $j \neq k$, ε_k are zero-mean noises, F_k are trigonometric polynomials

$$
F_k(\phi_k, \phi_j, \mathbf{a}_k) = w_k + \sum_{(m,n)\in\Omega_k} \left(\alpha_{k,m,n} \cos(m\phi_k - n\phi_j) + \beta_{k,m,n} \sin(m\phi_k - n\phi_j)\right),
\tag{12.8}
$$

$\mathbf{a}_k = (w_k, \{\alpha_{k,m,n}, \beta_{k,m,n}\}_{(m,n)\in\Omega_k})$ are vectors of their coefficients and Ω_k are summation ranges, i.e. sets of pairs (m, n) defining which monomials are contained in F_k. The terms with $m = n = 1$ can be induced by a linear coupling of the form kx_j or $k(x_j - x_k)$ in some "original equations" for the oscillators. The terms with $n = 2$ can be due to a driving force, which is quadratic with respect to the coordinate of the driving oscillator, e.g. kx_j^2. Various combinations are also possible so that the couplings in the phase dynamics equations (12.7) can be described with a set of monomials of different orders with $n \neq 0$. The strongest influence arises from the so-called "resonant terms", which correspond to the ratios $m/n \approx \omega_j/\omega_k$ in the equation for the kth oscillator phase. However, non-resonant terms can also be significant.

Intensity of the influence $j \to k$ can be reasonably defined via the mean-squared value of the partial derivative $\partial F_k(\phi_k, \phi_j, \mathbf{a}_k)/\partial\phi_j$ (Rosenblum and Pikovsky, 2001; Smirnov and Bezruchko, 2003):

$$
c_{j\to k}^2 = \frac{1}{2\pi^2} \int_0^{2\pi} \int_0^{2\pi} \left(\partial F_k(\phi_k, \phi_j, \mathbf{a}_k)/\partial\phi_j\right)^2 d\phi_j \, d\phi_k.
\tag{12.9}
$$

Indeed, the value of $c^2_{j \to k}$ depends only on the terms with $n \neq 0$ and reads as (Smirnov and Bezruchko, 2003)

$$c^2_{j \to k} = \sum_{(m,n) \in \Omega_k} n^2 \left(\alpha^2_{k,m,n} + \beta^2_{k,m,n} \right). \tag{12.10}$$

This is a theoretical coupling characteristic which can be computed if the polynomial coefficients in Eq. (12.8) are known.

In practice, one has only a time series of observed quantities x_1 and x_2 representing two oscillatory processes. So, one first extracts a time series of the phases $\phi_1(t)$ and $\phi_2(t)$ from the observed data with any of the existing techniques (Sect. 6.4.3). Then, one estimates the coupling characteristics by fitting the phase dynamics equations (12.7) with the functions (12.8) to the time series of the phases. For that, one can use the *ordinary least-squares technique* (Sect. 7.1.1) to get the estimates $\hat{\mathbf{a}}_k$ of the coefficients (Rosenblum and Pikovsky, 2001), i.e. one minimises the values of

$$\hat{\sigma}^2_k(\mathbf{a}_k) = \frac{1}{N - \tau/\Delta t} \sum_{n=\tau/\Delta t+1}^{N} \left(\phi_k(n\Delta t + \tau) - \phi_k(n\Delta t) - F_k(\phi_k(n\Delta t), \phi_j(n\Delta t), \mathbf{a}_k) \right)^2,$$

where $k, j = 1, 2, j \neq k$. The estimates can be written as $\hat{\mathbf{a}}_k = \arg\min_{\mathbf{a}_k} \hat{\sigma}^2_k(\mathbf{a}_k)$. The minimal value $\hat{\sigma}^2_k = \min_{\mathbf{a}_k} \hat{\sigma}^2_k(\mathbf{a}_k)$ characterises the noise level. The most direct way to estimate the coupling strengths $c_{j \to k}$ is to use the expression (12.10) and replace the true values $\mathbf{a}_k$ with the estimates $\hat{\mathbf{a}}_k$. Thereby, one gets the estimator

$$\hat{c}^2_{j \to k} = \sum_{(m,n) \in \Omega_k} n^2 \left(\hat{\alpha}^2_{k,m,n} + \hat{\beta}^2_{k,m,n} \right).$$

Sensitivity of the technique to weak couplings was demonstrated numerically (Rosenblum and Pikovsky, 2001). The estimator $\hat{c}_{j \to k}$ appears "good" for long and stationary signals, whose length should be about several hundreds of basic periods under moderate noise level. The technique has already given interesting results for a complex real-world process, where such data are available, namely for the interaction between human respiratory and cardio-vascular systems (Rosenblum et al., 2002). It appears that the character of the interaction in infants changes with their age from an almost symmetric coupling to a predominant influence of the respiratory system on the cardio-vascular one.

Application of the technique in practice encounters essential difficulties when time series are non-stationary. For instance, it is important to characterise an interaction between different brain areas from EEG recordings. However, their quasi-stationary intervals last for about a dozen of seconds, i.e. comprise not more than 100 basic periods for pathological (epileptic or Parkinsonian) oscillatory behaviour. Then, one could divide a long time series into quasi-stationary segments and compute coupling characteristics from each segment separately. However, for a time series of such a moderate length, the estimators $\hat{c}_{j \to k}$ turn out to be typically biased.

The reasons are described in Smirnov and Bezruchko (2003), where corrected estimators $\gamma_{j \to k}$ for the quantities $c_{j \to k}^2$ are suggested:

$$\gamma_{j \to k} = \sum_{(m,n) \in \Omega_k} n^2 \left(\hat{\alpha}_{k,m,n}^2 + \hat{\beta}_{k,m,n}^2 - 2\hat{\sigma}_{\hat{\alpha}_{k,m,n}}^2 \right),$$

the estimates of the variances $\hat{\sigma}_{\hat{\alpha}_{k,m,n}}^2$ of the coefficient estimates $\hat{\alpha}_{k,m,n}$ are derived in the form

$$\hat{\sigma}_{\hat{\alpha}_{k,m,n}}^2 = \frac{2\hat{\sigma}_k^2}{N - \tau/\Delta t} \left\{ 1 + 2 \sum_{l=1}^{\tau/\Delta t} \left(1 - \frac{l}{\tau/\Delta t} \right) \cos \left[\frac{l \left(m\hat{w}_k + n\hat{w}_j \right)}{\tau/\Delta t} \right] \exp \left[-\frac{l \left(m^2 \hat{\sigma}_k^2 + n^2 \hat{\sigma}_j^2 \right)}{2\tau/\Delta t} \right] \right\}.$$

95% confidence bands for the coupling strengths $c_{j \to k}^2$ are derived in Smirnov and Bezruchko (2003) for the case of trigonometric polynomials F_k of the third order (namely, for the set Ω_k, which includes the pairs of indices $m = n = 1$, $m = 1, n = -1$, $m = 1, n = 0$, $m = 2, n = 0$ and $m = 3, n = 0$) in the form $[\gamma_{j \to k} - 1.6\hat{\sigma}_{\gamma_{j \to k}}, \gamma_{j \to k} + 1.8\hat{\sigma}_{\gamma_{j \to k}}]$, where the estimates of the standard deviations $\hat{\sigma}_{\gamma_{j \to k}}$ are computed from the same short time series as

$$\hat{\sigma}_{\gamma_{j \to k}}^2 = \begin{cases} 2 \sum_{m,n} n^4 \hat{\sigma}_{\hat{\alpha}_{k,m,n}}^2 & , \gamma_{j \to k} \geqslant 5 \sqrt{\sum_{m,n} 2n^4 \hat{\sigma}_{\hat{a}_{1,m,n}}^2}, \\ \sum_{m,n} n^4 \hat{\sigma}_{\hat{a}_{k,m,n}}^2 & , \text{otherwise}, \end{cases}$$

and the estimate of the variance of the squared coefficient estimate is given as

$$\hat{\sigma}_{\hat{\alpha}_{k,m,n}^2}^2 = \begin{cases} 2\hat{\sigma}_{\hat{\alpha}_{k,m,n}}^4 + 4 \left(\hat{\alpha}_{k,m,n}^2 - \hat{\sigma}_{\hat{\alpha}_{k,m,n}}^2 \right) \hat{\sigma}_{\hat{\alpha}_{k,m,n}}^2, & \hat{\alpha}_{k,m,n}^2 - \hat{\sigma}_{\hat{\alpha}_{k,m,n}}^2 \geqslant 0, \\ 2\hat{\sigma}_{\hat{\alpha}_{k,m,n}}^4, & \text{otherwise}. \end{cases}$$

The value of $\gamma_{j \to k,c} = 1.6\hat{\sigma}_{\gamma_{j \to k}}$ represents a 0.975 quantile for the distribution of the estimator $\gamma_{j \to k}$ in the case of uncoupled processes. Hence, the presence of the influence can be inferred at the significance level 0.025 (i.e. with a probability of random error not more than 0.025) if it appears that $\gamma_{j \to k} > \gamma_{j \to k,c}$. The technique has been compared to other non-linear coupling analysis techniques in Smirnov and Andrzejak (2005) and Smirnov et al. (2007), where its superiority is shown for sufficiently regular oscillatory processes.

If directional couplings between processes are expected to be time-delayed, the technique can be readily generalised (Cimponeriu et al., 2004). Namely, one constructs the phase dynamics model in the form

$$\phi_k(t + \tau) - \phi_k(t) = F_k(\phi_k(t), \phi_j(t - \Delta_{j \to k}), \mathbf{a}_k) + \varepsilon_k(t), \tag{12.11}$$

where $k, j = 1, 2, \ j \neq k$, and $\Delta_{j \to k}$ is a trial *delay time* in the influence $j \to k$. One gets coupling estimates and their standard deviations depending on the trial delay: $\gamma_{j \to k}(\Delta_{j \to k})$ and $\hat{\sigma}_{\gamma_{j \to k}}(\Delta_{j \to k})$. Then, one selects the trial delay corresponding to the largest value of $\gamma_{j \to k}$, which significantly exceeds zero (if such a value of $\gamma_{j \to k}$ exists), i.e. exceeds $\gamma_{j \to k, c}(\Delta_{j \to k})$. Thereby, one also gets an estimate of the delay time.

The phase dynamics modelling technique is applicable if couplings are not very strong so that the degree of synchrony between the oscillators is low. This condition can be checked, e.g., via the estimation of the *phase synchronisation index* (Sect. 6.4.5) also called *mean phase coherence* (Mormann et al. (2000): $\rho(\Delta) = |\langle \exp(i(\varphi_1(t) - \varphi_2(t + \Delta))) \rangle_t|$. This quantity ranges from zero to one. The estimators $\gamma_{2 \to 1}(\Delta_{2 \to 1})$ and $\gamma_{1 \to 2}(\Delta_{1 \to 2})$ with their confidence bands can be considered reliable if the values of $\rho(-\Delta_{2 \to 1})$ and $\rho(\Delta_{1 \to 2})$ are less than 0.45 (Mokhov and Smirnov, 2006). The second condition of applicability is a sufficient length of the time series: not less than 40 basic periods (Mokhov and Smirnov, 2006; Smirnov and Bezruchko, 2003). Finally, the autocorrelation function of the residual errors for the models (12.7) or (12.11) should decrease down to zero over the interval of time lags $(0, \tau)$ to confirm appropriateness of the basic model (12.6) with white noises.

The corrected estimators $\gamma_{j \to k}$ are used for the analysis of two-channel EEG in a patient with temporal lobe epilepsy in Smirnov et al. (2005). Their further real-world applications are described in Sects. 12.3, 12.5 and 13.2.

12.3 Brain – Limb Couplings in Parkinsonian Resting Tremor

Many neurological diseases including epilepsy and Parkinson's disease are related to pathological synchronisation of large groups of neurons in the brain. Synchronisation of neurons in nuclei of thalamus and basal ganglia is a hallmark of Parkinson's disease (Nini et al., 1995). However, as yet its functional role in the generation of *Parkinsonian tremor* (involuntary regular oscillations of limbs at a frequency ranging from 3 to 6 Hz) is a matter of debate (Rivlin-Etzion et al., 2006). In particular, the hypothesis that the neural synchronisation drives the tremor has not yet got a convincing empirical confirmation (Rivlin-Etzion et al., 2006). The standard therapy for medically refractory Parkinson's disease is permanent electrical deep brain stimulation (DBS) at high frequencies (greater than 100 Hz) (Benabid et al., 1991). Standard DBS has been developed empirically, its mechanism of action is unclear (Benabid et al., 2005) and it has relevant limitations, e.g. side effects (Tass et al., 2003; Tass and Majtanik, 2006). It has been suggested to specifically counteract the pathological cerebral synchrony by desynchronising DBS (Tass, 1999), e.g. with coordinated reset stimulation (Tass, 2003). The verification of the tremor being generated by synchronised neural activity in the thalamus and the basal ganglia will further justify and strengthen the desynchronisation approach (Tass, 1999; 2003) and help to develop therapies, which may presumably be milder and lead to less side effects. Therefore, to detect couplings between limb oscillations and activity of different brain areas in Parkinsonian patients is a topical problem.

We have analysed more than 40 epochs of spontaneous Parkinsonian tremor recorded in three patients with Parkinson's disease (Bezruchko et al., 2008; Smirnov et al., 2008). Limb oscillations are represented by accelerometer signals recorded at sampling frequencies 200 Hz or 1 kHz. Information about the brain activity is represented by the recordings of local field potentials (LFPs) from a depth electrode implanted into the thalamus or the basal ganglia. The data are obtained at the Department of Stereotaxic and Functional Neurosurgery, University of Cologne, and at the Institute of Neuroscience and Biophysics – 3, Research Centre Juelich, Germany.

Accelerometer and LFP signals during an interval of strong Parkinsonian tremor are presented in Fig. 12.1 along with their power spectra. One can see oscillations in the accelerometer signal, which correspond to a peak in the power spectrum at the frequency of 5 Hz. The peak at the tremor frequency is seen in the LFP spectrum as well, even though it is wider. The phases of both signals can be unambiguously defined in the frequency band around the tremor frequency (e.g. 3–7 Hz). As a result of the phase dynamics modelling (Sect. 12.2), we have found statistically significant influence of the limb oscillations on the brain activity with a delay time not more than several dozens of milliseconds. The influence of the brain activity on the limb oscillations is present as well and is characterised by a delay time of 200–400 ms, i.e. one to two basic tremor periods (Fig. 12.2). The results are well reproduced, both qualitatively and quantitatively, for all three patients (Fig. 12.3). Some details are given in Sect. 13.2.

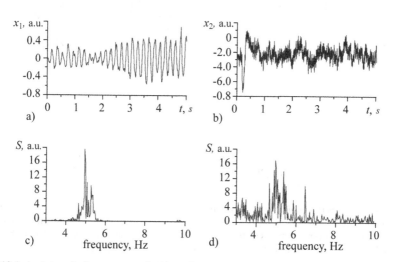

Fig. 12.1 An interval of spontaneous Parkinsonian tremor (total duration of 36 s, only the starting 5 s are shown): (**a, c**) an accelerometer signal in arbitrary units and its power spectrum estimate; (**b, d**) an LFP recording from one of the electrodes and its power spectrum estimate

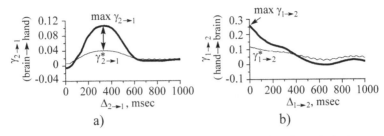

Fig. 12.2 Coupling estimates for the tremor epoch shown in Fig. 12.1 (dimensionless) versus a trial delay time: (**a**) brain → hand; (**b**) hand → brain. The phases are defined in the frequency band 3–7 Hz. *Thin lines* show the threshold values $\gamma^*_{j\to k} = 1.6\hat{\sigma}_{\gamma_{j\to k}}$. The values of $\gamma_{j\to k}$ exceeding $\gamma^*_{j\to k}$ differ from zero statistically significantly (at an overall significance level of $p < 0.05$). Thus, one observes an approximately zero delay time for the hand → brain influence and a delay time of about 335 ms for the brain → hand driving

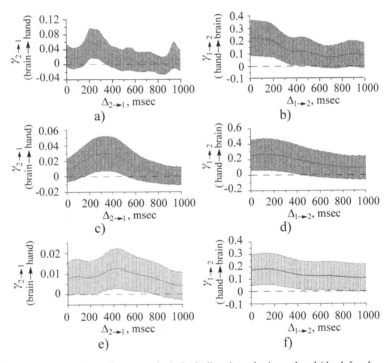

Fig. 12.3 Estimates of coupling strengths in both directions: brain → hand (the *left column*) and vice versa (the *right column*) for the three patients (*three rows*) versus a trial delay time. Coupling estimates, averaged over ensembles of 10–15 intervals of strong tremor, are shown along with their averaged 95% confidence bands (Smirnov et al., 2008)

Surrogate data tests (Dolan and Neiman, 2002; Schreiber and Schmitz, 1996) confirm statistical significance of our conclusions as well. Moreover, they show that linear techniques cannot reveal the influence of the thalamus and basal ganglia activity on the limbs.

Influence of the limb on the brain has been detected earlier with the linear Granger causality (Eichler, 2006; Wang et al., 2007). However, the phase dynamics modelling provides a new result: the brain → hand influence is detected and its delay time is estimated. This delay is quite big as compared to the conduction time of the neural pulses from the brain to the muscles. Therefore, it is interpreted (Smirnov et al., 2008) as a sign of indirect (after processing of the signals in the cortex) influence of the thalamus or the basal ganglia activity on the limb oscillations. Besides, it means that nuclei of the thalamus and the basal ganglia are elements of "feedback loops", which determine limb oscillations, rather than being just passive receivers of cerebral or muscle signals.

We have also estimated non-linear Granger causality for broadband accelerometer and LFP signals, rather than for the band-pass-filtered versions. It detects bidirectional couplings as well but does not give reliable estimates of the delay times. One reason can be that different time delays may correspond to different frequency bands leading to unclear results of the combined analysis.

An important area of further possible applications of the presented directionality analysis might be functional target point localisation diagnosis for an improvement of the depth electrode placement.

12.4 Couplings Between Brain Areas in Epileptic Rats

Over the years, electroencephalography is widely used in clinical practice for the investigation, classification and diagnosis of epileptic disorders. The EEG provides valuable information in patients with typical and atypical epileptic syndromes and offers important prognostic information. *Absence epilepsy*, previously known as petit mal, is classically considered as non-convulsive generalised epilepsy of unknown aetiology. Clinically, absence seizures occur abruptly, last from several seconds up to a minute and are accompanied by a brief decrease of consciousness that interrupts normal behaviour. Absences may either have or not have facial automatisms, e.g. minimal jerks and twitches of facial muscles, and eye blinks. In humans, EEGs during typical absence seizures are characterised by the occurrence of generalised 3–5-Hz spike – and - wave complexes which have an abrupt onset and offset (Panayiotopoulos, 1997). Similar EEG paroxysms, spike-and-wave discharges (SWDs) appear in rat strains with a genetic predisposition to absence epilepsy, such as WAG/Rij (Wistar Albino Glaxo from Rijswijk) (Coenen and van Luijtelaar, 2003). The EEG waveform and duration (1–30 s, mean 5 s) of SWD in rats and in humans are comparable, but the frequency of SWD in rats is higher, 8–11 Hz (Midzianovskaia et al., 2001; van Luijtelaar and Coenen, 1986).

EEG coherence was used previously to measure neuronal synchrony between populations of thalamic and cortical neurons (Sitnikova and van Luijtelaar, 2006).

The onset of SWD was characterised by area-specific increase of coherence that supported the idea that the cortico-thalamo-cortical circuitry is primarily involved in the initiation and propagation of SWD (Meeren et al., 2005; Steriade, 2005). However, the exact mechanism is unknown. A useful information to uncover it would be characteristics of directional couplings between different brain areas. Below, we describe our results on the estimation of interdependencies between local field potentials recorded simultaneously from the specific thalamus and the frontal cortex before, during and after SWD (Sitnikova et al., 2008).

Experiments were performed in five male 11–12-month-old WAG/Rij rats. The recordings are done at the Department of Biological Psychology, Radboud University of Nijmegen. EEGs were recorded from brain areas in which seizure activity is known to be the most robust: in the frontal cortex and in the ventroposteromedial (VPM) thalamic nucleus (Vergnes et al., 1987). EEG recordings were made in freely moving rats in a Faraday cage. Each recording session lasted from 5 to 7 h during the dark period of the day – night cycle. SWDs appeared in EEG as trains of stereotypic repetitive 7–10-Hz spikes and waves with high amplitude exceeding the background more than three times. SWDs lasted longer than 1 s (Midzianovskaia et al., 2001; van Luijtelaar and Coenen, 1986). In total, 53, 111, 34, 33 and 63 epileptic discharges in five rats were detected and analysed.

As it is mentioned in Sect. 11.1, non-stationarity is an intrinsic feature of the EEG signal. Since the above coupling estimation techniques require stationary data, we divided the EEG recordings into relatively short epochs in which the EEG signal revealed quasi-stationary behaviour. Time window lasting for 0.5 s seems to be a good choice. This duration corresponds to four spike-wave cycles. We report only results of the Granger causality estimation, since phase dynamics modelling gave no significant conclusions due to the shortness of quasi-stationary segments. Introduction of non-linearity (such as polynomials of the second and the third order) has no significant influence on the prediction quality of AR models before and after SWD. It suggests a predominance of the linear causal relations in non-seizure EEG. In contrast, the seizure activity (SWD) exhibits a non-linear character. However, the construction of non-linear models for seizure-related processes is quite non-trivial. Thus, we present only the results of the linear analysis.

Prediction improvements are computed using EEG data from the frontal cortex (x_1) and from the VPM (x_2). The linear AR models (12.1) and (12.3) are used to calculate the coupling characteristics $PI_{1\rightarrow2}$ (FC $\rightarrow$ VPM) and $PI_{2\rightarrow1}$ (VPM $\rightarrow$ FC). EEG recordings during a typical SWD are shown in Fig. 12.4. Figure 12.5 shows a typical dependence of the prediction error $\sigma_{1|2}^2$ on the dimensions d_1 and $d_{2\rightarrow1}$ at $\Delta_{2\rightarrow1} = 0$ for a 0.5-s interval of SWD. The error decreases when d_1 and $d_{2\rightarrow1}$ rise from 1 to 5. It reaches its saturation point for the values of $d_1 = d_{2\rightarrow1} = 5$, which are taken as optimal. The same dependence is observed for the error $\sigma_{2|1}^2$. Introduction of non-zero delays $\Delta_{j\rightarrow k}$ makes predictions worse, therefore, only zero delay times are used in the further analysis.

The first and the last spike in spike-and-wave sequences are used to mark the onset and the offset of seizure activity. Estimation of the thalamus-to-cortex and cortex-to-thalamus influences is performed for the EEG epochs covering a seizure

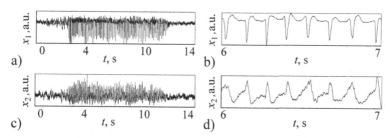

Fig. 12.4 EEG recordings of a spike-and-wave discharge in the frontal cortex (**a, b**) and in the specific ventroposteromedial thalamic nucleus (**c, d**). The panels (**b**) and (**d**) are magnified segments of the panels (**a**) and (**c**), respectively

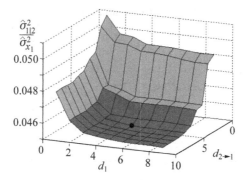

Fig. 12.5 The prediction error of model (12.3) for the frontal cortex EEG recording (fitted to the middle half a second in Fig. 12.4b) versus d_1 and $d_{2\to1}$

(SWD), 5 s before a seizure (pre-SWD) and 5 s after a seizure (post-SWD). The averaged results are illustrated in Fig. 12.6.

Before the onset of SWD, couplings are weak and remain constant until SWD begins. The first SWD-related disturbances of $PI_{j\to k}$ are observed about half a second before SWD onset. This effect is provoked by the seizure itself because the 0.5-s time window starts to capture the seizure activity. Still, the obtained values of $PI_{1\to2}$ and $PI_{2\to1}$ are statistically significantly greater than zero for majority of analysed epochs both before and during SWD at least at the level of $p = 0.05$ according both to F-test and surrogate data test (Schreiber and Schmitz, 1996). No changes in $PI_{j\to k}$ are found earlier than 0.5 s before the SWD onset, suggesting that the quantities $PI_{j\to k}$ are not capable of seizure prediction. The immediate onset of SWD is associated with a rapid growth in $PI_{1\to2}$ and $PI_{2\to1}$. The Granger causality characteristics reach their maximum within half a second after a seizure onset and remain high during the first 5 s of a seizure. The increase in couplings in both directions during an SWD as compared to pre-SWD epochs is significant. The ascending influence thalamus $\to$ cortex tends to be always stronger in terms of the PI values compared to the descending influence cortex $\to$ thalamus. Moreover, the occurrence of an SWD is associated with a tendency for a larger increase in the thalamus $\to$ cortex as compared to the cortex $\to$ thalamus influence. The results are similar for

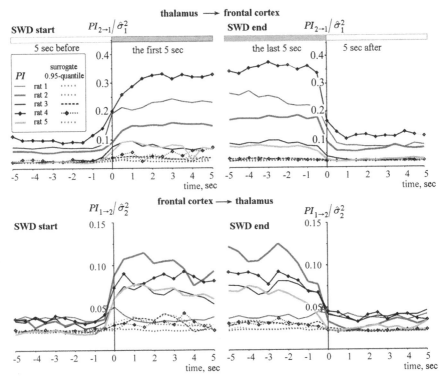

Fig. 12.6 Normalised prediction improvements averaged over all accessible SWDs for each animal versus the starting time instant of the moving window (0.5 s length). The onset and the offset of SWD are shown by *vertical lines*. The presence of SWD is associated with significant (and reversible) changes in the Granger causality in both directions. Surrogate data tests (*dotted lines*) are performed for each animal and confirm statistical significance of the Granger causality estimation results

all five rats analysed. Thus, bidirectional couplings between FC and VPM are always present, but the cortico-thalamo-cortical associations are reinforced during SWD.

Our results suggest that a reinforcement of predominant thalamus → cortex coupling accompanies the occurrence of an SWD, which can be interpreted as follows. In the described study, the EEG records were made in the areas where seizure activity is known to be the most robust (the frontal cortex and the VPM). It is important that direct anatomic connections between these structures are virtually absent, but both structures densely interconnect with the somatosensory cortex (Jones, 1985). As discussed in Meeren et al. (2002), the somatosensory cortex (the peri-oral region) contains an "epileptic focus" that triggers an SWD in WAG/Rij rats. The frontal EEGs are recorded rather far away from the "epileptic focus". Several groups report similar results of their investigations with other methods: the cortex, indeed, does not lead the thalamus when the cortical electrode is relatively far from the peri-oral area of the somatosensory cortex (Inoue et al., 1993; Polack et al., 2007; Seidenbecher et al., 1998).

12.5 El Niño – Southern Oscillation and North Atlantic Oscillation

El Niño – Southern Oscillation (ENSO) and *North Atlantic Oscillation* (NAO) represent the leading modes of interannual climate variability for the globe and the Northern Hemisphere (NH), respectively (CLIVAR, 1998; Houghton et al., 2001). Different tools have been used for the analysis of their interaction, in particular, cross-correlation function (CCF) and Fourier and wavelet coherence for the sea surface temperature (SST) and sea-level pressure (SLP) indices (Jevrejeva et al., 2003; Pozo-Vazquez et al., 2001; Rogers, 1984; Wallace and Gutzler, 1981).

One often considers a NAO index defined as the normalised SLP difference between Azores and Iceland (Rogers, 1984; http://www.cru.uea.ac.uk). It is further denoted as $NAOI_{cru}$. Alternatively, in http://www.ncep.noaa.gov, NAO is characterised as the leading decomposition mode of the field of 500 hPa geopotential height in the NH based on the "rotated principal component analysis" (Barnston and Livezey, 1987). It is denoted further as $NAOI_{ncep}$. Hence, $NAOI_{ncep}$ is a more global characteristic than $NAOI_{cru}$. ENSO indices T(Niño-3), T(Niño-3,4), T(Niño-4) and T(Niño-1+2) characterise SST in the corresponding equatorial regions of the Pacific Ocean (see, e.g., Mokhov et al., 2004). Southern oscillation index (SOI) is defined as the normalised SLP difference between Tahiti and Darwin. All the signals are rather short, which makes confident inference about the character of interaction difficult. We have investigated interaction between ENSO and NAO in Mokhov and Smirnov (2006) with non-linear Granger causality and phase dynamics modelling. The results are described below.

Mainly, the period 1950–2004 (660 monthly values) is analysed. The indices $NAOI_{cru}$ and $NAOI_{ncep}$ for NAO and T(Niño-3), T(Niño-3,4), T(Niño-4), T(Niño-1+2) and SOI for ENSO are used. Longer time series for $NAOI_{cru}$ (1821–2004), T(Niño-3) (1871–1997) and SOI (1866–2004) are also considered.

12.5.1 Phase Dynamics Modelling

Figure 12.7 demonstrates individual characteristics of the indices $NAOI_{ncep}$ (Fig. 12.7a) and T(Niño-3,4) (Fig. 12.7d). Wavelet analysis of each signal $x(t)$ is based on the wavelet transform

$$W(s, t) = \frac{1}{\sqrt{s}} \int\limits_{-\infty}^{\infty} x(t')\Phi^* \left((t - t')/s\right) dt', \qquad (12.12)$$

where $\Phi(\eta) = \pi^{-1/4} \left[\exp(-i\omega_0\eta) - \exp\left(-\omega_0^2/2\right)\right] \exp(-\eta^2/2)$ is the Morlet wavelet (see also Eq. (6.23) in Sect. 6.4.3), an asterisk means complex conjugate and s is the timescale. Global wavelet spectra S of the climatic signals, obtained by integration of Eq. (12.12) over time t at each fixed s, exhibit several peaks

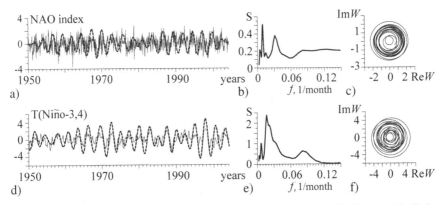

Fig. 12.7 Characteristics of NAOI$_{ncep}$ and T(Nino-3,4). (**a**) NAOI$_{ncep}$ (*thin line*) and ReW for $s = 32$ months (*dashed line*); (**b**) global wavelet spectrum of NAOI$_{ncep}$ ($f = 1/s$); (**c**) an orbit $W(t)$ for NAOI$_{ncep}$, $s = 32$ months; (**d–f**) the same as (**a–c**) for the index T(Niño-3,4)

(Fig. 12.7b, e). One can assume that the peaks correspond to oscillatory processes for which the phases can be adequately introduced. To get the phases of "different rhythms" in NAO and ENSO, we try several values of s in Eq. (12.12) corresponding to different spectral peaks. The phase is defined as an argument of the respective complex signal $W(s, t)$ at fixed s. For $\omega_0 = 6$ used below, this is tantamount to band-pass filtering of a signal x around the frequency $f = 1/s$ with a relative bandwidth $1/4$ and subsequent use of the Hilbert transform (see Sect. 6.4.3). Then, we estimate couplings between all the "rhythms" pairwise. The only case when substantial conclusions about the presence of coupling are inferred is the "rhythm" with $s = 32$ months for both signals (Fig. 12.7a, d, dashed lines). The phases are sufficiently well defined for both signals, since clear rotation around the origin takes place on the complex plane (Fig. 12.7c, f).

The results of the phase dynamics modelling are shown in Fig. 12.8 for $s = 32$ months and model (12.11) with $\tau = 32$ months, where ϕ_1 stands for the phase of NAO and ϕ_2 for ENSO. Figure 12.8a shows that the technique is applicable only for $\Delta_{2\to1} < 30$, where $\rho(-\Delta_{2\to1}) < 0.4$. The influence ENSO $\to$ NAO is pointwise significant for $0 < \Delta_{2\to1} < 30$ and maximal for $\Delta_{2\to1} = 24$ months (Fig. 12.8b). From here, we infer the presence of the influence ENSO $\to$ NAO at an overall significance level $p = 0.05$ as discussed in Mokhov and Smirnov (2006). Most probably, the influence ENSO $\to$ NAO is delayed by 24 months; however, this conclusion is not as reliable. No signs of the influence NAO $\to$ ENSO are detected (Fig. 12.8c).

Large ρ for $\Delta_{2\to1} > 30$ does not imply strong coupling. For such a short time series and close basic frequencies of the oscillators, the probability to get $\rho(\Delta) > 0.4$ for uncoupled processes is greater than 0.5 as observed in numerical experiments with exemplary oscillators.

All the reported results remain the same for any s in the range 29–34 months and relative bandwidths 0.2–0.4. Phase calculation based directly on a band-pass

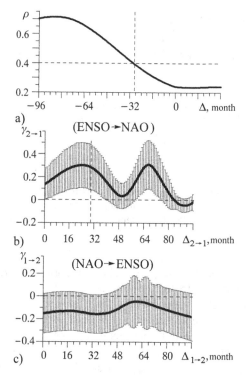

Fig. 12.8 Coupling between NAOI$_{ncep}$ and T(Niño-3,4) (over the period 1950–2004) in terms of the phase dynamics modelling: (**a**) mean phase coherence; (**b**, **c**) strengths of the influences ENSO → NAO and NAO → ENSO, respectively, with their 95% pointwise confidence bands

filtering and Hilbert transform leads to similar results, e.g., for the second-order Butterworth filter (Hamming, 1983) with the same bandwidths. The use of the other ENSO indices instead of T(Niño-3,4) gives almost the same results as in Fig. 12.8. Coupling is not pronounced only for T(Niño-1+2). Analysis of the other rhythms in NAOI$_{ncep}$ and T(Niño-3,4) does not lead to significant conclusions about the presence of interaction. For NAOI$_{cru}$ the width of the peak corresponding to $s = 32$ months is greater than that for NAOI$_{ncep}$. It leads to stronger phase diffusion of the 32-month rhythm as quantified by the mean-squared residual errors of the model (12.11) (Smirnov and Andrzejak, 2005). As a result, we have not observed significant coupling between NAOI$_{cru}$ and any of the ENSO indices for the period 1950–2004 as well as for the longer recordings (1871–1997 and 1866–2004).

12.5.2 Granger Causality Analysis

Cross-correlations between NAOI$_{ncep}$ (x_1) and T(Niño-3,4) (x_2) are not significant at $p < 0.05$. More interesting results are obtained from the non-linear Granger causality analysis based on the polynomial AR models like Eq. (12.4). Figure 12.9a

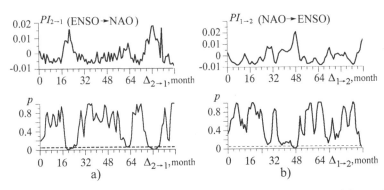

Fig. 12.9 Coupling between NAOI$_{ncep}$ and T(Niño-3,4) (1950–2004) in terms of the non-linear Granger causality. Prediction improvements are normalised by variances of the signals: (**a**) PI$_{2\to1}$/var[x_1], (**b**) PI$_{1\to2}$/var[x_2]. Pointwise significance level p estimated via F-test is shown below each panel

shows the normalised quantity PI$_{2\to1}$/var[x_1] for the parameters $d_1 = 0$, $d_{2\to1} = 1$ and $P_1 = 2$. It is about 0.015 for the time delays $19 \leq \Delta_{2\to1} \leq 21$ or $80 \leq \Delta_{2\to1} \leq 83$ months. Each of these PI values is pointwise significant at $p = 0.01$. Taking into account strong correlations of $PI_{2\to1}$ separated by $\Delta_{2\to1}$ less than 4 months, one can infer that the influence ENSO $\to$ NAO is present at the overall level $p < 0.05$ (Mokhov and Smirnov, 2006). Analogously, Fig. 12.9b shows the quantity PI$_{1\to2}$/var[x_2] for $d_2 = 0$, $d_{1\to2} = 1$ and $P_2 = 2$. Its pointwise significant values at $48 \leq \Delta_{1\to2} \leq 49$ months do not allow confident detection of the influence NAO $\to$ ENSO.

If d_1 and $d_{2\to1}$ are increased up to 2, no changes in PI values presented in Fig. 12.9a are observed. So, the reported PI is not achieved via complication of the individual model. Simultaneous increase in d_1 up to 3, P_1 up to 3 and $d_{2\to1}$ up to 2 leads to the absence of any confident conclusions due to large variance of the estimators.

Similar results are observed if T(Nino-3,4) is replaced with T(Niño-3), T(Niño-4) or SOI. However, the conclusion about the presence of the influence ENSO $\to$ NAO becomes less confident: $p \approx 0.1$. The use of T(Niño-1+2) leads to even less significant results. Analogous to the phase dynamics modelling, replacement of NAOI$_{ncep}$ with NAOI$_{cru}$ does not lead to reliable coupling detection neither for the period 1950–2004 nor for longer periods.

Finally, to reveal trends in coupling during the last decade, couplings between NAOI$_{ncep}$ and T(Niño-3,4) are estimated in a moving window of the length of 47 years. Namely, we start with the interval 1950–1996 and finish with the interval 1958–2004. PI values reveal an increase in the strength of the influence ENSO $\to$ NAO. The value of PI$_{2\to1}$ for $19 \leqslant \Delta_{2\to1} \leqslant 20$ months rises almost monotonously by 150% (Fig. 12.10). Although it is difficult to assess statistical significance of the conclusion, the monotone character of the increase indicates that it can hardly be an effect of random fluctuations. To a certain extent, it can be attributed to the strong 1997–1998 ENSO event.

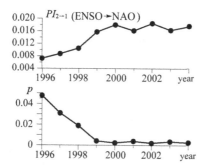

Fig. 12.10 Influence ENSO → NAO for NAOI$_{ncep}$ and T(Niño-3,4) in a 47-year moving window; the value max{PI$_{2→1}$($\Delta_{2→1}$ = 19), PI$_{2→1}$($\Delta_{2→1}$ = 20)}/var[x_1] is shown versus the last year of the moving window

Thus, the presence of coupling between ENSO and NAO is revealed by the use of two non-linear techniques and different climatic indices. Consistent results are observed in all cases. The influence ENSO → NAO is detected with confidence probability of 0.95 from the data for NAOI$_{ncep}$ (1950–2004). Estimate of its delay time ranges from several months up to 3 years with the most probable value of 20–24 months. Besides, an increase in the strength of the influence during the last decade is observed. Possible physical mechanisms underlying oscillations and interactions as slow and even slower than those reported here are considered, e.g., in Jevrejeva et al. (2003); Latif (2001); Pozo-Vazquez et al. (2001). The influence ENSO → NAO is not detected with the index NAOI$_{cru}$, which is a "more local" characteristic than the index NAOI$_{ncep}$. The influence NAO → ENSO is not detected with confidence for any indices.

12.6 Causes of Global Warming

A key global problem is related to the determination of the relative role of natural and anthropogenic factors in climate variations. Forecasts of the future climate change due to anthropogenic forcing depend on the present estimates of the impact of different factors on the climate. Thus, an impact of solar activity variations is quantified in Mokhov and Smirnov (2008); Mokhov et al. (2006); Moore et al. (2006) via the analysis of different reconstructions and measurement data for solar irradiance and *global surface temperature* (GST) of the Earth. A variable character of the solar activity impact in connection with its overall increase in the second half of the twentieth century is noted. Moreover, the use of a global climate model in three dimensions has led to the conclusion that solar activity influence can determine only a relatively small portion of the global warming observed in the last decades. A significant influence of the anthropogenic factor on the GST is noted in Verdes (2005). However, the question about the relative role of different factors is still not answered convincingly on the basis of the observation data analysis.

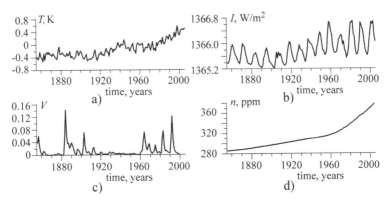

Fig. 12.11 The data: (**a**) mean GST (anomaly from the base period 1961–1990); (**b**) solar constant (irradiance in the range from infrared to ultraviolet wavelengths inclusively); (**c**) volcanic activity (optical depth of volcanic aerosol, dimensionless); (**d**) carbon dioxide atmospheric content in ppm (parts per million)

Here, we report our estimates of the influences of different factors on the GST (Mokhov and Smirnov, 2009) based on the analysis of the following data: annual values T of the mean GST anomaly in 1856–2005 (http://www.cru.uea.ac.uk), reconstructions and measurements of the annual *solar irradiance variations* I in 1856–2005 (http://www.cru.uea.ac.uk), *volcanic activity* V in 1856–1999 (Sato et al., 1993) and *carbon dioxide atmospheric content* n in 1856–2004 (Conway et al., 1994) (Fig. 12.11).

Firstly, we construct univariate AR models for the GST and then analyse the influences of different factors with bivariate and multivariate AR models. Since the main question is about the causes of the GST rise, we compute two characteristics for the different models: (i) the expectation of the value of T in 2005 denoted as T_{2005} and (ii) the expectation of the angular coefficient $\alpha_{1985-2005}$ of a straight line approximating the time profile $T(t)$ over the interval 1985–2005 in the least-squares sense (i.e. a characteristic of the recent trend). These two quantities for the original GST data take the values $T_{2005} = 0.502K$ and $\hat{\alpha}_{1985-2005} = 0.02K/$year.

The AR models are fitted to the intervals [1856 – L] for different L, rather than only for the largest possible $L = 2005$. Checking different L allows one to select a time interval, where each influence is most pronounced, and to determine a minimal value of L for which an influence can be revealed.

12.6.1 Univariate Models of the GST Variations

The mean-squared prediction error of a linear model (12.1) obtained from the interval [1856–2005] saturates at $d_T = 4$ (Fig. 12.12). Incorporation of any non-linear terms does not lead to statistically significant improvements (not shown). Thus, an optimal model reads

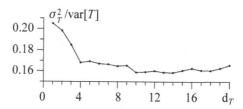

Fig. 12.12 Univariate AR models of the GST: the normalised prediction error variance versus the model order

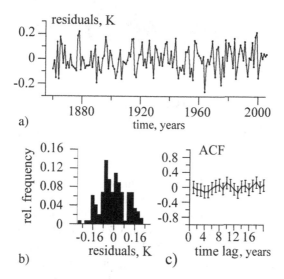

Fig. 12.13 Residual errors for the univariate AR model (12.13) at $d_T = 4$: (**a**) the time realisation; (**b**) the histogram; (**c**) the autocorrelation function with the 95% confidence interval estimates

$$T(t) = a_0 + \sum_{i=1}^{d_T} a_i T(t-i) + \xi(t), \tag{12.13}$$

where $d_T = 4$, $a_0 = -0.01 \pm 0.10\,K$, $a_1 = 0.58 \pm 0.08$, $a_2 = 0.03 \pm 0.09$, $a_3 = 0.11 \pm 0.09$, $a_4 = 0.29 \pm 0.08$. The intervals present standard deviation estimates coming from the least-squares routine (Sect. 7.4.1). The model prediction error is $\sigma_T^2 = 0.01\,K^2$, while the sample variance of the GST over the interval [1856–2005] is equal to $\mathrm{var}[T] = 0.06K^2$. In relative units $\sigma_T^2/\mathrm{var}[T] = 0.17$, i.e. 17% of the GST variance is not explained by the univariate AR model. Residual errors for the AR model with $d_T = 4$ look stationary (Fig. 12.13a) and their histogram exhibits maximum around zero (Fig. 12.13b). Their delta correlatedness holds true (Fig. 12.13c). The latter is the main condition for the F-test applicability to the further Granger causality estimation.

Time realisations of the obtained model (12.13) over 150 years at fixed initial conditions (equal to the original GST values in 1856–1859) look very similar to the original time series (Fig. 12.14a). For a quantitative comparison, Fig. 12.14b

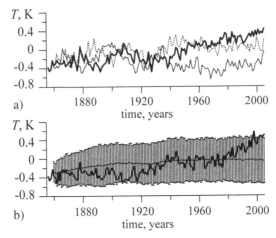

a)

b)

Fig. 12.14 Behaviour of the model (12.13) fitted to the interval [1856–2005]: (**a**) three time realisations taken randomly from an ensemble of 100 realisations; (**b**) mean values over the ensemble
(*thin line*) and the 95% intervals of the distributions (*error bars*) with the superimposed original
data for GST (*thick line*)

shows mean values and 95% intervals for the distributions of model values $T(t)$
computed from an ensemble of 100 simulated model time realisations. The original
time series does not come out of the intervals most of the time, i.e. the model quality
is sufficiently high. However, this is violated for the GST values in 2001–2005.
Thus, one may suspect that model (12.13) with constant parameters and constant
$\sigma_T^2 = 0.01\,K^2$ is not completely adequate, e.g. it may not take into account some
factors determining the essential GST rise over the last years.

The hypothesis finds a further confirmation under a more strict test. We check
whether the univariate model (12.13) fitted to the interval [1856–1985] can predict
the GST rise over the interval [1985–2005]. The results of model fitting are similar
to those for the interval [1856–2005]. Coefficient estimates differ to some extent:
$a_0 = -0.01 \pm 0.16K$, $a_1 = 0.56 \pm 0.09$, $a_2 = 0.05 \pm 0.10$, $a_3 = 0.02 \pm 0.10$, $a_4 =
0.29 \pm 0.09$. Prediction error is again $\sigma_T^2 = 0.01\,K^2$. However, the original GST
values over the last 16 years do not fall within the 95% intervals (Fig. 12.15). Thus,
one may assume that something has changed in the GST dynamics over the last
decades, e.g., as a result of external influences.

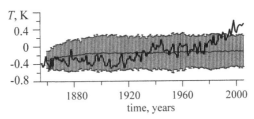

Fig. 12.15 The original GST (*thick line*) and the 95% "corridor" for the model (12.13) fitted to the
interval [1856–1985]

This is further analysed with bi- and multivariate AR models for the GST. We take $d_T = 4$ and select $d_{I \to T}, d_{n \to T}, d_{V \to T}$ and $\Delta_{I \to T}, \Delta_{n \to T}, \Delta_{V \to T}$ so as to provide the greatest GST prediction improvement and qualitative similarity between the model behaviour and the original GST time profile.

12.6.2 GST Models Including Solar Activity

An optimal choice of parameters is $d_{I \to T} = 1$ and $\Delta_{I \to T} = 0$. The influence $I \to T$ is most clearly seen when the interval [1856–1985] is used for model fitting (Fig. 12.16a). The model reads

$$T_t = a_0 + a_1 T_{t-1} + a_4 T_{t-4} + b_I I_{t-1} + \eta_t, \qquad (12.14)$$

where $a_1 = -93.7 \pm 44.4\,K$, $a_1 = 0.52 \pm 0.09$, $a_4 = 0.27 \pm 0.09$ and $b_I = 0.07 \pm 0.03\,K/(W/m^2)$. The prediction improvement is $PI_{I \to T}/\sigma_T^2 = 0.028$ and its positivity is statistically significant at $p < 0.035$. The model fitted to the interval [1856–2005] detects no influence $I \to T$ significant at $p < 0.05$. It may evidence that the impact of other factors, not related to solar activity, has increased during the interval [1985–2005]. Simulations with model (12.14) indirectly confirm this

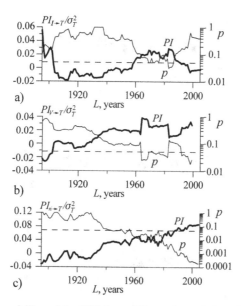

Fig. 12.16 Bivariate modelling of the GST from different time windows [1856 – L]. PI-values and significance levels for (**a**) the models taking into account solar activity; (**b**) the models taking into account volcanic activity; (**c**) the models taking into account CO_2 atmospheric content. The numerical values of PI (*thick lines*) are indicated on the left y-axes and significance levels (*thin lines*) on the right y-axes. The *dashed lines* show the level of $p = 0.05$

assumption. Figure 12.17a shows an ensemble of simulated realisations when an original time series $I(t)$ is used as input. The 95% intervals are narrower than those for the univariate model (cf. Fig. 12.14b), i.e. the incorporation of solar activity into the model allows better description of the GST in 1856–1985. However, the GST rise in 1985–2005 is not predicted by the bivariate model as well.

To assess the long-term effect of the solar activity trend on the GST rise, we simulate an ensemble of time realisations of model (12.14) when a detrended signal $I(t)$ (Lean et al., 2005) is used as input. The result is visually indistinguishable from the plot in Fig. 12.17a (not shown). Thus, the removal of the solar activity trend does not affect the model GST values. Quantitatively, we get $\langle T_{2005} \rangle = 0.0 \pm 0.02K$ and angular coefficients $\langle \alpha_{1985-2005} \rangle \leq 0.002K/\text{year}$ in both cases. The original trend $\hat{\alpha}_{1985-2005} = 0.02K/\text{year}$ is not explained by any of the bivariate models (12.14). Thus, despite it is detected that the solar activity variations affect the GST, the long-term analysis suggests that they are not the cause of the GST rise in the last years.

Introduction of non-linearity into the models does not improve their predictions so that the linear models seem optimal. This is the case for all models below as well. Therefore, all the results are presented only for the linear models.

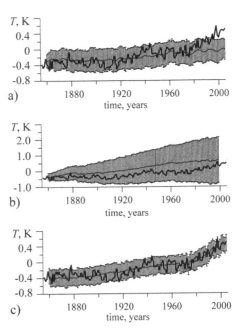

Fig. 12.17 The original GST values (*thick line*) and 95% "corridors" for the bivariate models of the GST: (a) model (12.14) with solar activity fitted to the interval [1856–1985]; model (12.15) with volcanic activity fitted to the interval [1856–1999]; (c) model (12.16) with CO_2 atmospheric content fitted to the interval [1856–2005]

12.6.3 GST Models Including Volcanic Activity

The influence of the volcanic activity appears of the same order of magnitude as that of the solar activity. An optimal choice is $d_{V \to T} = 1$ and $\Delta_{V \to T} = -1$, i.e. a model

$$T_t = a_0 + a_1 T_{t-1} + a_4 T_{t-4} + b_V V_t + \eta_t. \qquad (12.15)$$

The influence is detected most clearly from the entire interval [1856–1999] of the available data for $V(t)$ (Fig. 12.16b). For that interval $\mathrm{PI}_{V \to T}/\sigma_T^2 = 0.029$ and positivity of $\mathrm{PI}_{V \to T}$ is statistically significant at $p < 0.03$. Model coefficients are $a_0 = 0.25 \pm 0.14 K$, $a_1 = 0.55 \pm 0.08$, $a_4 = 0.29 \pm 0.08$, $b_V = -0.92 \pm 0.41 K$.

However, even if the original data for $V(t)$ are used as input, the model predicts only strong fluctuations of the GST around the mean value, e.g., in 1999 – around the value of $\langle T_{1999} \rangle = 0.7 \pm 0.14 K$ (Fig. 12.17b), rather than the rise in the GST during the last years. According to model (12.15), there is no trend in the GST on average: $\langle \alpha_{1985-2005} \rangle \leq 0.001 K/\text{year}$. If the signal $V(t) = 0$ is used as input, then the model predicts even greater values of the GST: $\langle T_{1999} \rangle = 1.5 \pm 0.16 K$. Indeed, the long-term effect of volcanic eruptions is to limit the GST values. Volcanic activity is relatively high in 1965–1995 (Fig. 12.17c), which should contribute to a decrease in the GST. Therefore, explaining the GST rise during the last decades by the volcanic activity influence is also impossible.

12.6.4 GST Models Including CO_2 Concentration

An optimal choice is $d_{n \to T} = 1$ and $\Delta_{n \to T} = 0$. Apart from highly significant prediction improvement, it provides a model which behaves qualitatively similar to the original data (in contrast to the models with $d_{n \to T} > 1$). The model reads as

$$T_t = a_0 + a_1 T_{t-1} + a_4 T_{t-4} + b_n n_{t-1} + \eta_t. \qquad (12.16)$$

The influence of CO_2 appears much more considerable than that of the other factors. It is detected most clearly from the entire available interval [1856–2005] (Fig. 12.16c), where $\mathrm{PI}_{n \to T}/\sigma_T^2 = 0.087$ and its positivity is significant at $p < 0.0002$. The coefficients of this model are $a_0 = -1.10 \pm 0.29 K$, $a_1 = 0.46 \pm 0.08$, $a_4 = 0.20 \pm 0.08$, $b_n = 0.003 \pm 0.001 \ K/\text{ppm}$.

An ensemble of time realisations (Fig. 12.17c) shows that the model (12.16) with the original data $n(t)$ used as input describes the original data $T(t)$ much more accurately than do the models taking into account the solar or the volcanic activity. Moreover, the model (12.16) fitted to a narrower interval, e.g. [1856–1960], exhibits practically the same time realisations as in Fig. 12.17c, i.e. it correctly predicts the GST rise despite the data over an interval [1960–2004] are not used for the model fitting. The model (12.16) fitted to any interval [1856 – L] with $L > 1935$ gives almost the same results.

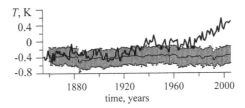

Fig. 12.18 The original GST values (*thick line*) and the 95% "corridor" for the bivariate model (12.16) if the signal $n(t) = \text{const} = n(1856)$ is used as input

If the artificial signal $n(t) = \text{const} = n(1856)$ is used as input for the model (12.16) fitted to the interval [1856–1985], then one observes just fluctuations of T about the level of T_{1856} (Fig. 12.18) and no trend, i.e. $\langle \alpha_{1985-2005} \rangle = 0$. If the original data for $n(t)$ are used as input, one gets the model characteristics $\langle T_{2005} \rangle \approx 0.5 K$ and $\langle \alpha_{1985-2005} \rangle = 0.17 K/\text{year}$, which are close to the observed ones. Thus, according to the model (12.16), the rise in the atmospheric CO_2 content explains a major part of the recent rise in the GST.

The results of the multivariate AR modelling confirm the above results of the bivariate analysis (the corresponding plots are not shown).

Thus, the Granger causality estimation and the investigation of the AR models' long-term behaviour allow to assess an effect of the solar activity, volcanic activity and carbon dioxide atmospheric content on the global surface temperature. The Granger causality shows that the three factors determine about 10% of the quantity σ_T^2, which is the variance of the short-term GST fluctuations unexplained by the univariate AR model. The impact of CO_2 is the strongest one, while an effect of the other two factors is several times weaker. The long-term behaviour of the models reveals that the CO_2 content is a determinative factor of the GST rise. According to the empirical AR models, the rise in the CO_2 concentration determines at least 75% of the GST trend over 1985–2005, while the other two factors are not the causes of the *global warming*. In particular, if the CO_2 concentration remained at the level of 1856 year, the GST would not rise at all during the last century. In contrast, model variations in the solar and volcanic activity do not lead to significant changes in the GST trend.

References

Akaike, H.: A new look at the statistical identification model. IEEE Trans. Autom. Control. **19**, 716–723 (1974)

Ancona, N., Marinazzo, D., Stramaglia, S.: Radial basis function approach to nonlinear Granger causality of time series. Phys. Rev. E. **70**, 056221 (2004)

Arnhold, J., Lehnertz, K., Grassberger, P., Elger, C.E.: A robust method for detecting interdependences: application to intracranially recorded EEG. Physica D. **134**, 419–430 (1999)

Balanov, A., Janson, N., Postnov, D., Sosnovtseva, O.: Synchronization: From Simple to Complex. Springer, Berlin, 2008

Barnston, A.G., Livezey, R.E.: Classification, seasonality, and persistence of low frequency atmospheric circulation patterns. Mon. Wea. Rev. **115**, 1083–1126 (1987)

Benabid, A.L., Pollak, P., Gervason, C., Hoffmann, D., Gao, D.M., Hommel, M., Perret, J.E., De Rougemont, J.: Long-term suppression of tremor by chronic stimulation of the ventral intermediate thalamic nucleus. The Lancet. **337**, 403–406 (1991)

Benabid, A.L., Wallace, B., Mitrofanis, J., Xia, R., Piallat, B., Chabardes, S., Berger, F. A putative generalized model of the effects and mechanism of action of high frequency electrical stimulation of the central nervous system. Acta Neurol. Belg. **105**, 149–157 (2005)

Bezruchko, B.P., Ponomarenko, V.I., Prokhorov, M.D., Smirnov, D.A., Tass, P.A.: Modeling nonlinear oscillatory systems and diagnostics of coupling between them using chaotic time series analysis: applications in neurophysiology. Physics – Uspekhi. **51**, 304–310 (2008)

Bezruchko, B.P., Ponomarenko, V.I., Rosenblum, M.G., Pikovsky, A.S.: Characterizing direction of coupling from experimental observations. Chaos. **13**, 179–184 (2003)

Boccaletti, S., Kurths, S., Osipov, G.,, Valladares, D., Zhou, C.: The synchronization of chaotic systems. Phys. Rep. **366**, 1–52 (2002)

Brea, J., Russell, D.F., Neiman, A.B.: Measuring direction in the coupling of biological A case study for electroreceptors of paddlefish. Chaos. **16**, 026111 (2006)

Cimponeriu, L., Rosenblum, M., Pikovsky, A.: Estimation of delay in coupling from time series. Phys. Rev. E. **70**, 046213 (2004)

CLIVAR Initial Implementation Plan. WCRP No.103. WMO/TD No.869. ICPO No.14. 313pp. http://www.clivar.dkrz.de/hp.html. (1998)

Coenen, A.M.L., van Luijtelaar, E.L.J.M.: Genetic Animal models for absence epilepsy: a review of the WAG/Rij strain of rats. Behavioral Genetics. **33**, 635–655 (2003)

Conway, J. et al. Evidence of interannual variability of the carbon cycle from the NOAA/CMDL global air sampling network. J. Geophys. Res. **99**, 22831–22855 (1994)

Dolan, K.T., Neiman, A.: Surrogate analysis of coherent multichannel data. Phys. Rev. E. **65**, 026108 (2002)

Eichler, M.: Graphical modelling of dynamic relationships in multivariate time series. In: Winterhalder, M., Schelter, B., Timmer, J. (eds) Handbook of Time Series Analysis, pp. 335–367. Wiley-VCH, Berlin (2006)

Faes, L., Porta, A., Nollo, G.: Mutual nonlinear prediction as a tool to evaluate coupling strength and directionality in bivariate time series: Comparison among different strategies based on k nearest neighbors. Phys. Rev. E. **78**, 026201 (2008)

Feldmann, U., Bhattacharya, J.: Predictability improvement as an asymmetrical measure of interdependence in bivariate time series. Int. J. Bif. Chaos. **14**, 505–514 (2004)

Friston, K.J., Harrison, L., Penny, W.: Dynamic causal modelling. NeuroImage. **19**, 1273–1302 (2003)

Granger, C.W.J.: Investigating causal relations by econometric models and cross-spectral methods. Econometrica. **37**, 424–438 (1969)

Hamming, R.W.: Digital Filters. 2nd edn. Prentice-Hall, Englewood Cliffs, NJ (1983)

Hlavackova-Schindler, K., Palus, M., Vejmelka, M., Bhattacharya, J.: Causality detection based on information-theoretic approaches in time series analysis. Phys. Rep. **441**, 1–46 (2007)

Houghton, J.T., Ding, Y., Griggs, D.J., Noguer, M., et al. (eds). Climate Change 2001: The Scientific Basis. Intergovernmental Panel on Climate Change. Cambridge University Press, Cambridge (2001)

Hramov, A.Ye., Koronovskii, A.A.: An approach to chaotic synchronization. Chaos. **14**, 603–610 (2004)

Inoue, M., Duysens, J., Vossen, J.M., Coenen, A.M.: Thalamic multiple-unit activity underlying spike-wave discharges in anesthetized rats. Brain Res. **612**, 35–40 (1993)

Jevrejeva, S., Moore, J., Grinsted, A. Influence of the Arctic Oscillation and El Nino – Southern Oscillation (ENSO) on ice conditions in the Baltic Sea: The wavelet approach. J. Geophys. Res. **108**(D21), 4677, doi:10.1029/2003JD003417 (2003)

Jones, E.G.: The Thalamus. Plenium Press, New York (1985)

Kiemel, T., Gormley, K.M., Guan, L., Williams, T.L., Cohen, A.H.: Estimating the strength and direction of functional coupling in the lamprey spinal cord. J. Comput. Neurosci. **15**, 233–245 (2003)

Kreuz, T., Mormann, F., Andrzejak, R.G., Kraskov, A., Lehnertz, K., Grassberger, P.: Measuring synchronization in coupled model systems: A comparison of different approaches. Physica D. **225**, 29–42 (2007)

Kreuz, T., Mormann, F., Andrzejak, R.G., Kraskov, A., Lehnertz, K., Grassberger, P.: Measuring synchronization in coupled model systems: A comparison of different approaches. Phys. D. **225**, 29–42 (2007)

Kuramoto, Y.: Chemical Oscillations, Waves and Turbulence. Springer, Berlin (1984)

Latif, M.: Tropical Pacific/Atlantic Ocean interactions at multidecadal time scales. Geophys. Res. Lett. **28**, 539–542 (2001)

Lean, J., Rottman, G., Harder, J., Kopp, G.: Source contributions to new understanding of global change and solar variability. Solar Phys. **230**, 27–53 (2005)

Le Van Quyen, M., Martinerie, J., Adam, C., Varela F.: Nonlinear analyses of interictal EEG map the brain interdependences in human focal epilepsy. Physica D. **127**, 250–266 (1999)

Maraun, D., Kurths, J.: Epochs of phase coherence between El Nino/Southern Oscillation and Indian monsoon. Geophys. Res. Lett. **32**, L15709 (2005)

Meeren, H., van Luijtelaar, G., Lopes da Silva, F., Coenen, A.: Evolving concepts on the pathophysiology of absence seizures: the cortical focus theory. Arch. Neurol. **62**, 371–376 (2005)

Meeren, H.K., Pijn, J.P., van Luijtelaar, E.L., Coenen, A.M., Lopes da Silva, F.H.: Cortical focus drives widespread corticothalamic networks during spontaneous absence seizures in rats. J. Neurosci. **22**, 1480–1495 (2002)

Midzianovskaia, I.S., Kuznetsova, G.D., Coenen, A.M., Spiridonov, A.M., van Luijtelaar, E.L.: Electrophysiological and pharmacological characteristics of two types of spike-wave discharges in WAG/Rij rats. Brain Res. 911, 62–70 (2001)

Mokhov, I.I., Bezerkhny, V.A., Eliseev, A.V., Karpenko, A.A.: Interrelation between variations in the global surface air temperature and solar activity based on observations and reconstructions. Doklady Earth Sci. **409**, 805–809 (2006)

Mokhov, I.I., Khvorostyanov, D.V., Eliseev, A.V.: Decadal and longer term changes in El Nino – Southern Oscillation characteristics. Intern. J. Climatol. **24**, 401–414 (2004)

Mokhov, I.I., Smirnov, D.A.: Diagnostics of a cause – effect relation between solar activity and the earth's global surface temperature. Izvestiya, atmos. Oceanic Phys. **44**, 263–272 (2008)

Mokhov, I.I., Smirnov, D.A.: El Nino Southern Oscillation drives North Atlantic Oscillation as revealed with nonlinear techniques from climatic indices. Geophys. Res. Lett. **33**, L0378 (2006)

Mokhov, I.I., Smirnov, D.A.: Empirical estimates of the influence of natural and anthropogenic factors on the global surface temperature. Doklady Acad. Sci. **426**(5), 679–684, (in Russian) (2009)

Moore, J., Grinsted, A., Jevrejeva, S.: Is there evidence for sunspot forcing of climate at multi-year and decadal periods?. Geophys. Res. Lett. **33**, L17705 (2006)

Mormann, F., Lehnertz, K., David, P., Elger, C.E.: Mean phase coherence as a measure for phase synchronization and its application to the EEG of epilepsy patients. Phys. D. **144**, 358–369 (2000)

Mosedale, T.J., Stephenson, D.B., Collins, M., Mills, T.C.: Granger causality of coupled climate processes: ocean feedback on the North Atlantic Oscillation. J. Climate. **19**, 1182–1194 (2006)

Mosekilde, E., Maistrenko Yu., Postnov, D.: Chaotic Synchronization. Applications to Living Systems. World Scientific, Singapore (2002)

Nini, A., Feingold, A., Slovin, H., Bergman, H.: Neurons in the globus pallidus do not show correlated activity in the normal monkey, but phase-locked oscillations appear in the MPTP model of parkinsonism. J. Neurophysiol. **74**, 1800–1805 (1995)

Osipov, G.V., Kurths, J., Zhou, C.: Synchronization in Oscillatory Networks. Springer, Berlin (2007)

Osterhage, H., Mormann, F., Wagner, T., Lehnertz, K.: Measuring the directionality of coupling: phase versus state space dynamics and application to EEG time series. Int. J. Neural Syst. **17**, 139–148 (2007)

Palus, M., Novotna, D.: Quasi-biennial oscillations extracted from the monthly NAO index and temperature records are phase-synchronized. Nonlin. Processes Geophys. **13**, 287–296 (2006)

Palus, M., Stefanovska, A.: Direction of coupling from phases of interacting oscillators: An information-theoretic approach. Phys. Rev. E. **67**, 055201(R) (2003)

Panayiotopoulos, C.P.: Absence epilepsies. In: Engel, J.J., Pedley, T.A. (eds.). Epilepsy: A Comprehensive Textbook, pp. 2327–2346. Lippincott-Raven Publishers, Philadelphia. (1997)

Pereda, E., Quian Quiroga, R., Bhattacharya, J.: Nonlinear multivariate analysis of neurophysiological signals. Progr. Neurobiol. **77**, 1–37 (2005)

Pikovsky, A.S., Rosenblum, M.G., Kurths, J.: Synchronisation. A Universal Concept in Nonlinear Sciences. Cambridge University Press, Cambridge (2001)

Polack, P.O., Guillemain, I., Hu, E., Deransart, C., Depaulis, A., Charpier, S.: Deep layer somatosensory cortical neurons initiate spike-and-wave discharges in a genetic model of absence seizures. J. Neurosci. **27**, 6590–6599 (2007)

Pozo-Vazquez, D., Esteban-Parra, M.J., Rodrigo, F.S., Castro-Diez, Y.: The association between ENSO and winter atmospheric circulation and temperature in the North Atlantic region. J. Climate. **14**, 3408–3420 (2001)

Prokhorov, M.D., Ponomarenko, V.I., Gridnev, V.I., Bodrov, M.B., Bespyatov, A.B.: Synchronization between main rhythmic processes in the human cardiovascular system. Phys. Rev. E. **68**, 041913 (2003)

Prusseit, J., Lehnertz, K.: Measuring interdependences in dissipative dynamical systems with estimated Fokker-Planck coefficients. Phys. Rev. E. **77**, 041914 (2008)

Rivlin-Etzion, M., Marmor, O., Heimer, G., Raz, A., Nini, A., Bergman, H.: Basal ganglia oscillations and pathophysiology of movement disorders. Curr. Opin. Neurobiol. **16**, 629–637 (2006)

Rogers, J.C.: The association between the North Atlantic Oscillation and the Southern Oscillation in the North Hemisphere. Mon. Wea. Rev. **112**, 1999–2015 (1984)

Romano, M.C., Thiel, M., Kurths, J., Grebogi, C.: Estimation of the direction of the coupling by conditional probabilities of recurrence. Phys. Rev. E. **76**, 036211 (2007)

Rosenblum, M.G., Cimponeriu, L., Bezerianos, A., et al. Identification of coupling direction: Application to cardiorespiratory interaction. Phys. Rev. E. **65**, 041909 (2002)

Rosenblum, M.G., Pikovsky, A.S., Kurths, J., Schaefer, C., Tass, P.A.: Phase synchronization: from theory to data analysis. In: Moss, F., Gielen, S. (eds.) Neuro-Informatics. Handbook of Biological Physics, vol. 4, pp. 279–321. Elsevier Science, New York (2001)

Rosenblum, M.G., Pikovsky, A.S.: Detecting direction of coupling in interacting oscillator. Phys. Rev. E. **64**, 045202(R) (2001)

Sato, M., Hansen, J.E., McCormick, M.P., Pollack, J.B.: Stratospheric aerosol optical depths, 1850–1990. J. Geophys. Res. **98**, 22987–22994 (1993)

Schelter, B., Winterhalder, M., Eichler, M., Peifer, M., Hellwig, B., Guschlbauer, B., Luecking, C.H., Dahlhaus, R., Timmer, J.: Testing for directed influences among neural signals using partial directed coherence. J. Neurosci. Methods. **152**, 210–219 (2006)

Schiff, S.J., So, P., Chang, T., Burke, R.E., Sauer, T.: Detecting dynamical interdependence and generalized synchrony through mutual prediction in a neural ensemble. Phys. Rev. E. **54**, 6708–6724 (1996)

Schreiber, T., Schmitz, A.: Improved surrogate data for nonlinearity tests. Phys. Rev. Lett. **77**, 635–638 (1996)

Schwarz, G.: Estimating the order of a model. Ann. Stat. **6**, 461–464 (1978)

Seidenbecher, T., Staak, R., Pape, H.C.: Relations between cortical and thalamic cellular activities during absence seizures in rats. Eur. J. Neurosci. **10**, 1103–1112 (1998)

Sitnikova E and van Luijtelaar, G.: Cortical and thalamic coherence during spike-wave seizures in WAG/Rij rats. Epilepsy Res. **71**, 159–180 (2006)

Sitnikova, E., Dikanev, T., Smirnov, D., Bezruchko, B., van Luijtelaar, G.: Granger causality: Cortico-thalamic interdependencies during absence seizures in WAG/Rij rats. J. Neurosci. Meth. **170**, 245–254 (2008)

Smirnov, D., Barnikol, U.B., Barnikol, T.T., Bezruchko, B.P., Hauptmann, C., Buehrle, C., Maarouf, M., Sturm, V., Freund, H.-J., Tass, P.A.: The generation of Parkinsonian tremor as revealed by directional coupling analysis. Europhys. Lett. **83**, 20003 (2008)

Smirnov, D., Schelter, B., Winterhalder, M., Timmer, J.: Revealing direction of coupling between neuronal oscillators from time series: Phase dynamics modeling versus partial directed coherence. Chaos. **17**, 013111 (2007)

Smirnov, D.A., Andrzejak, R.G.: Detection of weak directional coupling: phase dynamics approach versus state space approach. Phys. Rev. E. **71**, 036207 (2005)

Smirnov, D.A., Bezruchko, B.P.: Estimation of interaction strength and direction from short and noisy time series. Phys. Rev. E. **68**, 046209 (2003)

Smirnov, D.A., Bodrov, M.B., Perez Velazquez, J.L., Wennberg, R.A., Bezruchko, B.P.: Estimation of coupling between oscillators from short time series via phase dynamics modeling: limitations and application to EEG data. Chaos. **15**, 024102 (2005)

Staniek, M., Lehnertz, K.: Symbolic transfer entropy. Phys. Rev. Lett. **100**, 158101 (2008)

Steriade, M.: Sleep, epilepsy and thalamic reticular inhibitory neurons. Trends Neurosci. **28**, 317–324 (2005)

Tass, P.A. A model of desynchronizing deep brain stimulation with a demand-controlled coordinated reset of neural subpopulations. Biol. Cybern. **89**, 81–88 (2003)

Tass, P.A., Fieseler, T., Dammers, J., Dolan, K., Morosan, P., Majtanik, M., Boers, F., Muren, A., Zilles, K., Fink, G.R. Synchronization tomography: a method for three-dimensional localization of phase synchronized neuronal populations in the human brain using magnetoencephalography. Phys. Rev. Lett. **90**, 088101 (2003)

Tass, P.A., Majtanik, M.: Long-term anti-kindling effects of desynchronizing brain stimulation: a theoretical study. Biol. Cybern. **94**, 58–66 (2006)

Tass, P.A.: Phase Resetting in Medicine and Biology. Springer, Berlin (1999)

van Luijtelaar, E.L.J.M., Coenen, A.M.L.: Two types of electrocortical paroxysms in an inbred strain of rats. Neurosci. Lett. **70**, 393–397 (1986)

Verdes, P.F.: Assessing causality from multivariate time series. Phys. Rev. E. **72**, 026222 (2005)

Vergnes, M., Marescaux, C., Depaulis, A., Micheletti, G., Warter, J.M.: Spontaneous spike and wave discharges in thalamus and cortex in a rat model of genetic petit mal-like seizures. Exp. Neurol. **96**, 127–136 (1987)

Wallace, J.M., Gutzler, D.S.: Teleconnections in the geopotential height field during the northern hemisphere winter. Mon. Wea. Rev. **109**, 784–812 (1981)

Wang, S., Chen, Y., Ding, M., Feng, J., Stein, J.F., Aziz, T.Z., Liu, X.J.: Revealing the dynamic causal interdependence between neural and muscular signals in Parkinsonian tremor. J. Franklin Inst. **344**, 180–195 (2007)

Wang, W., Anderson, B.T., Kaufmann, R.K., Myneni, R.B. The relation between the North Atlantic Oscillation and SSTs in North Atlantic basin. J. Climate. **17**, 4752–4759 (2004)

Chapter 13
Outdoor Examples

In this final chapter, we illustrate different steps of the procedure of modelling from time series in more detail. For that, we use examples from the fields of electronics (Sect. 13.1), physiology (Sect. 13.2) and climatology (Sect. 13.3). They are presented in the order of decreasing amount of a priori information about an object: an appropriate model structure is completely known and only model parameters are estimated from data (Sect. 13.1); an appropriate model structure is partly known (Sect. 13.2); no specific ideas about suitable model equations are available (Sect. 13.3). For the sake of unity, we formulate the same purpose of empirical modelling in all the three cases, namely identification of directional couplings between the processes under study. This task has also been considered in the previous chapter, where a compact and more technical description of several techniques and applications has been given.

13.1 Coupled Electronic Generators

13.1.1 Object Description

An experimental object is a system of two self-sustained generators (Fig. 13.1) similar to that described in Dmitriev and Kislov (1989); Dmitriev et al. (1996). Both generators are constructed according to the same scheme and contain an RC low-pass filter (an element 1 in Fig. 13.1a), an RLC filter (an oscillatory circuit 2) and a non-linear element (an element 3) connected in a ring (Ponomarenko et al., 2004). The non-linear element with a quadratic transfer characteristic $U_{out} = A - B \cdot U_{in}^2$ consists of an electronic multiplier, which performs the operation of taking a squared value, and a summing amplifier, which adds the parameter A with a necessary sign to an output signal of the multiplier (Fig. 13.1b). Here, U_{out} is a voltage at the output of the non-linear element, B is a dimensional coefficient whose value is determined by the parameters of the electronic multiplier. A is used as a governing parameter. Under variations in A, one observes a transition to chaos via the cascade of period-doubling bifurcations in each generator.

B.P. Bezruchko, D.A. Smirnov, *Extracting Knowledge From Time Series*, Springer Series in Synergetics, DOI 10.1007/978-3-642-12601-7_13,
© Springer-Verlag Berlin Heidelberg 2010

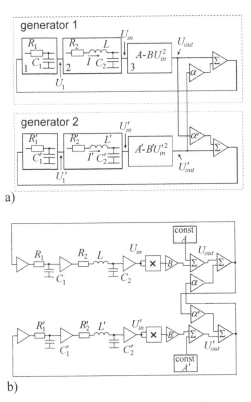

a)

b)

Fig. 13.1 Experimental set-up: (**a**) a more vivid block scheme; (**b**) a more detailed version. The parameters are $R_1 = 1000\,\Omega$, $R_2 = 60\,\Omega$, $C_1 = C_2 = 0.022\,\mu\text{F}$, $L = 6\,\text{mH}$, $B = 0.2\,\text{V}^{-1}$. Parameters of the generator 2 are the same up to an error of about 10%. A and A' control non-linear transfer functions of the non-linear elements, α and α' are coupling coefficients and Σ are summing amplifiers

Further, we denote the quantities relating to the second generator by a prime, while those for the first generator are not supplied with a prime. Interaction between the generators is possible due to the summing amplifiers Σ and the amplifiers with controlled gain factors α and α'. The latter ones serve to set the "interaction strength". By specifying different values of α and α', one provides bidirectional, unidirectional or zero coupling between the generators. Below, we describe the cases of uncoupled generators ($\alpha = \alpha' = 0$) and unidirectional coupling $2 \rightarrow 1$ ($\alpha \neq 0, \alpha' = 0$) for different values of A and A'. The settings considered are summarised in Table 13.1, where the corresponding dynamical regimes are identified as described below.

In what follows, the values of the parameters (such as inductance, capacity, resistance) in both generators are not regarded as a priori known. Only the equivalent electric scheme of Fig. 13.1 is considered as a priori known. Coupling character is revealed from data only with the use of the latter information.

Table 13.1 Parameters values and corresponding dynamical regimes considered. T is a characteristic timescale of the autonomous self-sustained oscillations (it is about 0.08 ms as described in Sect. 13.1.2). Namely, T corresponds to the highest peak in the power spectrum

| Trial no. | Parameters | | | | Dynamical regimes (Ponomarenko et al., 2004) |
	A, V	A', V	α	α'	
1	4.8	3.8	0	0	Cycle of the period $1T$
2	4.8	3.8	0.05	0	Torus in the generator 1, cycle of the period $1T$ in the generator 2
3	4.8	3.8	0.9	0	Cycle of the period $2T$ in the generator 1, cycle of the period $1T$ in the generator 2
4	4.8	5.8	0.1	0	Chaos in both generators
5	4.8	5.8	0	0	Cycle of the period $1T$ in the generator 1, chaos in the generator 2
6	8.8	7.4	0	0	Chaos in both generators
7	8.8	7.4	0.1	0	Chaos in both generators

13.1.2 Data Acquisition and Preliminary Processing

Observed quantities x and x' are linearly dependent on the voltages U_{in} and U'_{in} : $x = a + U_{in}/b$ and $x' = a' + U'_{in}/b'$. The shift and scaling parameters a, b, a', b' are some constants adjusted so as to provide a maximally efficient usage of the dynamics range of a 12-bit analogue-to-digital converter exploited for the measurements. The values of a, b, a', b' are not measured, since it would complicate the experimental set-up. The sampling frequency is 1.2 MHz which corresponds approximately to 100 data points per characteristic oscillation period.

Time series of x and x' is shown in Fig. 13.2a, b for the case of unidirectional coupling $2 \rightarrow 1$ (Table 13.1, trial 4). The driving generator 2 exhibits a chaotic regime (see below). The generator 1 without coupling would demonstrate a periodic regime. The quantities x and x' are presented in arbitrary units. Namely, the raw numerical values at the output of the ADC are integers and cover the range approximately from -2000 to 2000. For the sake of convenience, we divide them by 1000 and get the signals x and x' with the oscillation amplitudes of the order of 1. A rationale behind such scaling is that making typical values of all the analysed quantities of the order of unity allows the reduction of the computational errors (induced by the truncation) during the model fitting.

Estimates of the power spectra (Sect. 6.4.2) and the autocorrelation functions (Sect. 4.1.2) are shown in Fig. 13.2c–h. The highest peak is observed at the frequency of 12.36 kHz for the generator 2 (Fig. 13.2d). It corresponds to the timescale of 0.08 ms (about 100 data points), which is the distance between successive maxima in Fig. 13.2b. The peak is quite well pronounced, but it is somewhat wider than that for a quasi-periodic regime shown in Fig. 6.14b and, especially, for a periodic regime in Fig. 6.14a. Additional peaks are seen at 6.2 kHz (the first subharmonic, which is manifested as the alternation of higher and lower maxima in Fig. 13.2b), 18.6, 24.7, 30.9 and 37.1 kHz (overtones induced by the non-linearity of the

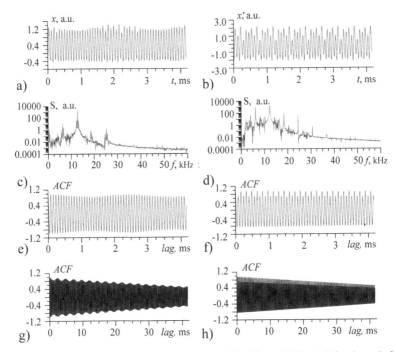

Fig. 13.2 Time series and their basic characteristics (Table 13.1, trial 4): the left column is for the generator 1; the right column is for the generator 2. Observed time series are shown in the panels (**a**) and (**b**), their periodograms in (**c**) and (**d**); their autocorrelation functions in (**e**) and (**f**) and, at different timescale, in (**g**) and (**h**). Both ACFs and periodograms are estimated from signals of length of 10^5 data points, rather than from the short pieces consisting of 5000 data points shown in the panels (**a**) and (**b**)

system and manifested as deviations of the temporal profile from a sinusoidal wave). The corresponding ACF exhibits a strong periodic component with a characteristic period of 0.08 ms and its subharmonic with a period twice as large. The ACF decreases with the time lag (Fig. 13.2h), but the rate of its decay is low: a linear envelope of the ACF reveals that the autocorrelations decrease from 1 to 0.5 over 40 ms (500 characteristic periods).

Projections of the orbits onto the plane of time-delayed coordinates are given in Fig. 13.3a, b. From Fig. 13.3b one can suspect a complex structure similar to a projection of a chaotic attractor e.g., like in Fig. 6.15. The regime observed in the generator 2 is, indeed, identified as chaotic, since we have evidenced its birth by tracing the evolution of the phase orbit of the generator under the parameter change (Ponomarenko et al., 2004). The regime under consideration has been established after a cascade of period-doubling bifurcations. We do not go into further details of the dynamical regime identification, since detection of chaos, estimation of fractal dimensions and similar questions are redundant for the formulated purpose of the model construction and the coupling estimation. In this example, a model dimension and even a complete structure of model equations are specified from physical considerations as presented below.

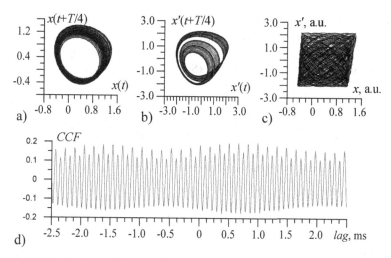

Fig. 13.3 Additional characteristics of the dynamics illustrated in Fig. 13.2: projections of the data onto different planes are shown in the panels (**a**), (**b**) and (**c**); cross-correlation function between $x(t)$ and $x'(t)$ is shown in panel (**d**)

The power spectrum and the ACF for the generator 1 are shown in Fig. 13.2c, e, g. A clear peak is observed at the frequency of 12.72 kHz (a natural frequency of the periodic oscillations in the generator 1). The satellite peaks observed at the close frequencies of 12.36 and 13.08 kHz are induced by the influence $2 \rightarrow 1$ and manifested as a periodic component of the ACF envelope with a period corresponding to the frequency mismatch of 0.36 kHz. The peaks at 6.2, 18.6 and 24.7 kHz are also seen in Fig. 13.2c. The ACF would not decrease for the non-driven generator 1; it decreases here due to the influence of the chaotic generator 2.

Traditional characteristics of the interdependence between the signals $x(t)$ and $x'(t)$ are illustrated in Fig. 13.3c, d. Figure 13.3c shows that a projection of the data onto the plane (x, x') fills almost a rectangular region. It means that even if the interdependence is present, it is not strong. This is confirmed by the cross-correlation function (Sect. 6.4.5) which takes absolute values less than 0.2, rather than close to 1, as illustrated in Fig. 13.3d.

13.1.3 Selection of the Model Equation Structure

Let us consider a single generator in the case of zero couplings. From Kirchhoff's laws, one can write down the following set of three ordinary first-order differential equations to model variations in currents and voltages:

$$
\begin{aligned}
C_1 \, dU_1/dt &= (U_{out} - U_1)/R_1, \\
C_2 \, dU_{in}/dt &= I, \\
L \, dI/dt &= U_1 - U_{in} - R_2 I.
\end{aligned}
\tag{13.1}
$$

All the variables entering the model are indicated in Fig. 13.1. Since only the values of U_{in} are observed experimentally, it is desirable to rewrite the equations in terms of this quantity and its derivatives. Thereby, one avoids coping with hidden variables, which is often a difficult task (Sect. 8.2). Via some algebraic manipulations, equation (13.1) can be equivalently rewritten as

$$\frac{d^3 U_{in}}{dt^3} = -\left(\frac{1}{R_1 C_1} + \frac{R_2}{L}\right)\frac{d^2 U_{in}}{dt^2} - \frac{1}{LC_2}\left(1 + \frac{R_2 C_2}{R_1 C_1}\right)\frac{dU_{in}}{dt} + \frac{A - U_{in} - BU_{in}^2}{LC_2 R_1 C_1}.$$

$$(13.2)$$

For two generators, one gets the set of six first-order equations:

$$C_1 \frac{dU_1}{dt} = \left(U_{out} - U_1 - \alpha U'_{out}\right)/R_1, \; C'_1\frac{dU'_1}{dt} = \left(U'_{out} - U'_1 - \alpha' U_{out}\right)/R'_1,$$

$$C_2 \frac{dU_{in}}{dt} = I, \qquad\qquad\qquad C'_2\frac{dU'_{in}}{dt} = I',$$

$$L\frac{dI}{dt} = U_1 - U_{in} - R_2 I, \qquad\qquad L'\frac{dI'}{dt} = U'_1 - U'_{in} - R'_2 I',$$

$$(13.3)$$

where the terms $\alpha U'_{out}$ and $\alpha' U_{out}$ describe the couplings between the generators. The equations can be rewritten in terms of the variables U_{in}, U'_{in} as

$$\frac{d^3 U_{in}}{dt^3} = -\left(\frac{1}{R_1 C_1} + \frac{R_2}{L}\right)\frac{d^2 U_{in}}{dt^2} - \frac{1}{LC_2}\left(1 + \frac{R_2 C_2}{R_1 C_1}\right)\frac{dU_{in}}{dt}$$

$$+ \frac{A - U_{in} - BU_{in}^2}{LC_2 R_1 C_1} + \frac{\alpha\left(A' - B'U_{in}'^2\right)}{LC_2 R_1 C_1},$$

$$\frac{d^3 U'_{in}}{dt^3} = -\left(\frac{1}{R'_1 C'_1} + \frac{R'_2}{L'}\right)\frac{d^2 U'_{in}}{dt^2} - \frac{1}{L'C'_2}\left(1 + \frac{R'_2 C'_2}{R'_1 C'_1}\right)\frac{dU'_{in}}{dt}$$

$$+ \frac{A' - U'_{in} - B'U_{in}'^2}{L'C'_2 R'_1 C'_1} + \frac{\alpha'\left(A - BU_{in}^2\right)}{L'C'_2 R'_1 C'_1}.$$

$$(13.4)$$

To estimate model parameters from a time series of $x(t)$ and $x'(t)$, we specify the model structure as

$$\frac{d^3 x}{dt^3} = c_1 \frac{d^2 x}{dt^2} + c_2 \frac{dx}{dt} + c_3 + c_4 x + c_5 x^2 + c_6 x' + c_7 x'^2, \qquad (13.5)$$

$$\frac{d^3 x'}{dt^3} = c'_1 \frac{d^2 x'}{dt^2} + c'_2 \frac{dx'}{dt} + c'_3 + c'_4 x' + c'_5 x'^2 + c'_6 x + c'_7 x^2, \qquad (13.6)$$

where the terms $c_6 x'$, $c'_6 x$ absent from equation (13.4) are introduced to allow for the scaling and shift parameters a, b, a', b' in the measurement procedure. The unknown values of a, b, a', b' are not important, since they affect only the numerical values of the model coefficients c_k, c'_k, rather than model behaviour and relative approximation error considered below. One can infer the presence of the influence $2 \rightarrow 1$ and $1 \rightarrow 2$ by comparing the models (13.5) and (13.6) with the models which do not allow for the coupling terms and read as

$$\frac{d^3 x}{dt^3} = c_1 \frac{d^2 x}{dt^2} + c_2 \frac{dx}{dt} + c_3 + c_4 x + c_5 x^2 \tag{13.7}$$

and

$$\frac{d^3 x'}{dt^3} = c_1' \frac{d^2 x'}{dt^2} + c_2' \frac{dx'}{dt} + c_3' + c_4' x' + c_5' x'^2. \tag{13.8}$$

More precisely, we use two model-based approaches to reveal couplings. Firstly, we compare approximation errors for the individual models (13.7) or (13.8) to the errors for the joint models (13.5) or (13.6), respectively. This is similar to the characterisation of the Granger causality (Sect. 12.1). Secondly, we use a criterion which unites the analysis of couplings with the model validation and can be called a "free-run version" of the Granger causality. Namely, if an empirical model of the form (13.7) cannot reproduce the observed dynamics of x, while a model of the form (13.5) can do it, then one infers the presence of the influence $2 \rightarrow 1$ and its considerable effect on the dynamics. If already the model (13.5) describes the dynamics of x satisfactorily and the model (13.7) does not improve anything, then the influence $2 \rightarrow 1$ is insignificant. A similar comparison of the models (13.8) and (13.6) applies to the detection of the influence $1 \rightarrow 2$. If even the models (13.5) and (13.6) cannot adequately describe the observed dynamics, then the physical ideas behind the model equations are invalid and one must seek for other model structures.

13.1.4 Model Fitting, Validation and Usage

13.1.4.1 Individual Models

Let us start with the modelling of the generator 2 with the aid of the "individual" equation (13.8). We use the chaotic time series $x'(t)$ shown in Fig. 13.2b as a training time series. It is of a moderate length: $N = 5000$ data points, i.e. about 50 basic periods. The model (13.8) should be appropriate, since the generator 2 is not influenced by the generator 1. However, a complete validity of the model structure is not trivial, since the basic equation (13.2) is derived under the assumptions of strictly quadratic non-linearity (though the transfer characteristic follows a quadratic parabola with errors of about 1%), a constant inductance L (though it is realised with a ferrite core so that nonlinear properties might be observed at big oscillation amplitudes), etc.

According to the procedure described in the beginning of Chap. 8 and in Sect. 9.1, one first performs numerical differentiation of the signal $x'(t)$ to get the estimates

$$\frac{d\hat{x}'}{dt}, \frac{d^2\hat{x}'}{dt^2}, \frac{d^3\hat{x}'}{dt^3}$$

of the derivatives

$$\frac{dx'}{dt}, \frac{d^2x'}{dt^2}, \frac{d^3x'}{dt^3}$$

entering the model equation (13.8). We use a digital smoothing polynomial (Sect. 7.4.2): the estimates

$$\frac{d\hat{x}'(t_n)}{dt}, \frac{d^2\hat{x}'(t_n)}{dt^2}, \frac{d^3\hat{x}'(t_n)}{dt^3}$$

and a smoothed signal $\hat{x}'(t_n)$ at a time instant t_n are obtained from an algebraic polynomial of an order L fitted to the time window $\{t_n - m\Delta t; t_n + m\Delta t\}$ consisting of $2m + 1$ data points. Let us note the following purely technical detail. For the sake of convenience, we rescale the time units similar to the above scaling of the variables x and x': the time units are selected so as to provide the values of the derivatives of the order of 1. This is achieved if the angular frequency of the oscillations becomes of the order of 1. Thus, we define the time units so that sampling interval (which equals $1/1.2$ μs in the physical units) gets equal to $\Delta t = 0.1$.

Secondly, the model coefficients are estimated via minimisation of the error

$$\varepsilon^2 = \frac{1}{N - 2m} \sum_{i=m+1}^{N-m} \left[\frac{d^3\hat{x}'(t_i)}{dt^3} - c_1' \frac{d^2\hat{x}'(t_i)}{dt^2} - c_2' \frac{d\hat{x}'(t_i)}{dt} - c_3' - c_4'\hat{x}'(t_i) - c_5'\hat{x}'^2(t_i) \right]^2$$

The minimal relative approximation error is

$$\varepsilon_{rel} = \sqrt{\min_{\{c_k'\}} \varepsilon^2 \Big/ \text{var}[d^3\hat{x}'/dt^3]},$$

where "var" stands for the sample variance (Sect. 2.2.1) of the argument. Next, a free-run behaviour of an obtained model is simulated. A projection of a model phase orbit onto the plane $(x', dx'/dt)$ is compared to the corresponding projection of an observed data. For quantitative comparison, we compute a prediction time, i.e. a time interval τ_{pred}, over which the prediction error $\sigma(\tau) =$

$$\sqrt{\left\langle \left(x'(t_0 + \tau) - x'_{pred}(t_0 + \tau) \right)^2 \right\rangle_{t_0}}$$ (Sect. 2.2.4) rises up to $0.05\sqrt{\text{var}[x']}$. Here,

$x'_{pred}(t_0 + \tau)$ is a prediction at a time instant $t_0 + \tau$ obtained by simulation of a model orbit from the initial condition $\hat{x}'(t_0), d\hat{x}'(t_0)/dt, d^2\hat{x}'(t_0)/dt^2$ at the initial time instant t_0. A test time series of the length of 10000 data points (which is a continuation of the data segment shown in Fig. 13.2b) is used both for the qualitative comparison and for the prediction time estimation.

Table 13.2 Characteristics of the models (13.8) for different m and L

m	L	$\hat{c}'_1$	$\hat{c}'_2$	$\hat{c}'_3$	$\hat{c}'_4$	$\hat{c}'_5$	ε_{rel}	τ_{pred}
2	3	−15.9	−1.71	−2.20	−6.09	1.09	0.45	$0.05T$
5	3	−0.40	−0.44	−0.19	−0.15	0.12	0.78	$0.25T$
12	3	−0.42	−0.42	−0.17	−0.15	0.10	0.07	$0.25T$
40	3	−0.30	−0.20	−0.04	−0.05	0.03	0.11	$0.07T$
40	7	−0.43	−0.42	−0.18	−0.17	0.11	0.03	$0.71T$

We try different values of the parameters $L \geq 3$ and $m \geq L/2$. Model quality essentially depends on m and L, since these parameters determine the errors in the derivative estimates and, hence, the errors in the estimates of the model coefficients. The model coefficients and some characteristics are shown in Table 13.2. Thus, small values of m are not sufficient to reduce the noise influence so that the derivative estimates are very noisy as seen in Fig. 13.4 for $m = 2$ and $L = 3$. Both the second and the third derivatives are strongly noise corrupted. The corresponding model is completely invalid (Fig. 13.5); it exhibits a stable fixed point instead of chaotic oscillations (Fig. 13.5b, c), a very large approximation error $\varepsilon_{rel} = 0.45$ and a small prediction time $\tau_{pred} = 0.05T$, where the characteristic period is $T \approx 100\Delta t$ (Fig. 13.5a, d). Greater values of m at fixed $L = 3$ allow some noise reduction. The value of $m = 5$ gives a reasonable temporal profile of the second derivative but still noisy fluctuations in the third derivative (Fig. 13.6). The resulting model is better, but still invalid, since it exhibits a periodic behaviour rather than a chaotic one (Fig. 13.7). The value of $m = 12$ gives the best results: reasonable profiles of all the derivatives (Fig. 13.8), a much smaller approximation error $\varepsilon_{rel} = 0.07$ and a chaotic dynamics which is qualitatively very similar to the observed one (Fig. 13.9). A further increase in m worsens the results. Thus, at $m = 40$ the derivative estimates look even "smoother" (Fig. 13.10b–d) and the signal itself becomes somewhat distorted (Fig. 13.10a shows the difference between the original data and a smoothed signal $\hat{x}'(t)$). Hence, the random errors in the derivative estimates are smaller, but

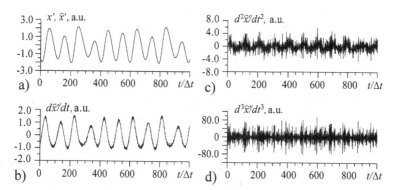

Fig. 13.4 Numerical differentiation of the signal $x'(t)$ with a digital smoothing polynomial at $m = 2$, $L = 3$: (**a**) an original signal $x'(t)$ and its smoothed estimate $\hat{x}'(t)$ fully coincide at this scale; (**b**)–(**d**) the derivatives of increasing order

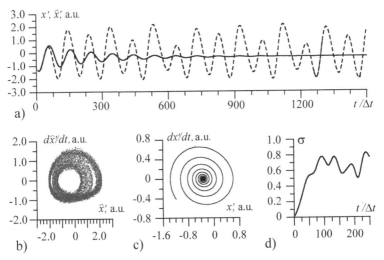

Fig. 13.5 Behaviour of an empirical model (13.8) obtained at $m = 2$, $L = 3$: (**a**) the test time series $x'(t)$ (the *dashed line*) and the model time realisation at the same initial conditions (the *solid line*); (**b**) a projection of the original data onto the plane $(\hat{x}', d\hat{x}'/dt)$; (**c**) a projection of the model phase orbit onto the same plane; (**d**) a model prediction error σ for the prediction $t/\Delta t$ sampling intervals ahead

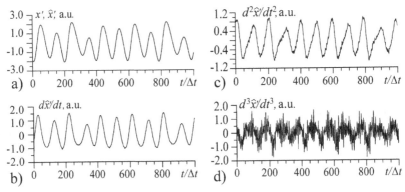

Fig. 13.6 Numerical differentiation of the signal $x'(t)$ at $m = 5$, $L = 3$. Notations are the same as in Fig. 13.4

there appears a significant bias. The corresponding model exhibits an approximation error greater than at $m = 12$ and a periodic behaviour (Fig. 13.11), which is not similar to the observed dynamics.

Big values of m can be used in combination with big values of L to allow the smoothing polynomial to reproduce the signal waveform in a wider window $\{t_n - m\Delta t; t_n + m\Delta t\}$. In particular, the pair $m = 40$ and $L = 7$ gives a model with chaotic behaviour and the best prediction time of $0.7T$ (even this prediction time is quite moderate, which is not surprising for a chaotic regime). However, a chaotic attractor of this model is less similar to the observed dynamics (it does not

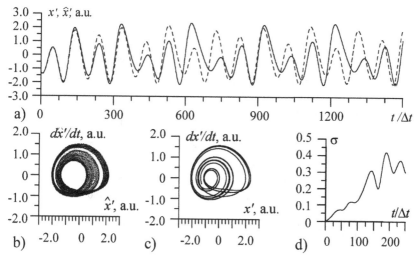

Fig. 13.7 Behaviour of an empirical model (13.8) obtained at $m = 5$, $L = 3$. Notations are the same as in Fig. 13.5

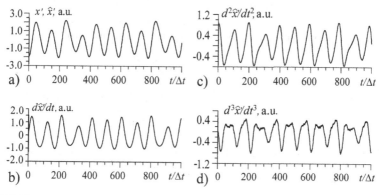

Fig. 13.8 Numerical differentiation of the signal $x'(t)$ at $m = 12$, $L = 3$. Notations are the same as in Fig. 13.4

exhibit a gap between the two bands seen in Fig. 13.9b, c) than that for the model obtained at $m = 12$, $L = 3$. However, the coefficients of both models are close to each other (Table 13.2). Below, we describe modelling of the coupled generators at $m = 12$, $L = 3$, since this choice has provided the best qualitative description of the chaotic dynamics of the generator 2. Still, both $m = 12$, $L = 3$ and $m = 40$, $L = 7$ appear to give very similar results of the coupling analysis.

13.1.4.2 Modelling of the Coupled Generators

The derivatives of $x(t)$ for the trial 4 are estimated also at $m = 12$, $L = 3$. Fitting the model equations (13.6), which involve the data from the generator 1 to describe

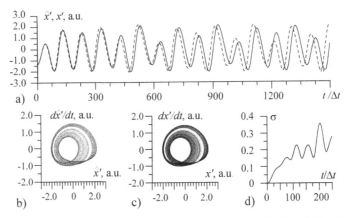

Fig. 13.9 Behaviour of an empirical model (13.8) obtained at $m = 12$, $L = 3$. Notations are the same as in Fig. 13.5

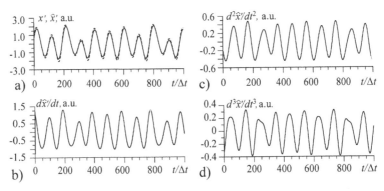

Fig. 13.10 Numerical differentiation of the signal $x'(t)$ at $m = 40$, $L = 3$. Notations are the same as in Fig. 13.4. An original signal $x'(t)$ in the panel a (the *dashed line*) somewhat differs from its smoothed estimate $\hat{x}'(t)$ (the *solid line*)

the time series of the generator 2, gives an approximation error $\varepsilon_{rel} = 0.07$, i.e. the same as for the individual model (13.8). The model coefficients responsible for the influence $1 \rightarrow 2$ appear close to zero: $\hat{c}'_6 = -0.001$ and $\hat{c}'_7 = 0.001$. The other coefficients are very close to the corresponding coefficients of the individual model. Thus, no quantitative improvement is observed under the use of the model (13.6) as compared to the model (13.8), i.e. one cannot see any signs of the influence $1 \rightarrow 2$.

As for the modelling of the generator 1, the results for the individual model (13.7) and the joint model (13.5) are reported in Table 13.3. One can see the reduction in the approximation error by 20% and the coupling coefficient estimate $\hat{c}_7$, which is not as small as that for the influence $1 \rightarrow 2$. Thus, some signs of coupling $2 \rightarrow 1$ are observed already from these characteristics. The presence of coupling and its unidirectional character becomes completely obvious when a free-run behaviour of different models is considered.

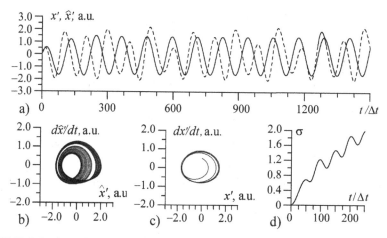

Fig. 13.11 Behaviour of an empirical model (13.8) obtained at $m = 40$, $L = 3$. Notations are the same as in Fig. 13.5

Table 13.3 Characteristics of models (13.5) and (13.7) for the generator 1

Model	$\hat{c}_1$	$\hat{c}_2$	$\hat{c}_3$	$\hat{c}_4$	$\hat{c}_5$	$\hat{c}_6$	$\hat{c}_7$	ε_{rel}
(13.7)	−0.28	−0.43	0.03	−0.23	0.13	–	–	0.15
(13.5)	−0.32	−0.43	0.03	−0.25	0.13	−0.002	0.006	0.12

Projections of the original and model phase orbits onto different planes are presented in Fig. 13.12. The observed dynamics of the generator 1 is illustrated in Fig. 13.12a. The individual model (13.7) gives a periodic regime (Fig. 13.12b), i.e. it is inadequate. The model allowing for unidirectional coupling $2 \rightarrow 1$, i.e. Eqs. (13.5) and (13.8), exhibits a chaotic attractor very similar to the observed behaviour (Fig. 13.12c). A model with bidirectional coupling, i.e. Eqs. (13.5) and (13.6), does not give any further improvement (Fig. 13.12d). The results for the generator 2 are presented similarly in the second row. Already the individual model (13.8) adequately reproduces the dynamics (cf. Fig. 13.12e and f). A model with a unidirectional coupling $1 \rightarrow 2$, i.e. Eqs. (13.7) and (13.6), and a model with a bidirectional coupling demonstrate the same behaviour (Fig. 13.12g, h). Similar conclusions are made from the projections onto the plane (x, x'): the model with unidirectional coupling $1 \rightarrow 2$ is insufficient to reproduce the dynamics qualitatively (cf. Fig. 13.12i and j), while the model with unidirectional coupling $2 \rightarrow 1$ (Fig. 13.12k) and the model with bidirectional coupling (Fig. 13.12l) exhibit the same dynamics similar to the observed one.

Thus, via the analysis of a free-run behaviour of the empirical models and their approximation errors, we infer from data that the unidirectional coupling scheme $2 \rightarrow 1$ is realised in the trial 4. This is a correct conclusion. In this manner, the global modelling helps both to get an adequate model, when parameters of the circuits are unknown, and to reveal the coupling scheme. Similar results are observed

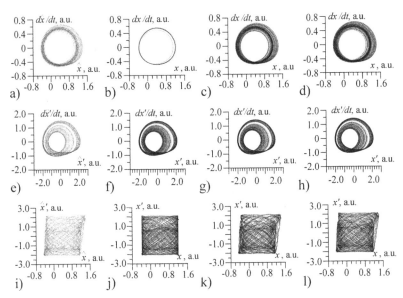

Fig. 13.12 Comparison of the original data and a behaviour of different empirical models obtained at $m = 12$, $L = 3$: (**a**), (**e**), (**i**) projections of the original data for the test time series; (**b**) a model (13.7), no coupling; (**c**), (**k**) models (13.5) and (13.8), i.e. a unidirectional coupling $2 \to 1$ is allowed for; (**d**), (**h**), (**l**) models (13.5) and (13.6), i.e. a bidirectional coupling is allowed for; (**f**) a model (13.8), no coupling; (**g**), (**j**) models (13.7) and (13.6), i.e. a unidirectional coupling $1 \to 2$ is allowed

for all seven trials (Table 13.1). The results in terms of the approximation error are summarised in Fig. 13.13, where we use $m = 40$, $L = 7$, since this choice always gives a good description of the dynamics, while the choice of $m = 12$, $L = 3$ is the best one for the trial 4 and a couple of other cases (anyway, the results are very

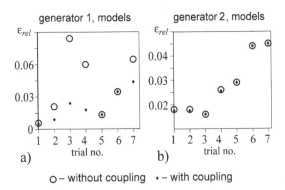

Fig. 13.13 Relative approximation errors for the models without coupling (*circles*) and allowing for a unidirectional coupling (rhombs) at $m = 40$, $L = 7$: (**a**) the models (13.7) and (13.5) for the generator 1; (**b**) the models (13.8) and (13.6) for the generator 2

similar for these two choices of the pair m,L). A unidirectional coupling $2 \rightarrow 1$ is correctly detected in the trials 2, 3, 4 and 7: the reduction in the approximation error is observed for the model (13.5), since the rhombs are located considerably lower than the circles. No coupling is detected in the other three trials, which is also correct. The same conclusions are obtained when we compare the prediction times or consider the projections of the model phase orbits (not shown).

Finally, we note that the trial 3 corresponds to a phase synchronisation regime: the phases of the generators are locked, but variations in their instantaneous amplitudes are a bit different due to the non-identity of the circuits (a slight parameter mismatch). Coupling character cannot be revealed from the locked phases with the phase dynamics modelling technique (Sect. 12.2). The Granger causality estimation may also face difficulties, since an inappropriate model structure in the case of a strong unidirectional coupling may lead to a spurious detection of a bidirectional coupling. Fitting the models (13.5) and (13.6) allows one to reveal the coupling character correctly even in this case due to the adequate model structure (and due to non-identity of the amplitude variations). This is an advantage of the situation when detailed a priori information about an object is available: a further useful knowledge can be extracted then from minimally informative experimental data.

Thus, with the electronic example we have illustrated the full procedure of constructing an empirical model under the "transparent box" setting and revealing a coupling character with its help.

13.2 Parkinsonian Tremor

13.2.1 Object Description

As mentioned in Sect. 12.3, one of the manifestations of Parkinson's disease is a strong resting tremor, i.e. regular high-amplitude oscillations of limbs. The mechanism of the parkinsonian tremor is still widely discussed. It is known that populations of neurons located in the thalamus and the basal ganglia fire in a synchronised and periodical manner at a frequency similar to that of the tremor (Lenz et al., 1994; Llinas and Jahnsen, 1982; Pare et al., 1990) as a result of local pathophysiology. Also, there is an important peripheral mechanism involved in the generation of these abnormal oscillations: the receptor properties in the muscle spindles. They contribute to a servo control (closed-loop control of position and velocity) and amplify synchronised input from central nervous structures by strongly synchronised feedback (Stilles and Pozos, 1976). The resulting servo loop oscillations are regarded as a basic mechanism for tremor generation (Stilles and Pozos, 1976). Although subcortical oscillations are not strictly correlated with the tremor (Rivlin-Etzion et al., 2006), it was shown that limb oscillations influence subcortical activity by the proprioceptive feedback from muscle spindles (Eichler, 2006). As yet, it was difficult to reveal empirically whether the parkinsonian tremor is affected by subcortical oscillations (Brown, 2003; Rivlin-Etzion et al., 2006).

For deeper understanding of the mechanism of the parkinsonian tremor generation, it is important to reveal the character of interaction between different brain areas and oscillating limbs. Due to the regularity of the parkinsonian tremor, application of the phase dynamics modelling technique (Sect. 12.2) has appeared fruitful (Bezruchko et al., 2008; Smirnov et al., 2008). Namely, a bidirectional coupling between limb and subcortical oscillations has been detected and time delays in both directions have been estimated. Below, we describe such an analysis of a single tremor epoch in a patient with bilateral resting tremor. The patient had more pronounced pathological oscillations of the left hand.

13.2.2 Data Acquisition and Preliminary Processing

Local field potentials (LFPs) from the subthalamic nucleus (STN) and accelerometer signals, assessing the hand tremor, were recorded simultaneously. It was done by the group of Prof. P. Tass (Institute of Neuroscience and Biophysics – 3, Research Centre Juelich, Germany) and their colleagues at the Department of Stereotaxic and Functional Neurosurgery, University of Cologne. Recordings were performed during or after deep brain stimulation electrode implantation. Intraoperative recordings from the right STN were performed with the ISIS MER system (Inomed, Teningen, Germany). The latter is a "Ben's gun" multi-electrode for acute basal ganglia recordings during stereotaxic operations (Benabid et al., 1987), i.e. an array consisting of four outer electrodes separated by 2 mm from a central one. Proper electrode placement was confirmed by effective high-frequency macro-stimulations, intraoperative X-ray controls (Treuer et al., 2005), postoperative CT scans and intraoperative micro-recordings. The LFP recordings represented voltages on the depth electrodes against a remote reference. The recordings were performed after overnight withdrawal of antiparkinsonian medication. The study was approved by the local ethical committee. The patient gave a written consent.

Accelerometer and LFP signals are denoted further as $x_1(t)$ and $x_2(t)$, respectively, where $t = n\Delta t$, $n = 1, 2, \ldots$, the sampling interval is $\Delta t = 5$ ms. An accelerometer signal from the left hand during an epoch of strong resting tremor (of the length of 83.5 s or 16700 data points) is shown in Fig. 13.14a. The simultaneous LFP recording performed via the central depth electrode is shown in Fig. 13.14b. We describe only the central electrode, since all the results are very similar for all depth electrodes. The accelerometer signal displays a sharp peak in the power spectrum (Fig. 13.14c) at the frequency of 5 Hz. The corresponding spectral peak in the power spectrum of LFP is also observed (Fig. 13.14d). A spectral peak at the tremor frequency is often manifested in the power spectrum of the LFP recorded from the depth electrode contralateral to (i.e. at the opposite side of) the tremor (Brown, 2003; Deuschl et al., 1996; Rivlin-Etzion et al., 2006; Zimmermann et al., 1994).

The parkinsonian resting tremor is highly regular: the peak at the tremor frequency is rather narrow for the accelerometer signal. The peak in the LFP power spectrum is wider. This regularity is further illustrated by the autocorrelation

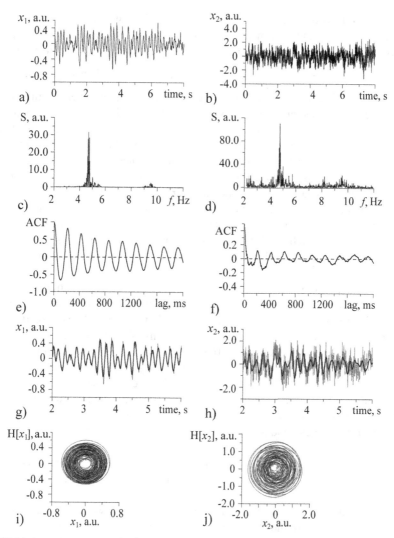

Fig. 13.14 A spontaneous epoch of parkinsonian tremor of length 83.5 s. The left column relates to an accelerometer signal and the right one to a simultaneous LFP signal: (**a**), (**b**) the time series at the beginning of the epoch; (**c**), (**d**) their power spectra estimates (periodograms); (**e**), (**f**) their autocorrelation functions; (**g**), (**h**) magnified segments of the original signals (*grey lines*) and their versions band-pass filtered (*black lines*) in the frequency band 2–9 Hz; (**i**), (**j**) the orbits on the plane "a band-pass filtered signal (2–9 Hz) versus its Hilbert transform"

functions in Fig. 13.14e, f: the ACF of x_1 (Fig. 13.14e) decays much slower than that of x_2 (Fig. 13.14f). A signal with a strong periodic component is often called "phase coherent", since the phases of two orbits with initially the same phase diverge very slowly in time (see, e.g., Pikovsky et al., 2001). In such a case, one speaks also of weak phase diffusion, which is manifested as weak variations in the instantaneous period (a time interval between successive local maxima in a time series). The phase diffusion can be characterised by the coefficient of variation of the instantaneous period $k = \sqrt{\langle (T_i - \langle T_i \rangle)^2 \rangle} / \langle T_i \rangle$, where T_i are the intervals between successive maxima ($i = 1, 2, \ldots$) and angle brackets denote averaging over the time index i. For the accelerometer signal shown in Fig. 13.14a, the value of k is equal to 0.05 (local maxima and the distances T_i are determined from a band-pass filtered signal, shown by the black line Fig. 13.14g, to avoid fast fluctuations inappropriate for the determination of the tremor period). Since the instantaneous period varies only by 5% as compared to its mean value of $\langle T_i \rangle \approx 200$ ms, the process can be regarded as rather phase coherent. Our analysis of 41 tremor epochs from three different patients (Smirnov et al., 2008) shows that the coefficient k for the accelerometer signals typically takes the values of 0.05–0.1. For such regular signals, it is most reasonable to introduce phases and estimate couplings via the phase dynamics modelling as justified in Smirnov and Andrzejak (2005) and Smirnov et al. (2007).

Before the phase dynamics modelling, it is necessary to select a frequency band to define the phases. The results may differ for different bands. Inclusion of the frequencies below 2–3 Hz may shift the focus to slow processes like the heart beat or technical trends, which are not related to the parkinsonian tremor generation. Inclusion of the frequencies higher than 9–10 Hz implies a description of fast fluctuations, in particular, higher harmonics of the basic frequency. Such fluctuations may play a role of noise in the modelling of the tremor and make confident conclusions about the coupling presence more difficult. However, all that is not known in advance. Hence, one must try different frequency bands. We present the results only for a rather wide frequency band around the tremor frequency and then briefly comment what differs if other bands are used.

Both signals $x_1(t)$ and $x_2(t)$ are filtered in the relatively wide frequency band of 2–9 Hz (Fig. 13.14g, h). Their Hilbert transforms (Sect. 6.4.3) are illustrated in Fig. 13.14i, j, where rotation about a clearly defined centre is seen for both signals. Thus, the phases $\phi_1(t)$ and $\phi_2(t)$ are defined reasonably well. Ten characteristic periods at both edges of the phase time series are removed from the further analysis, since the corresponding phase values may be strongly distorted due to the edge effects as discussed in Pikovsky et al. (2000). The resulting phase time series of length 15900 data points (approximately 400 oscillation periods) is used for model fitting.

The cross-correlation function between the LFP and the contralateral hand acceleration is shown in Fig. 13.15a. Within a range of time lags, the CCF significantly differs from zero. Thus, the presence of coupling can be inferred reliably already from the CCF. The CCF exhibits some asymmetry: local maximum of its absolute value, closest to zero time lag, is observed at -25 ms. It could be a sign that the signal $x_2(t)$ "leads". However, somewhat higher peaks are observed at positive time lags, more distant from zero. Thus, directional coupling characteristics cannot be extracted from the CCF unambiguously, which is a typical case.

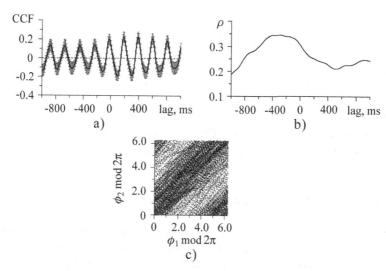

Fig. 13.15 Simple characteristics of interdependence between the accelerometer and LFP signals, shown in Fig. 13.14: (**a**) the cross-correlation function, error bars show the 95% confidence bands estimated via Bartlett's formula (Bartlett, 1978); (**b**) phase synchronisation index for the phases defined in the frequency band 2–9 Hz; (**c**) the data points on the plane of wrapped phases

13.2.3 Selection of the Model Equation Structure

At variance with the electronic example (Sect. 13.1), model equations for the parkinsonian tremor and subcortical activity cannot be written down from physical or physiological considerations. Thus, the structure of model equations cannot be regarded as completely known. On the other hand, it has appeared possible to introduce the phases of the oscillations related to the tremor frequency band. As discussed in Sect. 12.2, an adequate description of the phase dynamics for a wide range of oscillatory processes is achieved with the phase oscillator model, i.e. a first-order stochastic differential equation (Kuramoto, 1984; Rosenblum et al., 2001). For two processes, the model reads as

$$
\begin{aligned}
d\phi_1(t)/dt &= \omega_1 + G_1(\phi_1(t), \phi_2(t - \Delta_{2\to1})) + \xi_1(t), \\
d\phi_2(t)/dt &= \omega_2 + G_2(\phi_2(t), \phi_1(t - \Delta_{1\to2})) + \xi_2(t),
\end{aligned}
\tag{13.9}
$$

where ω_k is the angular frequency, the function G_k is 2π periodic with respect to both arguments, $\Delta_{2\to1}, \Delta_{1\to2}$ are time delays in couplings, ξ_k are zero-mean white noises with ACFs $\langle \xi_k(t)\xi_k(t')\rangle = \sigma_{\xi_k}^2 \delta(t - t')$, $\sigma_{\xi_k}^2$ characterises noise intensity. Empirical model is convenient to be sought for in the form of the corresponding difference equations:

$$
\begin{aligned}
\phi_1(t + \tau) - \phi_1(t) &= F_1(\phi_1(t), \phi_2(t - \Delta_{2\to1}), \mathbf{a}_1) + \varepsilon_1(t), \\
\phi_2(t + \tau) - \phi_2(t) &= F_2(\phi_2(t), \phi_1(t - \Delta_{1\to2}), \mathbf{a}_2) + \varepsilon_2(t),
\end{aligned}
\tag{13.10}
$$

where τ is the fixed time interval equal to the basic oscillation period (200 ms in our case), ε_k is the zero-mean noise, F_k is the third-order trigonometric polynomial (12.8), $\mathbf{a}_k$ is the vector of its coefficients, $\Delta_{2\to1}$, $\Delta_{1\to2}$ are trial time delays.

Thus, a suitable model structure (or, at least, a good guess for it) for the phases of the processes $x_1(t)$ and $x_2(t)$ can be considered as partly known due to relatively high regularity of the processes. Indeed, the first-order difference equation (13.10) with low-order trigonometric polynomials is a sufficiently universal choice, but not as arbitrary as in the black box case (Chap. 10): one does not need to try different model dimensions and different types of approximating functions.

13.2.4 Model Fitting, Validation and Usage

The three conditions of applicability of the phase dynamics modelling technique (Smirnov and Bezruchko, 2003; 2009) are imposed on the time series length (not less than 40 characteristic periods), the synchronisation index (not greater than 0.45) and the autocorrelation function of the model residual errors (the ACF decreases down to 0 or, at least, gets less than 0.2, over the interval of time lags up to τ). In our example, the length of the considered time series is about 400 characteristic oscillation periods, which is sufficiently big. Phase synchronisation index

$$\rho(\Delta) = \left| \frac{1}{N} \sum_{n=1}^{N} e^{i(\phi_1(n\Delta t) - \phi_2(n\Delta t + \Delta))} \right|$$

is less than 0.4 for any time lag Δ (Fig. 13.15b) as required. Sufficiently weak interdependence between simultaneous values of ϕ_1 and ϕ_2 is also illustrated in Fig. 13.5c: the distribution of the observed values on the plane of the wrapped phases fills the entire square and exhibits only weak non-uniformity. The ACF of the residual errors must be checked after fitting the model (13.10) to the phase time series.

We have fitted the equations to the data as described in Sect. 12.2. Namely, we have fixed $\tau = 200$ ms and minimised mean-squared approximation errors

$$\hat{\sigma}_{k,\mathbf{a}}^2(\Delta_{j\to k}, \mathbf{a}_k) = \frac{1}{N - \tau/\Delta t} \sum_{n=\tau/\Delta t + 1}^{N} \left(\phi_k(n\Delta t + \tau) - \phi_k(n\Delta t) \right.$$

$$\left. - F_k(\phi_k(n\Delta t), \phi_j(n\Delta t - \Delta_{j\to k}), \mathbf{a}_k) \right)^2,$$

where $k, j = 1, 2$ ($j \neq k$). For a fixed value of the trial delay $\Delta_{j\to k}$, this is a linear problem so that the coefficient estimates $\hat{\mathbf{a}}_k(\Delta_{j\to k}) = \arg\min_{\mathbf{a}_k} \hat{\sigma}_{k,\mathbf{a}}^2(\Delta_{j\to k}, \mathbf{a}_k)$ are found by solving a linear set of algebraic equations. The minimisation is performed for different trial delays. Then, the quantity $\hat{\sigma}_k^2(\Delta_{j\to k}) = \min_{\mathbf{a}_k} \hat{\sigma}_{k,\mathbf{a}}^2(\Delta_{j\to k}, \mathbf{a}_k)$ is plotted versus $\Delta_{j\to k}$ (Fig. 13.16a, b). Its minimal value characterises the phase

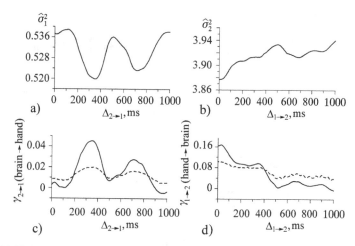

Fig. 13.16 Fitting the phase dynamics model (13.10) to the data of Fig. 13.14: (**a**), (**b**) approximation errors $\hat{\sigma}_1^2$ and $\hat{\sigma}_2^2$ (dimensionless) versus a trial time delay; (**c**), (**d**) coupling characteristics $\gamma_{2\to1}$ and $\gamma_{1\to2}$ (dimensionless) versus a trial time delay (*solid lines*) along with the pointwise 0.975 quantiles $\gamma_{2\to1,c}$ and $\gamma_{1\to2,c}$ (*dashed lines*)

diffusion intensity: $\hat{\sigma}_{k,\min}^2 = \min\limits_{\Delta_{j\to k}} \hat{\sigma}_k^2(\Delta_{j\to k})$. From the coefficient estimates $\hat{\mathbf{a}}_k$, the coupling strength $\gamma_{j\to k}(\Delta_{j\to k})$ is computed as their weighted sum. The formula for $\gamma_{j\to k}$ is given in Sect. 12.2. For the low-order trigonometric polynomials used, it appears that the maximum of $\gamma_{j\to k}(\Delta_{j\to k})$ corresponds to the minimum of $\hat{\sigma}_k^2(\Delta_{j\to k})$ (Fig. 13.16c, d).

The formula for the critical value $\gamma_{j\to k,c}(\Delta_{j\to k})$, which is a pointwise 0.975 quantile for the estimator $\gamma_{j\to k}(\Delta_{j\to k})$ in the case of uncoupled processes, is also available (Sect. 12.2). If one observes that $\gamma_{j\to k}(\Delta_{j\to k}) > \gamma_{j\to k,c}(\Delta_{j\to k})$ for a range of time delays wider than half a basic oscillation period (for an entire tried range of time delays covering five basic periods), then the presence of coupling can be inferred at the resulting significance level of 0.05 (i.e. with a probability of random error less than 0.05). This is the case in our example (Fig. 13.16c, d). Thus, a bidirectional coupling between the brain activity and the limb oscillations is detected.

Then, the location of the maximum of $\gamma_{j\to k}(\Delta_{j\to k})$ or the minimum of $\hat{\sigma}_k^2(\Delta_{j\to k})$ gives an estimate of the time delay: $\hat{\Delta}_{j\to k} = \arg\min\limits_{\Delta_{j\to k}} \hat{\sigma}_k^2(\Delta_{j\to k})$. In our case, the estimated time delay is $\hat{\Delta}_{2\to1} = 350$ ms for the influence $2 \to 1$ (brain to hand, $\gamma_{2\to1}(\hat{\Delta}_{2\to1}) = 0.045$) and $\hat{\Delta}_{1\to2} = 0$ ms for the influence $1 \to 2$ (hand to brain, $\gamma_{1\to2}(\hat{\Delta}_{1\to2}) = 0.16$).

Finally, the obtained model (13.10) with coefficients $\hat{\mathbf{a}}_k(\hat{\Delta}_{j\to k})$ should be validated including the properties of its residual errors. Our optimal model is specified by the phase diffusion intensities $\hat{\sigma}_{1,\min}^2 = 0.52$ and $\hat{\sigma}_{2,\min}^2 = 3.88$, and the following significantly non-zero coefficients (see their notations in Sect. 12.2): $w_1 = 6.03 \pm 0.07$, $\beta_{1,2,0} = 0.019 \pm 0.017$, $\alpha_{1,1,1} = 0.213 \pm 0.076$;

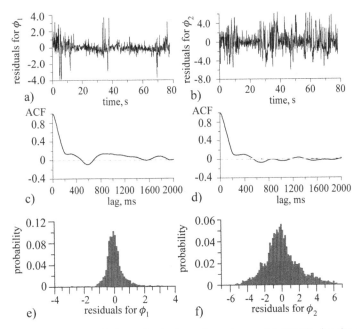

Fig. 13.17 Residual errors of the optimal phase dynamics model (13.10) for the epoch of Fig. 13.14. The left column shows residuals for the accelerometer signal phase, the right one is for the LFP signal phase; (**a**), (**b**) residual errors versus time; (**c**), (**d**) their autocorrelation functions; (**e**), (**f**) their histograms

$w_2 = 6.28 \pm 0.20$, $\beta_{2,1,1} = 0.44 \pm 0.21$. Time realisations of this model represent almost linear increase of the phases with some fluctuations. Qualitatively, they are very similar to the observed almost linearly increasing phases (not shown, since the plots are not informative). The model residuals are shown in Fig. 13.17a, b and their ACFs in Fig. 13.17c, d. The ACFs get less than 0.2 for time lags greater than $\tau = 200$ ms as required. Thus, applicability of the technique is confirmed.

The required properties of the residuals and proper behaviour of the time realisations validate the obtained model (13.10). The main usage of the model in our case is to make conclusions about the couplings between the processes. As already mentioned, it reveals a bidirectional coupling. Further, for the time series considered, the brain-to-hand influence appears delayed by more than a basic oscillation period, while the opposite influence is non-delayed. These results appear reproducible (another tremor epoch is illustrated in Figs. 12.1 and 12.2): the analysis of 41 tremor epochs from three different patients (Smirnov et al., 2008) in 30 cases has revealed the "coupling pattern" similar to that in Fig. 13.16, i.e. a bidirectional coupling which is significantly time delayed in the brain-to-hand direction and almost non-delayed in the opposite direction. In the other 11 epochs, no coupling has been detected. The cause seems to be that epochs of strong tremor occur intermittently in all the patients and the estimated curves $\gamma_{j \to k}(\Delta_{j \to k})$ fluctuate stronger for shorter epochs. It can be interpreted as an effect of noise. Averaging over all tremor epochs available for a given patient exhibits the plots $\gamma_{j \to k}(\Delta_{j \to k})$ shown in Fig. 12.3, which clearly confirm the coupling pattern observed in Fig. 13.16.

The results do not change under sufficiently strong variations in the frequency band used to define the phases. Namely, the results are practically the same if the lower cut-off frequency is not less than 2 Hz and not greater than $f_{\text{tremor}} - f_c$ and the upper cut-off frequency is not less than $f_{\text{tremor}} + f_c$ and not greater than $2 f_{\text{tremor}} - f_c$, where f_{tremor} is the basic tremor frequency and f_c is equal to 1 Hz. In our example, $f_{\text{tremor}} = 5\,\text{Hz}$ so that the acceptable values of the lower cut-off frequency range from 2 to 4 Hz and those of the upper cut-off frequency are from 6 to 9 Hz. Above, we have presented the results for a maximally wide acceptable frequency band. Its further enlargement or movement to higher frequencies gives less regular signals (a stronger phase diffusion) and insignificant conclusions about the coupling presence.

13.2.5 Validation of Time Delay Estimation

An analytic formula for the error in the time delay estimates is unavailable. To check correctness of the time delay estimation and assess its typical errors, we apply the same modelling procedure to a toy model consisting of the noisy van der Pol oscillator (an "analogue" of the hand oscillations) and a strongly dissipative linear oscillator (an "analogue" of the brain signal). Parameters of these oscillatory systems are selected so that they give stronger phase diffusion for the "LFP" signals and weaker one for the "tremor" signals as it is observed for the measurement data. Thus, the accelerometer $(a_1(t) = \mathrm{d}^2 y_1(t)/\mathrm{d}t^2)$ and LFP $(y_2(t))$ model oscillators read as

$$\frac{\mathrm{d}^2 y_1(t)}{\mathrm{d}t^2} - \left(\lambda - y_1^2(t)\right)\frac{\mathrm{d}y_1(t)}{\mathrm{d}t} + y_1(t) = k_{2 \to 1}(y_2(t - \tau_{2 \to 1}) - y_1(t)) + \xi_1(t),$$

$$\frac{\mathrm{d}^2 y_2(t)}{\mathrm{d}t^2} + 0.15\frac{\mathrm{d}y_2(t)}{\mathrm{d}t} + y_2(t) = k_{1 \to 2}(y_1(t - \tau_{1 \to 2}) - y_2(t)) + \xi_2(t), \quad (13.11)$$

where ξ_1, ξ_2 are independent white noises with ACFs $\langle \xi_k(t)\xi_k(t')\rangle = \sigma_{\xi_k}^2 \delta (t - t')$, $\sigma_{\xi_1} = \sigma_{\xi_2} = 0.1$, $\lambda = 0.05$ and $\tau_{2 \to 1}, \tau_{1 \to 2}$ are time delays. Angular frequencies of both oscillators are approximately equal to 1 so that their periods are about six time units. To generate a time series, the equations are integrated with the Euler technique (Sect. 4.5.2) at the step size of 0.01. The sampling interval is equal to 0.15, i.e. gives approximately 40 data points per basic period of oscillations.

The reasoning behind the simple model equation (13.11) is as follows. Firstly, the non-linear oscillator is chosen as a model of the accelerometer signal, since in a large number of tremor epochs our attempts to reconstruct a model equation from an accelerometer time series resulted in similar models (not shown). In fact, the spinal cord is able to produce self-sustained rhythmic neural and muscular activity due to its central pattern generators (Dietz, 2003). Moreover, similar models were previously obtained both for parkinsonian and essential tremor dynamics (Timmer et al., 2000). We consider the oscillator which is close to the point of the Andronov – Hopf bifurcation and demonstrates self-sustained oscillations (positive value of λ) perturbed by noise. However, the coupling estimation results are very similar for small negative λ, since the noise induces similar oscillations for small negative and small positive values of λ. Secondly, the linear oscillator is chosen as a model of the LFP signal, since construction of polynomial autoregressive models with different polynomial orders did not detect pronounced non-linearity (not shown).

The time series and power spectra estimates for the coupled oscillators (13.11) are shown in Fig. 13.18 (cf. Fig. 13.14).

We have analysed ensembles of time series generated by equation (13.11) with exactly the same procedure as applied to the experimental data above. In the numerical simulations, we used ensembles consisting of 100 time series of length of 100 basic periods. The results for a single epoch of Fig. 13.18 are shown in Figs. 13.19 and 13.20: they are qualitatively similar to the corresponding experimental results in Figs. 13.16 and 13.17. The averaged plots of coupling estimates for the observables $a_1(t)$ and $y_2(t)$ are shown in Fig. 13.21 in the same form as for the experimental results in Fig. 12.3. Without coupling (i.e. for $k_{2\to1} = k_{1\to2} = 0$), Fig. 13.21a, b

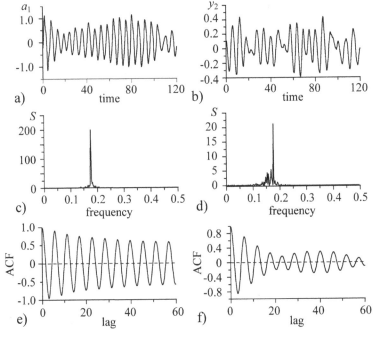

Fig. 13.18 A simulated time realisation of equation (13.11) of the duration 600 units of time t (4000 data points, about 100 basic periods) for the parameters $k_{2\to1} = 0.2$, $\tau_{2\to1} = 13.0$, $k_{1\to2} = 0.05$, $\tau_{1\to2} = 0$: (**a**) a signal $a_1(t) = d^2y_1(t)/dt^2$ at the beginning of the epoch considered, an analogue of the band-pass-filtered accelerometer signal $x_1(t)$ (Fig. 13.14g); (**b**) a simultaneous signal $y_2(t)$, an analogue of the band-pass-filtered LFP $x_2(t)$ (Fig. 13.14h); (**c**), (**d**) periodograms of $a_1(t)$ and $y_2(t)$, respectively; (**e**), (**f**) their autocorrelation functions

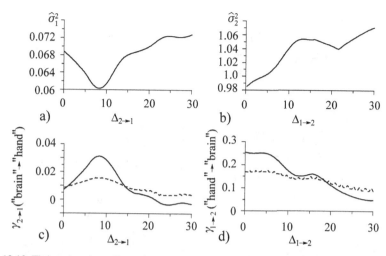

Fig. 13.19 Fitting the phase dynamics equation (13.10) to the time realisations of the model (13.11) illustrated in Fig. 13.18: (a), (b) the approximation errors $\hat{\sigma}_1^2$ and $\hat{\sigma}_2^2$ versus a trial time delay; (c), (d) the coupling characteristics $\gamma_{2\rightarrow1}$ and $\gamma_{1\rightarrow2}$ versus a trial time delay (*solid lines*) along with the pointwise 0.975 quantiles $\gamma_{2\rightarrow1,c}$ and $\gamma_{1\rightarrow2,c}$ (*dashed lines*)

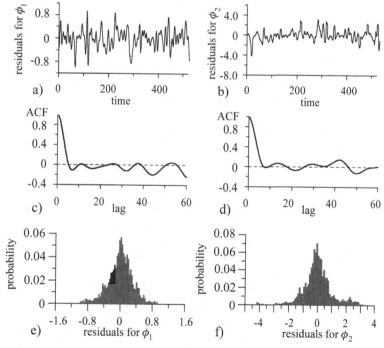

Fig. 13.20 Residual errors of the optimal phase dynamics model (13.10) for the time series of Fig. 13.18. The left column shows the residuals for the phase ϕ_1 and the right one is for the phase ϕ_2: (a), (b) the residual errors versus time; (c), (d) their autocorrelation functions; (e), (f) their histograms

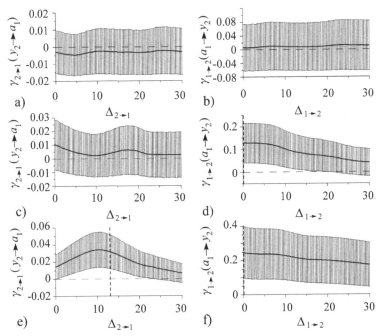

Fig. 13.21 The coupling characteristics averaged over ensembles of the time series from the system (13.11). The *error bars* indicate the averaged values of the analytic 95% confidence bands. The vertical *dashed lines* show true delay times: (**a**), (**b**) uncoupled oscillators; (**c**), (**d**) a unidirectional coupling with $k_{2\to1} = 0$, $k_{1\to2} = 0.07$, $\tau_{1\to2} = 0$; (**e**), (**f**) a bidirectional time-delayed coupling with $k_{2\to1} = 0.2$, $\tau_{2\to1} = 13.0$, $k_{1\to2} = 0.05$, $\tau_{1\to2} = 0$ (an example of a single time series for these parameters is given in Figs. 13.18, 13.19 and 13.20)

evidences that there is no false coupling detection on average. For a unidirectional "hand-to-brain" coupling ($k_{2\to1} = 0$, $k_{1\to2} = 0.07$, $\tau_{1\to2} = 0$), the unidirectional coupling pattern is observed in Fig. 13.21c, d. The experimental coupling pattern of Fig. 12.3 is qualitatively reproduced in our model (13.11) with a bidirectional time-delayed coupling, e.g., for $k_{2\to1} = 0.2$, $\tau_{2\to1} = 13.0$ (i.e. twice as large as the basic period of the oscillations), $k_{1\to2} = 0.05$, $\tau_{1\to2} = 0$, see Fig. 13.21e, f. At that, $\gamma_{2\to1}$ gets maximal for $\hat{\Delta}_{2\to1} = 9.0$, which is smaller than $\tau_{2\to1} = 13.0$ approximately by 0.6 of the basic period. For a range of true coupling coefficients and time delays, we have observed that the time delay estimate is less than the true time delay by half of the basic period on average.

Thus, with the numerical example we have qualitatively illustrated that reasonable estimates of the time delays are obtained with the phase dynamics modelling technique. Quantitatively, the time delay estimates may have an error about half a basic period and, hence, are not very accurate. Yet, we can conclude that the time delay in the brain-to-hand influence estimated from the parkinsonian tremor data is greater than the delay in the hand-to-brain direction.

To summarise, the results of the work (Smirnov et al., 2008), described above, fit to the assumption that the subcortical oscillations drive and synchronise premotor and motor cortex which activates the contralateral muscles via the spinal cord (Brown, 2003; Rivlin-Etzion et al., 2006; Tass et al., 1998; Timmermann et al., 2003). However, the long brain-to-tremor delay indicates a more complex mechanism compared to a simple forward transmission. In contrast, the short tremor-to-brain delay fits to a direct neural transmission time of a proprioceptive feedback loop (Eichler, 2006). These results provide a new picture of the old servo loop oscillation concept, where feedback and feed-forward are acting via straight transmission lines (Stilles and Pozos, 1976). Rather, one can suggest that the synchronised subcortical oscillatory activity feeds into a multistage re-entrant processing network, most likely involving cortico-subcortical and spinal reflex loops (see also Brown, 2003; Rivlin-Etzion et al., 2006, Stilles and Pozos, 1976).

13.3 El-Niño/Southern Oscillation and Indian Monsoon

13.3.1 Object Description

Major climatic processes in Asian – Pacific region, which are of global importance, are related with the phenomena of El-Niño/Southern Oscillation (ENSO) and Indian monsoon (Solomon et al., 2007). The strongest interannual variations in the global surface temperature depend on the intensity of the ENSO phenomenon. Two-thirds of the Earth population live in the monsoon-related regions with a key role of Indian monsoon (Zhou et al., 2008). Thus, investigation of the interaction between ENSO and Indian monsoon activity is of both regional and global interest.

The presence of interdependence between these processes has been reliably detected with different techniques in many works (Kripalani and Kulkarni, 1997; 2001, Krishnamurthy and Goswami, 2000; Kumar et al., 1999; Maraun and Kurths, 2005; Sarkar et al., 2004; Solomon et al., 2007; Walker and Bliss, 1932; Yim et al., 2008; Zubair and Ropelewski, 2006). Indeed, an increase in the sea surface temperature (SST) in equatorial Pacific during El-Nino along with the corresponding change in convective processes, the Walker zonal circulation, the Hadley meridional circulation and the displacement of the intertropical convergence zone, is accompanied by considerable seasonal anomalies of temperature and precipitation in many regions. At that, there are significant variations in the correlation between characteristics of ENSO and Indian monsoon, in particular, its noticeable decrease starting from the last quarter of the XX century (Solomon et al., 2007). Along with the characterisation of an overall coupling strength provided by the coherence and synchronisation analysis, climatologists are strongly interested in a quantitative estimation of directional couplings between ENSO and Indian monsoon along with tendencies of their temporal changes.

Below, we describe estimation of the directional couplings by using the empirical AR models, i.e. the Granger causality (Sect. 12.1), which gets more and more often used in the Earth sciences (Mokhov and Smirnov, 2006, 2008; Mosedale et al., 2006; Wang et al., 2004).

13.3.2 Data Acquisition and Preliminary Processing

We have analysed monthly values of the ENSO and Indian monsoon indices for the period 1871–2006 illustrated in Fig. 13.22. Indian monsoon is characterised by variations in all-India monthly precipitation (Mooley and Parthasarathy, 1984). The corresponding data of the Indian Institute of Tropical Meteorology are available at http://climexp.knmi.nl/data/pALLIN.dat. As the ENSO index, we use SST in the area Niño-3 (5S–5N, 150W–90W) in the Pacific Ocean. We take the UK Meteorological Office GISST2.3 data for the period 1871–1996 (Rayner et al., 2003), which are available at http://paos.colorado.edu/research/wavelets/nino3data.asc, and supplement them with the data of the Climate Prediction Center obtained via Reynolds' optimal interpolation (Reynolds and Smith, 1994) for the period 1997–2006, which are available at http://www.cpc.noaa.gov/data/indices/sstoi.indices. The concatenation of the data is done in analogy with the work of Torrence and Compo presented at http://atoc.colorado.edu/research/wavelets/wavelet1.html.

Seasonal variations in both processes are clearly seen in Fig. 13.22. They are related to the common external driving, i.e. to the insolation cycle. Common exter-

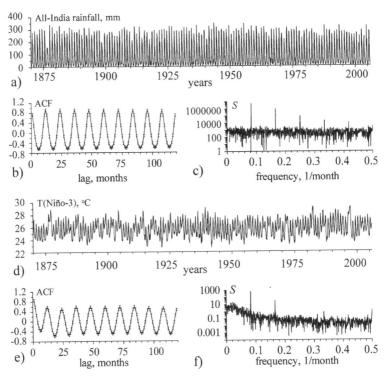

Fig. 13.22 Climatic data available and their characteristics: (**a**) an index of Indian monsoon, (**b**) its ACF estimate; the error bars show 95% confidence intervals according to Bartlett's formula (Bartlett, 1978); (**c**) its periodogram; (**d**) an ENSO index, (**e**) its ACF estimate, (**f**) its periodogram

nal driving can lead to prediction improvements (Sect. 12.1) and to erroneous conclusions about the presence of mutual influences. Therefore, we have removed the component with 12-month period and its higher harmonics from both signals. It is realised as follows. An averaged value of an observed quantity η is computed separately for each calendar month, e.g. for January. The averaging is performed over the entire interval 1871–2006. This averaged value is subtracted from all the January values of η. The values of η corresponding to each of the 12 months are processed analogously. Below, we deal only with such deseasonalised signals and denote the resulting monsoon index as $x_1(t)$ and the ENSO index as $x_2(t)$. The resulting time series are shown in Fig. 13.23a, d and their length is $N = 1632$ data points. Their power spectra and ACFs do not reveal any signs of a 12-month cycle (Fig. 13.23b, c, e, f) as desired.

The cross-correlation function for the signals $x_1(t)$ and $x_2(t)$ reaches the value of -0.22 for the 3-month time delay of the ENSO index relative to the monsoon index (Fig. 13.24). According to Bartlett's formula (Bartlett, 1978), the width of a pointwise 95% confidence band for the CCF estimate is ± 0.05. Therefore, although the absolute value of the CCF is not very large, its difference from zero at the time lags close to zero is highly statistically significant. The CCF indicates the presence of an

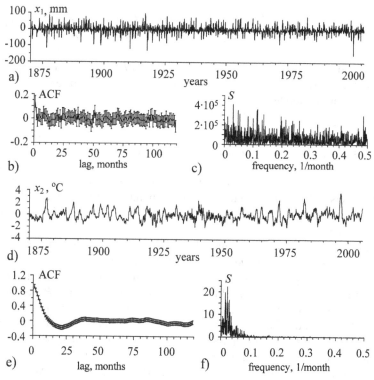

Fig. 13.23 Climatic data after the removal of the 12-month component: (**a**)–(**c**) an index of Indian monsoon with its ACF and periodogram; (**d**)–(**f**) an ENSO index with its ACF and periodogram

Fig. 13.24 The cross-correlation function between deseasonalised monsoon and ENSO indices

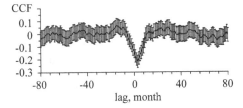

interdependence between the processes, but it does not allow to reveal whether the coupling is unidirectional (and, then, to find out its directionality) or bidirectional.

13.3.3 Selection of the Model Equation Structure

Since any a priori information about an appropriate model structure is absent, we use universal autoregressive models (Sect. 12.1) to describe the observed dynamics and reveal the character of coupling. Namely, individual (univariate) models are constructed in the form

$$x_1(t) = f_1(x_1(t-1), \ldots, x_1(t-d_1)) + \xi_1(t),$$
$$x_2(t) = f_2(x_2(t-1), \ldots, x_2(t-d_2)) + \xi_2(t), \qquad (13.12)$$

where f_1 and f_2 are polynomials of the orders P_1 and P_2, respectively, d_1 and d_2 are the model dimensions (orders), ξ_1 and ξ_2 are Gaussian white noises. Analogously, the joint (bivariate) model structure is

$$x_1(t) = f_{1|2}(x_1(t-1), \ldots, x_1(t-d_1), x_2(t-1), \ldots, x_2(t-d_{2\to1})) + \eta_1(t),$$
$$x_2(t) = f_{2|1}(x_2(t-1), \ldots, x_2(t-d_2), x_1(t-1), \ldots, x_1(t-d_{1\to2})) + \eta_2(t), \qquad (13.13)$$

where $f_{2|1}$ and $f_{1|2}$ are polynomials of the same orders P_1 and P_2 as for the individual models (13.12), $d_{2\to1}$ and $d_{1\to2}$ are the numbers of the values of the other process taken into account (they characterise inertial properties of couplings), η_1 and η_2 are Gaussian white noises.

Polynomial coefficients in the models (13.12) and (13.13) are estimated via the *ordinary least-squares technique*, i.e. via minimisation of the sums of the squared residual errors (Sect. 12.1). Since any a priori information about an appropriate model structure is absent, we try different values of d_k, $d_{j\to k}$ and P_k to find optimal ones. It is important to select the form of the non-linear functions properly. Due to relatively short time series at hand, we use low-order algebraic polynomials (Ishiguro et al., 2008; Mokhov and Smirnov, 2006) as a reasonable universal choice under the "black box" problem setting (Chap. 10).

Concretely, to select d_k, $d_{j\to k}$ and P_k, we proceed as follows. At a fixed P_k, the value of d_k is selected according to Schwarz's information criterion (see the discussion of the cost functions in Sect. 7.2.3), i.e. so as to minimise the value of

$$S_k = \frac{N}{2}\ln\hat{\sigma}_k^2 + \frac{\ln N}{2}P_k,$$

where $\hat{\sigma}_k^2$ is the minimal mean-squared prediction error of the individual AR model (13.12) at the given d_k and P_k (see the notations in Sect. 12.1). Then, we validate the univariate AR model obtained. Firstly, we check whether its residual errors are delta correlated to assure further applicability of the F-test for the coupling estimation (Sect. 12.1). Secondly, we check whether its time realisations are close to the observed time series $x_k(t)$ in a statistical sense: temporal profiles look similar; the ranges of probable values of the model and observed variables are almost the same. If all that is fulfilled, then the univariate model is regarded satisfactory, otherwise the value of d_k is increased.

Given d_k, we use Schwarz's criterion to select $d_{j\rightarrow k}$, i.e. we minimise

$$S_{j\rightarrow k} = \frac{N}{2}\ln\hat{\sigma}_{k|j}^2 + \frac{\ln N}{2}P_{k|j},$$

where $\hat{\sigma}_{k|j}^2$ is the minimal mean-squared prediction error of the bivariate AR model (13.13). However, for the purposes of coupling detection, another approach may be even more appropriate: one can select such value of $d_{j\rightarrow k}$, which maximises $\text{PI}_{j\rightarrow k} = \hat{\sigma}_k^2 - \hat{\sigma}_{k|j}^2$ or corresponds to the value of $\text{PI}_{j\rightarrow k}$, which exceeds zero at the smallest significance level p. We use the latter approach as well and compare the results of both approaches. An obtained bivariate AR model is validated in the same way as the univariate models.

Different values of P_k are tried. The above analysis, including the selection of d_k and $d_{j\rightarrow k}$, is performed for each P_k. The most appropriate value of P_k is selected both according to Schwarz's criterion and to the most significant $\text{PI}_{j\rightarrow k}$ and the results are compared. The trial values of d_k, $d_{j\rightarrow k}$ and P_k are varied within such a range that the number of coefficients in any fitted AR model remains much less than the time series length N, namely the number of model coefficients does not exceed $\sqrt{N}$, i.e. approximately 40 in our case.

13.3.4 Model Fitting, Validation and Usage

Firstly, we fit models and estimate couplings for the entire period 1871–2006. Secondly, the analysis is done in moving windows of length ranging from 10 to 100 years to get time-resolved coupling characteristics.

13.3.4.1 Univariate models

The number of coefficients in the linear models is equal to $P_k = d_k + 1$ so that d_k can be increased up to 39 when a model is fitted to the entire period 1871–2006. For the quadratic models, we get

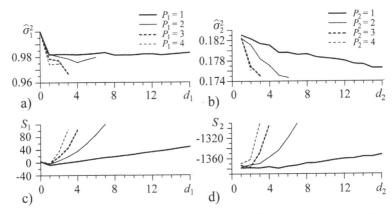

Fig. 13.25 Mean-squared prediction errors (*first row*) and Schwarz's criterion (*second row*) for the individual AR models of the Indian monsoon index (*left column*) and the ENSO index (*right column*)

$$P_k = \frac{(d_k + 1)(d_k + 2)}{2}$$

so that d_k may not exceed 7. It should be $d_k \leq 4$ for $P_k = 3$, $d_k \leq 3$ for $P_k = 4$, etc.

For the monsoon index, an optimal model is achieved at $d_1 = 1$ for any P_1 (Fig. 13.25a, c). Schwarz's criterion takes the smallest values for the linear models. Thus, an optimal model is linear with $d_1 = 1$. It gives a prediction error with the variance $\hat{\sigma}_1^2/\text{var}[x_1] = 0.98$, where $\text{var}[x_1]$ is the sample variance of x_1. The model explains only 2% of $\text{var}[x_1]$.

For the ENSO index, an optimal model dimension is $d_2 = 1$ at $P_2 = 2, 4$ and $d_2 = 2$ at $P_2 = 3$, but the best model is linear with $d_2 = 5$ (Fig. 13.25b, d). The normalised variance of its prediction error is $\hat{\sigma}_2^2/\text{var}[x_2] = 0.18$.

The obtained individual models appear valid: the residual errors of the optimal models (Fig. 13.26a, d) for both processes are delta correlated (Fig. 13.26b, e) and exhibit distributions with quickly decreasing tails (Fig. 13.26c, f); model time realisations are statistically close to the observed data (the plots are not shown, since they are similar to those for a bivariate model presented below).

13.3.4.2 ENSO-to-Monsoon Driving

To construct bivariate models for the monsoon index, we use $d_1 = 1$ at different values of P_1 based on the results shown in Fig. 13.25c. The value of $d_{2 \to 1} = 1$ appears optimal at $P_1 = 1$ and 3 (Fig. 13.27a). The linear model gives the smallest value of Schwarz' criterion. However, the model with $P_1 = 3$ gives greater and the most statistically significant prediction improvement (Fig. 13.27c, e). This is a sign of non-linearity in the ENSO-to-monsoon influence, which would be ignored if the linear model were used to estimate the coupling. To avoid such a negligence, we regard the model with $P_1 = 3$ as optimal. Its prediction improvement

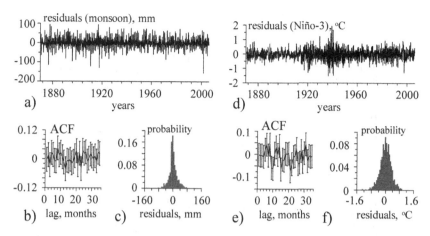

Fig. 13.26 Residual errors of the optimal individual models (13.12) with ACFs and histograms: (a)–(c) for the Indian monsoon index; (d)–(f) for the ENSO index

is $PI_{2\to1}/\hat\sigma_1^2 = 0.028$, i.e. it equals only 2.8% of the variance of all factors, which remain unexplained by the univariate model. Yet, the ENSO-to-monsoon influence is detected with high confidence ($p < 10^{-8}$).

The optimality of the value of $d_{2\to1} = 1$ means "inertialless" ENSO-to-monsoon influence. The model reads as

$$x_1(t) = a_{1,1}x_1(t-1) + b_{1,1}x_2(t-1) + c_{1,1}x_1^2(t-1)x_2(t-1) + c_{1,2}x_2^3(t-1)$$
$$+\eta_1(t),$$

(13.14)

where $\sigma_{\eta_1}^2 = 5.84 \times 10^2$ mm^2 and estimates of the coefficients and their standard deviations (see Sect. 7.4.1 and Seber, 1977) are the following: $a_{1,1} = 0.071 \pm 0.035$, $b_{1,1} = -4.65 \pm 1.11$ mm K^{-1}, $c_{1,1} = (-3.53 \pm 0.76) \cdot 10^{-3}$ mm^{-1} K^{-1} and $c_{1,2} = 1.53 \pm 0.38$ mm K^{-3}. We have shown only the terms whose coefficients differ from zero at least at the pointwise significance level of 0.05, i.e. the absolute value of a coefficient is at least twice as big as its standard deviation. The linear coupling coefficient $b_{1,1}$ is negative, which corresponds to the above-mentioned negative correlation between the signals x_1 and x_2.

13.3.4.3 Monsoon-to-ENSO Driving

A bivariate model for the ENSO index is optimal at $P_2 = 1$ and $d_{1\to2} = 3$ (Fig. 13.27b). It corresponds to the most significant prediction improvement $PI_{1\to2}/\hat\sigma_2^2 = 0.024$ exceeding zero at the significance level of $p < 10^{-8}$ (Fig. 13.27d, f). The model reads as

$$x_2(t) = a_{2,1}x_2(t-1) + a_{2,5}x_2(t-5) + b_{2,1}x_1(t-1) + b_{2,2}x_1(t-2)$$
$$+b_{2,3}x_1(t-3) + \eta_2(t),$$

(13.15)

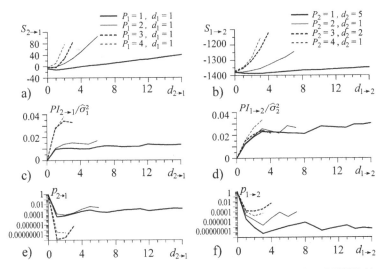

Fig. 13.27 Fitting the bivariate models to the monsoon (the *left column*) and ENSO (the *right column*) data: Schwarz's criterion (*first row*), the prediction improvement (*second row*) and the significance level (*third row*)

where $\sigma_{\eta_2}^2 = 0.11\,\mathrm{K}^2$, $a_{2,1} = 0.92 \pm 0.025$, $a_{2,5} = -0.083 \pm 0.025$, $b_{2,1} = (-1.44 \pm 0.34) \times 10^{-3}\,\mathrm{mm}^{-1}\,\mathrm{K}$, $b_{2,2} = (-1.04 \pm 0.35) \times 10^{-3}\,\mathrm{mm}^{-1}\,\mathrm{K}$ and $b_{2,3} = (-1.01 \pm 0.35) \times 10^{-3}\,\mathrm{mm}^{-1}\,\mathrm{K}$. The monsoon-to-ENSO influence is inertial, since the optimal value of $d_{1 \to 2} > 1$. Namely, the behaviour of the ENSO index depends on the values of the monsoon index for three previous months. The coupling coefficients $b_{2,1}, b_{2,2}, b_{2,3}$ are negative and also correspond to the observed anti-correlation between x_1 and x_2. All the three coupling coefficients are almost identical, i.e. the total contribution of the monsoon index to equation (13.15) $(b_{2,1}x_1(t-1) + b_{2,2}x_1(t-2) + b_{2,3}x_1(t-3))$ is approximately proportional to its average value over 3 months. No signs of non-linearity of the monsoon-to-ENSO influence are detected.

13.3.4.4 Validation of the Bivariate Model

ACFs and histograms of the residual errors for the bivariate models (13.14) and (13.15) are very similar to those for the individual models in Fig. 13.26; they exhibit delta correlatedness and quickly decreasing tails (not shown). The correlation coefficient between the residual errors for the monsoon and ENSO indices is 0.02 ± 0.05, i.e. equals zero within the estimation error. Thus, the noises η_1 and η_2 are considered independent when realisations of the bivariate model (13.14) and (13.15) are simulated. Time realisations of this optimal model with $P_1 = 3$, $d_1 = 1$, $d_{2 \to 1} = 1$, $P_2 = 1$, $d_2 = 5$, $d_{1 \to 2} = 3$ look similar to the observed time series (Fig. 13.28a,c). For a quantitative comparison, an ensemble of model realisations at the same initial conditions is generated and 95% intervals of the distributions

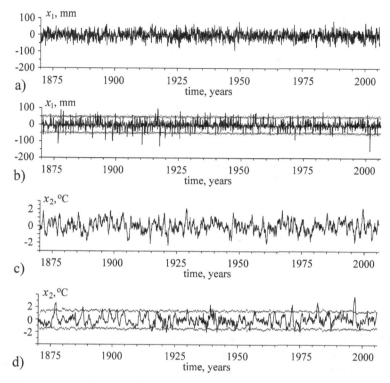

Fig. 13.28 Behaviour of the optimal AR model (13.14) and (13.15): (**a**), (**c**) model time realisations corresponding to the monsoon index and the ENSO index, respectively; (**b**), (**d**) 95% intervals of the model variables (*dotted lines*) and the observed data (*solid lines*), respectively

of the model variables are determined. It appears that 95% of the observed values of the ENSO and monsoon indices fall within those intervals (Fig. 13.28b,d), which confirms validity of the model.

13.3.4.5   Coupling Analysis in Moving Window

Finally, let us consider temporal variations in the coupling characteristics by using the moving window analysis, i.e. the intervals $[T - W, T]$, where W is the window length and T is a coordinate of the window endpoint (in years). At a fixed value of W (which is systematically changed from 10 years to 100 years with a step of 10 years), the Granger causality estimates are calculated for T ranging from $1871 + W$ till 2006.

To assess significance levels of the conclusions about the coupling present under the moving-window scenario, a multiple test correction (Lehmann, 1986) must be applied. Namely, according to the above procedure, one gets the estimates of the prediction improvement $PI_{j \to k}$ and the corresponding significance level p for each time window. This is the so-called pointwise significance level, i.e. a probability of

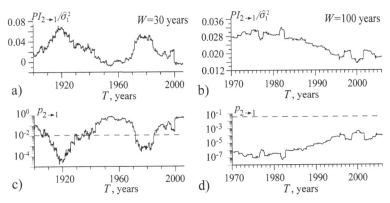

Fig. 13.29 Estimates of the ENSO-to-monsoon influence in a moving window $[T-W, T]$ versus the coordinate of the window endpoint T: (**a**), (**b**) prediction improvements; (**c**), (**d**) pointwise significance levels. Different panels show the results for the two different window lengths of 30 and 100 years. The *dashed lines* show the critical values p_c of the pointwise significance level corresponding to the resulting significance level of $p = 0.05$ (see the text)

a random erroneous conclusion for a single time window considered separately. The probability of a false positive conclusion at the pointwise level p for at least one of M non-overlapping time windows may reach the value of $p \cdot M$, because probability of a union of independent events is approximately equal to the sum of their individual probabilities (if the resulting value $p \cdot M$ is still much less than unity). Thus, one can establish the presence of coupling for a particular time window among M non-overlapping windows at a "true" significance level p if the pointwise significance level for this window equals p/M, where the multiplier $1/M$ is called the Bonferroni correction. The dashed lines in Figs. 13.29 and 13.30 show such a

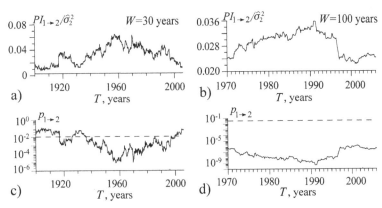

Fig. 13.30 Estimates of the monsoon-to-ENSO influence in a moving window $[T-W, T]$ versus the coordinate of the window endpoint T: (**a**), (**b**) prediction improvements; (**c**), (**d**) pointwise significance levels. Different panels show the results for the two different window lengths of 30 and 100 years. The *dashed lines* show the critical values p_c of the pointwise significance level corresponding to the resulting significance level of $p = 0.05$ (see the text)

threshold value of $p_c = 0.05/(N/W)$, where N/W estimates the number of non-overlapping windows; if the pointwise significance level p for any time window gets less than p_c, we infer coupling presence for this window at the resulting significance level less than 0.05.

Estimates of the ENSO-to-monsoon driving for the optimal non-linear model with the parameters $d_1 = d_{2 \to 1} = 1$, $P_1 = 3$ are presented in Fig. 13.29 for the window lengths of 30 years (Fig. 13.29a, c) and 100 years (Fig. 13.29b, d). The 100-year windows give highly significant results for any T. A long-term tendency consists of a weak rise of the ENSO-to-monsoon coupling strength at the beginning of the investigated period, reaching a maximum, and a subsequent decrease. The duration of the decrease period is longer than that of the rise period. When the time window length is decreased, the temporal resolution enhances at the expense of the significance of the results. Thus, the 30-year windows reveal coupling for $1910 \leq T \leq 1930$ and $1975 \leq T \leq 1985$, i.e. over the intervals 1880–1930 and 1945–1985. For shorter moving windows, the non-linear model gets relatively too "big" and gives less significant results. In total, the ENSO-to-monsoon driving is relatively weak before 1880, during the period 1930–1945 and after 1985.

One can see statistically significant influence of the monsoon on ENSO in a 100-year moving window for any T (Fig. 13.30). A long-term tendency is the same as for the ENSO-to-monsoon driving, but the monsoon-to-ENSO coupling strength starts to decrease later; a maximum of its temporal profile is closer to the year of 2006. A significant monsoon-to-ENSO influence in a 30-year window is observed for $1917 \leq T \leq 1927$ and, especially, for $1935 \leq T \leq 2000$. With a 20-year moving window, a significant influence is detected only over the interval 1930–1960; with a 10-year moving window, it is not detected at all (the plots are not shown). In total, the monsoon-to-ENSO influence is not seen only before 1890 and it is the most essential during the period 1930–1950.

Thus, the intervals of the strongest ENSO-to-monsoon and monsoon-to-ENSO influences do not coincide in time but follow each other. The coupling between both processes is approximately symmetric "in strength": the normalised prediction improvement is about 2–3% for the entire interval 1871–2006 and reaches about 7% in a 30-year moving window for both directions.

We note that in Maraun and Kurths (2005) the authors found intervals of 1:1 synchronisation between both signals, i.e. the intervals when the phase difference $\phi_1 - \phi_2$ is approximately constant. These are the intervals 1886–1908 and 1964–1980, which correspond to the strong ENSO-to-monsoon influence detected by the Granger causality. Next, the intervals of 1:2 synchronisation (when the difference $\phi_1 - 2\phi_2$ is approximately constant) appear during 1908–1921 (corresponds to a predominant monsoon-to-ENSO driving detected with the Granger causality), 1935–1943 (the strongest monsoon-to-ENSO driving and no significant ENSO-to-monsoon driving) and 1981–1991 (a predominant monsoon-to-ENSO driving). Thus, the 1:1 synchronisation coincides with the intervals of a stronger ENSO-to-monsoon influence, while the 1:2 synchronisation to a predominant monsoon-to-ENSO influence.

13.4 Conclusions

The investigation of coupling between the climatic processes considered here represents a black-box problem. With the universal model structure (polynomial AR models), we have obtained valid models and used them to characterise directional couplings between ENSO and Indian monsoon. The results complement previous knowledge of anti-correlation between these processes (Walker and Bliss, 1932) and their phase synchrony intervals (Maraun and Kurths, 2005). Namely, the empirical modelling has revealed bidirectional coupling between ENSO and Indian monsoon with high confidence. The ENSO-to-monsoon influence appears inertialless and non-linear. The monsoon-to-ENSO influence is linear and inertial; the values of the monsoon index for three months affect the future behaviour of the ENSO index. However, in some sense the coupling is symmetric; prediction improvement is about 2–3% in both directions. The moving window analysis has revealed an alternating character of the coupling. The monsoon-to-ENSO coupling strength rises since the end of the nineteenth century till approximately the period of 1930–1950, when it is maximal. This influence weakens in the last decade of the twentieth century. The opposite influence is strongest during the period of 1890–1920. It is also noticeable in 1950–1980 and not detected in 1920–1950 and after 1980.

To summarise, the three examples considered in Chap. 13 illustrate an empirical modelling procedure under three different settings: complete a priori information about a model structure, partial information and no information. The first setting takes place for the laboratory electronic systems, which is typical since laboratory experiments can be devised so as to control many properties of the objects under study. The second setting corresponds to a physiological problem, where partial information about an appropriate model structure is available due to specific properties (considerable regularity) of the observed signals. The third example of no specific knowledge about a model structure is taken from climatology. Estimates of couplings between the investigated processes provided by the empirical modelling can be regarded as most "valuable" in the latter case, where the results seem to be obtained practically "from nothing". However, under all the three settings, empirical modelling allows to get useful information such as validation of the physical ideas behind model equations and quantitative characterisation of individual dynamics and interactions.

References

Bartlett, M.S.: Stochastic Processes. Cambridge University Press: Cambridge (1978)

Benabid, A.L., Pollak, P., Louveau, A., Henry, S., de Rougemont, J. Combined (thalamotomy and stimulation) stereotactic surgery of the VIM thalamic nucleus for bilateral Parkinson's disease. Appl. Neurophysiol. **50**, 344–346 (1987)

Bezruchko, B.P., Ponomarenko, V.I., Prokhorov, M.D., Smirnov, D.A., Tass, P.A.: Modeling non-linear oscillatory systems and diagnostics of coupling between them using chaotic time series analysis: applications in neurophysiology. Physics – Uspekhi. **51**, 304–310 (2008)

Brown, P.: Oscillatory nature of human basal ganglia activity: relationship to the pathophysiology of Parkinson's disease. Mov. Disord. **18**, 357–363 (2003)

Deuschl, G., Krack, P., Lauk, M., Timmer, J.: Clinical neurophysiology of tremor. J. Clin. Neurophysiol. **13**, 110–121 (1996)

Dietz, V.: Spinal cord pattern generators for locomotion. Clin. Neurophysiol. **114**, 1379–1389 (2003)

Dmitriev, A.S., Kislov, V.Ya. Stochastic Oscillations in Radiophysics and Electronics. Nauka, Moscow, (in Russian) (1989)

Dmitriev, A.S., Panas, A.I., Starkov, S.O.: Ring oscillating systems and their application to the synthesis of chaos generators. Int. J. Bifurc. Chaos. **6**, 851–865 (1996)

Eichler, M.: Graphical modelling of dynamic relationships in multivariate time series. In: Winterhalder, M., Schelter, B., Timmer, J. (eds) Handbook of Time Series Analysis, pp. 335–367. Wiley-VCH, Berlin (2006)

Ishiguro, K., Otsu, N., Lungarella, M., Kuniyoshi, Y.: Detecting direction of causal interactions between dynamically coupled signals. Phys. Rev. E. **77**, 026216 (2008)

Kripalani, R.H., Kulkarni, A.: Monsoon rainfall variations and teleconnections over South and East Asia. Int. J. Climatol. **21**, 603–616 (2001)

Kripalani, R.H., Kulkarni, A.: Rainfall variability over Southeast Asia: Connections with Indian monsoon and ENSO extremes: New perspectives. Int. J. Climatol. **17**, 1155–1168 (1997)

Krishnamurthy, V., Goswami, B.N.: Indian monsoon-ENSO relationship on interdecadal timescale. J. Climate. **13**, 579–595 (2000)

Kuramoto, Y.: Chemical Oscillations, Waves and Turbulence. Springer, Berlin (1984)

Lehmann, E.: Testing Statistical Hypothesis. Springer, Berlin (1986)

Lenz, F.A., Kwan, H.C., Martin, R.L., Tasker, R.R., Dostrovsky, J.O., Lenz, Y.E.: Single unit analysis of the human ventral thalamic nuclear group. Tremor related activity in functionally identified cells. Brain. **117**, 531–543 (1994)

Llinas, R., Jahnsen, H.: Electrophysiology of mammalian thalamic neurones in vitro. Nature. **297**, 406–408 (1982)

Maraun, D., Kurths, J.: Epochs of phase coherence between El Nino/Southern Oscillation and Indian monsoon. Geophys. Res. Lett. **32**, L15709 (2005)

Mokhov, I.I., Smirnov, D.A.: Diagnostics of a cause – effect relation between solar activity and the earth's global surface temperature. Izvestiya, Atmos. Oceanic Phys. **44**, 263–272 (2008)

Mokhov, I.I., Smirnov, D.A.: El Nino Southern Oscillation drives North Atlantic Oscillation as revealed with nonlinear techniques from climatic indices. Geophys. Res. Lett. **33**, L0378 (2006)

Mooley, D.A., Parthasarathy, B.: Fluctuations in all-India summer monsoon rainfall during 1871–1978. Clim. Change. **6**, 287–301 (1984)

Mosedale, T.J., Stephenson, D.B., Collins, M., Mills, T.C.: Granger causality of coupled climate processes: ocean feedback on the North Atlantic Oscillation. J. Climate. **19**, 1182–1194 (2006)

Pare, D., Curro'Dossi, R., Steriade, M.: Neuronal basis of the parkinsonian resting tremor. Neuroscience. **217**, 217–226 (1990)

Pikovsky, A.S., Rosenblum, M.G., Kurths, J.: Phase synchronization in regular and chaotic systems. Int. J. Bif. Chaos. **10**, 2291–2305 (2000)

Pikovsky, A.S., Rosenblum, M.G., Kurths, J.: Synchronisation. A Universal Concept in Nonlinear Sciences. Cambridge: Cambridge University Press (2001)

Ponomarenko, V.I., Smirnov, D.A., Bodrov, M.B., Bezruchko, B.P.: Global reconstruction from time series in application to determination of coupling direction. In: Gulyaev, Yu.V., Sinitsyn, N.I., Anikin, V.M. Voprosy prikladnoy fiziki (Question of Applied Physics). Inter-University Scientific Transactions. Issue 11, pp. 192–200. College (2004)

Rayner, N.A., Parker, D.E., Horton, E.B., Folland, C.K., Alexander, L.V., Rowel, D.P., Kent, E.C., Kaplan, A.: Global analyses of sea surface temperature, sea ice, and night marine air temperature since the late nineteenth century. J. Geophys. Res. **108**(D14). doi:10.1029/2002JD002670 (2003)

Reynolds, R.W., Smith, T.M.: Improved global sea surface temperature analyses. J. Climate. **7**, 929–948 (1994)

Rivlin-Etzion, M., Marmor, O., Heimer, G., Raz, A., Nini, A., Bergman, H.: Basal ganglia oscillations and pathophysiology of movement disorders. Curr. Opin. Neurobiol. **16**, 629–637 (2006)

Rosenblum, M.G., Pikovsky, A.S., Kurths, J., Schaefer, C., Tass, P.A.: Phase synchronization: from theory to data analysis. In: Moss, F., Gielen, S. (eds.) Neuro-Informatics. Handbook of Biological Physics, vol. 4, pp. 279–321.Elsevier Science, New York (2001)

Sarkar, S., Singh, R.P., Kafatos, M.: Further evidences for the weakening relationship of Indian rainfall and ENSO over India. Geophys. Res. Lett. **31**, L13209. doi:10.1029/2004GL020259 (2004)

Seber, G.A.F.: Linear Regression Analysis. Wiley, New York (1977)

Smirnov, D., Barnikol, U.B., Barnikol, T.T., Bezruchko, B.P., Hauptmann, C., Buehrle, C., Maarouf, M., Sturm, V., Freund, H.-J., Tass, P.A.: The generation of Parkinsonian tremor as revealed by directional coupling analysis. Europhys. Lett. **83**, 20003 (2008)

Smirnov, D., Schelter, B., Winterhalder, M., Timmer, J.: Revealing direction of coupling between neuronal oscillators from time series: Phase dynamics modeling versus partial directed coherence. Chaos. **17**, 013111 (2007)

Smirnov, D.A., Andrzejak, R.G.: Detection of weak directional coupling: phase dynamics approach versus state space approach. Phys. Rev. E. **71**, 036207 (2005)

Smirnov, D.A., Bezruchko, B.P.: Estimation of interaction strength and direction from short and noisy time series. Phys. Rev. E. **68**, 046209 (2003)

Solomon, S., Qin, D., Manning, M., Marquis, M., Averyt, K., Tignor, M.M.B., LeRoy Miller, H., Chen, Z. (eds.): Climate Change 2007: The Physical Science Basis. Cambridge University Press: Cambridge (2007)

Stilles, R., Pozos, R. A mechanical-reflex oscillator hypothesis for parkinsonian hand tremor. J. Appl. Physiol. **40**, 990–998 (1976)

Tass, P.A.: Phase Resetting in Medicine and Biology. Springer, Berlin (1999)

Timmer, J., Haeussler, S., Lauk, M., Luecking, C.H.: Pathological tremors: deterministic chaos or nonlinear stochastic oscillators?. Chaos. 10, 278–288 (2000)

Timmermann, L., Gross, J., Dirks, M., Volkmann, J., Freund, H.J., Schnitzler, A.: The cerebral oscillatory network of parkinsonian resting tremor. Brain. **126**, 199–212 (2003)

Treuer, H., Klein, D., Maarouf, M., Lehrke, R., Voges, J., Sturm, V.: Accuracy and conformity of stereotactically guided interstitial brain tumour therapy using I-125 see. Radiother. Oncol. **77**, 202–209 (2005)

Walker, G.T., Bliss, E.W.: World weather V. Mem. R. Meteorol. Soc. **4**, 3–84 (1932)

Wang, W., Anderson, B.T., Kaufmann, R.K., Myneni, R.B. The relation between the North Atlantic Oscillation and SSTs in North Atlantic basin. J. Climate. **17**, 4752–4759 (2004)

Yim, S.-Y., Jhun, J.-G., Yeh, S.-W.: Decadal change in the relationship between east Asian – western North Pacific summer monsoons and ENSO in the mid-1990s. Geophys. Res. Lett. **35**, L20711, doi:10.1029/2008GL035751 (2008)

Zhou, T., Zhang, L., Li, H.: Changes in global land monsoon area and total rainfall accumulation over the last half century. Geophys. Res. Lett. **35**, L16707, doi:10.1029/2008GL034881 (2008)

Zimmermann, R., Deuschl, G., Hornig, A., Schulte-Munting, J., Fuchs, G., Lucking, C.H.: Tremors in Parkinson's disease: symptom analysis and rating. Clinical Neuropharmacology. 17, 303–314 (1994)

Zubair, L., Ropelewski, C.F.: The strengthening relationship between ENSO and Northeast Monsoon rainfall over Sri Lanka and Southern India. J. Climate. **19**, 1667–1575 (2006)

Summary and Outlook

Mathematical modelling is one of the main methods in scientific research. It is discussed in multitude of textbooks and monographs, each of them being devoted to selected aspects of the topic, such as models developed in a specific field of science, peculiarities of certain mathematical tools and technical applications. The main distinctive feature of the present book consists in the consideration of recently developed non-linear methods for dynamical modelling of complex (irregular, non-stationary, noise-corrupted) processes and signals. These properties are inherent to many real-world systems. The corresponding experimental data, recorded with modern digital devices, are often represented as a time series, i.e. a sequence of measured values of an observable. Such a form is "understandable" for a computer. If necessary, the data can be visualised on a screen or subjected to further mathematical transformations. In the book, we have considered opportunities of the time series-based construction of mathematical models, which can be used to predict future behaviour of an object in time or under parameter variations and solve a number of other practical tasks. It is not a rare case when fundamental laws of an object functioning are unknown and, therefore, such a way of constructing a model (i.e. its restoration from the observation data) appears the only possible.

The Main Points of Part I

Despite an overall applied character of the book, it starts with a general outlook, with a brief discussion of the experience obtained from centuries-old activity related to the scientific description of natural phenomena with mathematical models. Chapter 1 presents the most fruitful approaches to modelling and lists some "eternal questions" such as the causes of amazing efficiency of mathematics, the number of possible models for a given object and existence of a "true" model. History of mechanical models is exposed as an example of non-trivial evolution of model representations.

Two basic approaches to modelling, resulting from activity of many generations of scientists, are discussed in Chap. 2. The first one is dynamical (deterministic) approach associated mainly with the names of I. Newton and P. Laplace.

B.P. Bezruchko, D.A. Smirnov, *Extracting Knowledge From Time Series*, Springer Series in Synergetics, DOI 10.1007/978-3-642-12601-7, © Springer-Verlag Berlin Heidelberg 2010

Deterministic models (dynamical systems) allow precise forecast of their future behaviour if an initial state is given. The latest jump of interest to the dynamical description of natural phenomena lasts from the 1960s till now. This is the "Golden Age" of non-linear dynamics with its possibilities of a vivid representation of complex behaviour, the concept of dynamical chaos and a set of paradigmatic simple models with complex dynamics. Basic concepts of non-linear dynamics are described in the beginning of Chap. 2. Then, we discuss the grounds to use an alternative approach to modelling, which is called stochastic or probabilistic. There are several quite different ideas allowing to call a process "random" (stochastic). They complement each other and, sometimes, lead to mutually opposite statements. Non-trivial interrelations between the deterministic and the stochastic approaches are discussed and illustrated in the end of Chap. 2.

Chapter 3 concentrates on the mathematical tools used under the deterministic approach to describe the systems of different levels of complexity in respect of their dimensionality and spatial extension. These tools range from explicit functions of time and one-dimensional discrete maps to partial differential equations and networks with complex topology. The concepts of linearity and non-linearity are discussed in some detail. Also, we consider etalon systems of non-linear dynamics, which are useful in the study of a complex behaviour due to a wide set of possible oscillatory regimes and their non-trivial evolution under parameter variations.

Chapter 4 gives a general information on stochastic models. It comprises basic concepts and models used in the theory of random processes. Peculiarities of numerical integration of stochastic differential equations are highlighted. Currently, investigations in the field of "purely deterministic" non-linear dynamics have become less popular, since incorporation of random terms into model equations allows a more realistic description of many natural processes and a study of some fundamental noise-induced phenomena. In particular, we briefly touch on a possible constructive role of noise (stochastic and coherence resonances).

The Main Points of Part II

It is devoted to modelling from time series including its ideology, techniques and applications. This is only one of the four ways of model construction presented in Sect. 1.5. Experience of many researchers shows that the modelling procedure summarised in a vivid scheme in Chap. 5 often gives successful results if techniques and pitfalls described, e.g., in Chaps. 6, 7, 8, 9 and 10 are taken into account. Probably, such a technology as fitting models to data cannot be considered as a highest degree of perfection: ill-posedness of modelling problems and other difficulties are mentioned already in Chap. 5. However, its attractiveness rises under the realistic setting of a deficit of prior information about an object with complex behaviour.

The field of empirical modelling emerged long time ago due to the problems of approximation of experimental dependencies with smooth functions. Its essential development in mathematical statistics was associated with the theory of system identification. Then, the interest of scientists to specific problems of empirical modelling varied in time. Its rise during the 1980s–1990s was determined by the birth of the concept of dynamical chaos, proofs of the celebrated Takens' theorems and availability of fast computers. Indeed, if a simple quadratic map is capable of demonstrating a hierarchy of periodic regimes ending with chaos (Sect. 3.6.2), then what a potential could be expected from more complex non-linear equations, whose investigation became possible!

One could even think that modelling from data series would become a routine operation. However, researchers became disappointed soon at frequent failures of the developed universal approaches in practice. The models were often quite cumbersome, unstable to variations in parameters, etc. In particular, the difficulties in modelling rise with increasing dimensionality of a model, the so-called "curse of dimensionality". However, chances for a success can be increased if one refuses universal techniques and pays more attention to specific features of an object. We have tried to develop and illustrate this idea through Part II: *It is not fruitful to look for a "panacea", a universal model, and an omnipotent algorithm of its construction from data. It seems more reasonable to develop specialised techniques for modelling certain classes of systems.* At that, one must be accurate and attentive to the details of the modelling procedure at all stages of the scheme shown in Chap. 5, at any level of prior uncertainty ranging from complete knowledge of the form of appropriate equations to complete absence of any information about a suitable model structure.

Fruitful ideas for the model structure selection can be obtained from a preliminary analysis of the observed data as discussed in Chap. 6, where we describe peculiarities of getting experimental time series, their visual inspection and obtaining time courses of model variables from observed data. Chapter 7 considers construction of models in the simplest form of explicit functions of time. However, it includes basic ideas and techniques exploited in the construction of much more sophisticated models. Chaps. 8, 9 and 10 are devoted to the construction of dynamical model equations under different degrees of prior uncertainty. The case of complete knowledge of the model structure is presented in Chap. 8, where parameter estimation and dealing with hidden variables are described. Chapter 9 presents the case of partial uncertainty with the emphasis on approximation problems and restoration of non-linear characteristics. Chapter 10 considers the most difficult and intriguing "black box" problem, i.e. the case of complete prior uncertainty about the model structure.

Applications of empirical modelling are given in Chaps. 11, 12 and 13, where we present examples of different levels of complexity. Firstly, reconstruction of etalon dynamical systems from their numerical solutions is useful from illustrative and methodological point of view. Secondly, modelling of laboratory systems, which allow to select regimes of their functioning purposefully, is more complicated (since these are real systems subjected to random influences, etc.) but has mainly a methodological value too. Thirdly, the investigation of real-world processes, where possibilities of active manipulations with an object are quite restricted or even absent (e.g., in geophysics), is the most complex and practically important case. At that,

Chap. 13 gives more detailed illustrations of the different steps of a modelling procedure and intended for a wider audience.

General Concluding Remarks

Techniques of the time series analysis can be distinguished by convention into two groups: *"direct" processing* and *model-based approach.*

The former way is more traditional, e.g., a cardiologist considers an electrocardiogram by visual inspection of the form of PQRST complexes. Analogously, one can compute power spectra and correlation functions, estimate fractal dimensions and restore phase orbits directly from time series. In part, this way is concerned in Chap. 6, where we present also some newer techniques including wavelet analysis, empirical mode decomposition and recurrence analysis. Currently, one of the most popular directions in the field of "direct" techniques of time series analysis is the development of the concept and quantitative characteristics of *complexity* and methods for their estimation (see, for example, Badii and Politi, 1997; Boffetta et al., 2002; Shalizi, 2003).

Under the model-based approach, one creates and uses a necessary intermediate element for the analysis of a process. This element is a mathematical model, which describes the observed dynamics. We stress that, typically, it is a predictive model of the observed motion, rather than a model of the object structure or of the entire mechanism of its functioning. The model-based approach is required to solve the problems of the forecast of a future behaviour, prediction of changes under parameter variations, signal classification, data compression and storage, validation of ideas about an object, "measurement" of hidden variables, etc.

When reporting results of our investigations, we face some disappointment of listeners who expected a universal recipe of empirical modelling but learned a specific technique directed to a certain class of objects. Indeed, mathematical modelling remains and, probably, will always stay a kind of art to a significant extent or a kind of piecework requiring a preliminary preparation of unique tools. However, if a problem or an object under study is important enough, then it can be worth spending a lot of time and efforts to such a piecework. Furthermore, let us recall that making semiconductor elements was a kind of art 50 years ago when getting two identical transistors was almost impossible. Currently, technological achievements allow one to make identical processors consisting of millions of transistors. Not all specialists would share such an optimistic attitude to the field of empirical modelling. However, the results which are already achieved in this field deserve studying and application in practice.

Our teaching experience shows that the topic is of interest for university students and Ph.D. students. The talks devoted to modelling problems induce conceptual discussions at scientific conferences of different profiles, which have already led our group to fruitful collaboration with climatologists (Mokhov and Smirnov, 2006, 2008, 2009), neurophysiologists (Dikanev et al., 2005; Sitnikova et al., 2008;

Smirnov et al., 2005a, 2008; Stoop et al., 2006), and cardiologists (Prokhorov et al., 2003). We hope that this book will further contribute to the rise in such interest and attract attention of people from different fields of science, technology and medicine. Of course, it is only "an excursus into ...", rather than an exhaustive discussion of the eternal question about possibilities of mathematical description of real-world objects and phenomena.

References

Badii, R., Politi, A.: Complexity: Hierarchical Structures and Scaling in Physics. Cambridge University Press, Cambridge (1997)

Boffetta, G., Cencini, M., Falcioni, M., Vulpiani, A.: Predictability: a way to characterize complexity. Phys. Rep. **356**, 367–474 (2002)

Dikanev, T., Smirnov, D., Wennberg, R., Perez Velazquez, J.L., Bezruchko, B.: EEG nonstationarity during intracranially recorded seizures: statistical and dynamical analysis. Clin. Neurophysiol. **116**, 1796–1807 (2005)

Mokhov, I.I., Smirnov, D.A.: El Nino Southern Oscillation drives North Atlantic Oscillation as revealed with nonlinear techniques from climatic indices. Geophys. Res. Lett. **33**, L0378 (2006)

Mokhov, I.I., Smirnov, D.A.: Diagnostics of a cause – effect relation between solar activity and the earth's global surface temperature. Izvestiya, atmos. Ocean. Phys. **44**, 263–272 (2008)

Mokhov, I.I., Smirnov, D.A.: Empirical estimates of the influence of natural and anthropogenic factors on the global surface temperature. Doklady Acad. Sci. **426**(5), 679–684, (in Russian) (2009)

Prokhorov, M.D., Ponomarenko, V.I., Gridnev, V.I., Bodrov, M.B., Bespyatov, A.B. Synchronization between main rhythmic processes in the human cardiovascular system. Phys. Rev. E. **68**, 041913 (2003)

Shalizi, C.R.: Methods and techniques of complex systems science: an overview, vol. 3, arXiv:nlin.AO/0307015 (2003). Available at http://www.arxiv.org/abs/nlin.AO/0307015

Sitnikova, E., Dikanev, T., Smirnov, D., Bezruchko, B., van Luijtelaar, G., Granger causality: Cortico-thalamic interdependencies during absence seizures in WAG/Rij rats. J. Neurosci. Meth. **170**, 245–254 (2008)

Smirnov, D., Barnikol, U.B., Barnikol, T.T., Bezruchko, B.P., Hauptmann, C., Buehrle, C., Maarouf, M., Sturm, V., Freund, H.-J., Tass, P.A.: The generation of Parkinsonian tremor as revealed by directional coupling analysis. Europhys. Lett. **83**, 20003 (2008)

Smirnov, D.A., Bodrov, M.B., Perez Velazquez, J.L., Wennberg, R.A., Bezruchko, B.P.: Estimation of coupling between oscillators from short time series via phase dynamics modeling: limitations and application to EEG data. Chaos. **15**, 024102 (2005)

Stoop, R., Kern, A., Goepfert, M.C., Smirnov, D.A., Dikanev, T.V., Bezrucko, B.P.: A generalization of the van-der-Pol oscillator underlies active signal amplification in Drosophila hearing. Eur. Biophys. J. **35**, 511–516 (2006)

List of Mathematical Models

List of Real-World Examples

Index

Interfacial Wave Theory of Pattern Formation
Selection of Dentritic Growth and Viscous
Fingerings in Hele-Shaw Flow By Jian-Jun Xu

**Asymptotic Approaches in Nonlinear
Dynamics** New Trends and Applications
By J. Awrejcewicz, I. V. Andrianov,
L. I. Manevitch

Brain Function and Oscillations
Volume I: Brain Oscillations.
Principles and Approaches
Volume II: Integrative Brain Function.
Neurophysiology and Cognitive Processes
By E. Başar

**Asymptotic Methods for the Fokker-Planck
Equation and the Exit Problem in Applications**
By J. Grasman, O. A. van Herwaarden

**Analysis of Neurophysiological Brain
Functioning** Editor: Ch. Uhl

Phase Resetting in Medicine and Biology
Stochastic Modelling and Data Analysis
By P. A. Tass

Self-Organization and the City By J. Portugali

Critical Phenomena in Natural Sciences
Chaos, Fractals, Selforganization and Disorder:
Concepts and Tools 2nd Edition By D. Sornette

Spatial Hysteresis and Optical Patterns
By N. N. Rosanov

**Nonlinear Dynamics of Chaotic and Stochastic
Systems** Tutorial and Modern Developments
2nd Edition
By V. S. Anishchenko, V. Astakhov,
A. Neiman, T. Vadivasova,
L. Schimansky-Geier

Synergetic Phenomena in Active Lattices
Patterns, Waves, Solitons, Chaos
By V. I. Nekorkin, M. G. Velarde

Brain Dynamics
Synchronization and Activity Patterns
in Pulse-Coupled Neural Nets with Delays
and Noise By H. Haken

From Cells to Societies
Models of Complex Coherent Action
By A. S. Mikhailov, V. Calenbuhr

Brownian Agents and Active Particles
Collective Dynamics in the Natural and Social
Sciences By F. Schweitzer

**Nonlinear Dynamics of the Lithosphere
and Earthquake Prediction**
By V. I. Keilis-Borok, A.A. Soloviev (Eds.)

Nonlinear Fokker-Planck Equations
Fundamentals and Applications
By T.D. Frank

**Patterns and Interfaces
in Dissipative Dynamics**
By L.M. Pismen

Synchronization in Oscillatory Networks
By G. Osipov, J. Kurths, C. Zhou

Turbulence and Diffusion Scaling Versus
Equations
By O.G. Bakunin

Stochastic Methods A Handbook for the Natural
and Social Sciences
4th Edition
By C. Gardiner

**Permutation Complexity in Dynamical
Systems** Ordinal Patterns, Permutation Entropy
and All That
By J.M. Amigó

Quantum Signatures of Chaos
3rd Edition By F. Haake

Reaction–Transport Systems Mesoscopic
Foundations, Fronts, and Spatial Instabilities
By V. Méndez, S. Fedotov, W. Horsthemke

Extracting Knowledge From Time Series
An Introduction to Nonlinear Empirical
Modeling
By B.P. Bezruchko, D.A. Smirnov